Ruling *the* Waves

CYCLES OF DISCOVERY,
CHAOS, AND WEALTH
FROM THE COMPASS
TO THE INTERNET

Debora L. Spar

HARCOURT, INC.
New York San Diego London

www.harcourt.com

Library of Congress Cataloging-in-Publication Data
Spar, Debora L.
Ruling the waves: cycles of discovery, chaos, and wealth from the compass to the internet/Debora L. Spar.
p.   cm.
Includes index.
ISBN 0-15-100509-5
1. Technological innovations—Social aspects. 2. Technological innovations—Economic aspects. 3. Information technology—Social aspects. 4. Information technology—Economic aspects. 5. Internet—Social aspects. 6. Internet—Economic aspects. 7. Technology and state. I. Title.
HM846 .S63 2001
303.48'3—dc21   2001024256

Text set in Plantin
Designed by Kaelin Chappell
First edition
K J I H G F E D C B A
Printed in the United States of America

CONTENTS

The View from Partenia

Partenia is a lonely place. Strewn across the sands of the Sahara, it is formally located in Tunisia, or Algeria, or Libya, depending on to whom you talk or which way the winds are blowing. It is an ancient place, Partenia, a remnant of a world that hardly anyone can even remember. Yet in a very strange way Partenia is coming back.

In 1995, the Vatican dismissed an outspoken French bishop named Jacques Gaillot. Arguing that Gaillot had been far too liberal for the Church's doctrine, Vatican officials removed him from his diocese outside Paris and sent him to Partenia. Clearly it was a symbolic move, for the Church never expected Gaillot to preach to the empty drifts of the Sahara. They simply wanted to defrock him gently, pushing the unruly bishop to one of the several jurisdictions reserved for retired, aging, or unwanted

priests. Gaillot, however, wasn't prepared to go quietly, or to renounce the liberal views that had angered his superiors in
Rome. So he went to Partenia—virtually.

One year after his dismissal, Gaillot launched the world's
first "virtual diocese." Named Partenia, it is a site for liberal
Catholics, a "place of freedom," according to Gaillot, where
Catholics can discuss the issues that Gaillot had come to stand
for: the problem of homelessness, the spread of AIDS, the evils
of nuclear testing, and the wisdom of married priests. In the
first six weeks of 1996, Partenia registered 250,000 hits. The
Vatican, presumably, was not impressed with Gaillot's move
and spent a good deal of time trying to concoct a strategy for
dealing with this unsettling cyber-priest. But there really wasn't
much that they could do, so they left Gaillot and his liberal site
alone. Partenia had won.

In cyberspace, Partenia is everywhere. Dotted
along the Internet's web are millions of places where rebels like
Bishop Gaillot reside. There are pornography sites accessible to
straitlaced Singaporeans, Liberian gambling dens, and secluded
banking services run from the tiny island of Anguilla. There are
networks of Burmese dissidents, collecting information on the
dictatorial regime in Rangoon and e-mailing it to thousands of
supporters around the world. There are bootleg copies of academic papers and Snoop Doggy Dogg's latest hits. In cyberspace, even solemn corporations indulge their rebel side,
slipping around the real-world laws that govern things such as
export controls and truth in advertising.

If we look at cyberspace from the viewpoint of
Partenia, then, it looks very much like a frontier town—like
California of the 1890s, or the Indies to which Europe
scrambled in the seventeenth century. There are the usual
hordes of rebels and rogues, plus scores of pioneers and gold-
diggers, each scrambling to carve out new territories and stake
their claims in them. There are people like Marc Andreessen
and Jerry Yang (the respective founders of Netscape and
Yahoo!), who ventured west to test their mettle and made incredible fortunes virtually overnight. There are prophets who

scream of a brave new world and traveling salesmen hawking IPOs instead of snake oil. (The connection, of course, may not be that distant.) As on any good frontier, there are not a lot of rules or marshals in town, so justice is rough and the winners grab whatever they can. There are, to be sure, some remote authorities (the U.S. Federal Communications Commission, the European Commission's DG IV) who claim to be patrolling the area, but everyone knows that their guns are not loaded. For cyberspace, it seems, is a lawless realm, a place where unruly bishops can confound the Pope and Jerry Yang can start a multibillion-dollar industry before turning thirty.

This sense of anarchy permeates the farthest reaches of the Net. In Silicon Valley, along Route 128, and in the *samizdat* cafes of Beijing and Rangoon, there is a palpable sense of excitement, a prevailing belief that authority is dead and that digital technologies have killed it. And to some extent this is true. Digital technologies have created a revolution of sorts. They have allowed entrepreneurs to build empires out of fiber and thin air and to establish these empires in a realm without rules. They have challenged governments and their traditional authority—not by design or intent, but purely as a result of technological accident. Because digital technologies allow information to flow seamlessly and invisibly across national borders, they make it very difficult for governments to do many of the things to which they have grown accustomed. Governments can't patrol their physical territories in cyberspace; they can't easily enforce property rights over ephemeral ideas and rapidly moving bits; they can't control information flows; they even may not be able to collect taxes. Such is the nature of politics along the technological frontier.

Yet even in the midst of all this tumult, it is useful to maintain a sense of perspective, and of history. Cyberspace is indeed a brave new world, but it's not the only new world. There have been other moments in time that undoubtedly felt very much like the present era, other moments when technology raced faster than governments and called forth whole new markets and social structures. Other entrepreneurs sensed that they, too, were standing on the edge of history, bending authority to their will and reaping fabulous profits along the way. Some of them succeeded beyond their wildest dreams. Pioneers such as Thomas Edison and Guglielmo Marconi, for example, saw the fantastic opportunity of technology and ran nimbly along its curve. They built empires where none existed and wrote rules to serve their own advantage. Other pioneers, however, were far less successful. Even if they had path-breaking technology, and even if they flourished for some time in a period of blissful chaos, many entrepreneurs eventually found themselves caught by a system that bit back—by markets that reasserted their old ways or governments that outraced the technological frontier and claimed it for themselves. The new world in these cases fell back into the old, leaving the pioneers stranded on what once seemed to be the future.

This is a book that tries to yank the Internet out of the spotlight of the twenty-first century and back to its older and dimmer roots. It argues that while cyberspace is new and sparkling with opportunity, it is not that new and that much sparklier than other technologies were on the eve of their creation. We are undeniably living in a revolutionary period. We see this revolution every day and feel it crack the structures of our lives. We see it in the rush toward Silicon Valley; in the euphoria that drove Internet stocks to unbelievable heights; in the intrusion of e-mail and surfing and "dot.com" everything. At a more profound level, it is also clear that this revolution will seriously affect both business and politics. It will open vast new vistas for commerce and, in the process, will challenge relations between private firms and the governments that seek to regulate them. The information revolution is alive and well. It will change the way we work, they way we play, and the way in which we order our societies. It will change in particular how we think about governments, because cyberspace is a realm that seems inherently to ignore traditional authorities. Cyberspace, in fact, is a truly global phenomenon, something that spans borders irrepressibly and imperceptibly. Purely by accident, the Net shatters our notions of what a "state" does or what a "national economy" is. For cyberspace is bigger than any state and well beyond traditional powers of enforcement. What can the Pope do if Bishop Gaillot uses his site to condemn celibacy in the priesthood and encourage the use of condoms? Not much. How can Singapore stop its citizens from peeking at *Hustler* on their laptops? Or the U.S. government prevent American firms from using high-powered security software in their overseas affiliates? Again, they essentially can't. Silently, cyberspace challenges the power of government by going where it, by definition, cannot: across national borders.

Theoretically, this shift in geography should be a tremendous boon to firms, just as it is a rather terrifying prospect for states. Freed from governmental control, firms in cyberspace should be able to operate freely and without rules. They should be able to write their own terms and strike their own deals without having to pay any heed to bureaucratic whim or regulation. This, after all, is the political thrill of the Net. Yet this is also where history suggests a certain amount of prudence. Other technologies have challenged government's authority; other pioneers have gleefully declared the death of the state. What their stories show us, though, is that while technology can gravely wound governments, it rarely kills them. Instead, governments survive because, ironically, both society and entrepreneurs want them. Governments provide the property rights that entrepreneurs eventually want, the legal stability that commerce craves, and the stability that society demands. For in the end, even pirates and pioneers want order. Once they have staked their claim or claimed their loot, they want someone else to protect it. And that someone else is usually the state.

Consider what happened during the first round of the Information Revolution. Up until the fifteenth century, information was a highly guarded, tightly controlled commodity. The Catholic Church, which essentially performed a quasi-governmental function at that time, sat at the center of all information flows. Only priests and a handful of scholars could read; only monks were permitted to write and copy the manuscripts that formed the backbone of knowledge in the Middle Ages. All information then was written (aside, of course, from knowledge passed through the oral tradition) and the Church controlled the writing process. Indeed, one of the major tasks of the Catholic Church during this period was to act as a kind of labor-intensive, multinational publishing operation. Across Europe, monks spent their lives huddled over exquisite manuscripts, painstakingly redrawing each letter, each sketch of just a few texts. The technology of the period was pen and parchment, and it survived for hundreds of years.

And then a young silversmith named Johannes Gutenberg came along. In 1453, after years of tinkering with typography and dies cast from various minerals, Gutenberg devised a rudimentary printing press. The technology was rather straightforward: blocks etched with different letters were simply inserted into a shallow box, inked, and then pressed against sheets of paper to leave a written imprint. But the implications wreaked slow-boiling havoc on Europe's established social order. Initially, the Church welcomed Gutenberg's machine, describing it as a supernatural invention that would enable priests to read more easily to their congregations. As the medium spread, though, the Church grew wary, seeing in moveable type the makings of a revolution. The more Bibles were printed, the more people wanted to read them and to interpret the Church's doctrine for themselves. The more access people had to printing technologies, the more freedom they had to challenge the Church and to break its monopoly on priestly information. Thus the technology—a simple mix of ink and metal—became political. It created a space in which rebels could play and ignored the structure of existing rules.

Before too long this radical potential erupted. In 1517 a young German priest named Martin Luther used a Gutenberg-type press to publish copies of his "95 Theses," a list of arguments that directly challenged the Church and its practices. By themselves, the theses were not really all that radical; for centuries, disgruntled clerics had made similar arguments and compiled similar lists. But Luther had technology on his side. Rather than staying tacked to the door of a single local church, Luther's theses circulated widely around Europe, creating excitement just by virtue of their circulation and the renown that soon accompanied them. They stirred a priestly buzz, to be sure, but a buzz all the same. And the implications were dramatic. Luther's theses let loose the Reformation and the rise of Protestantism, the first sustained crack in Catholicism's religious control of Europe. Simultaneously, the expansion of printing technology made it easier and easier for laypersons to read, and thus to challenge the vested authority of the Church's priestly elite. By 1534, when King Henry VIII declared Catholicism illegal in Great Britain and made his subjects members of the newly created Church of England, the Catholic Church had been dealt a formidable blow. Printing technologies didn't cause this shift of power, of course, but they certainly facilitated it. Without Gutenberg, Luther would have remained a local hero—a thorn in the side of the Church, perhaps, but not the founder of an entirely separate structure. This, again, is the nature of power along the technological frontier.

But even Luther's victory was not complete. Technology challenged the Church, forcing it to change in some ways, but it certainly did not kill it. On the contrary, once Vatican officials realized the power of printing, they scrambled to use it for their own purposes. They established Catholic publishing houses to foster a Counter-Reformation, circulated the Bible and other key texts, and learned how to reach out to a growing mass of literate followers. While they undeniably lost the informal monopoly over printing that the monks had provided, they managed to control key forms of information and the power that accompanied them. Rules changed, and power shifted, but the patterns that had dominated before Gutenberg's time remained solidly in place.

An equally striking dynamic surrounded the development of radio, another major stage in the information revolution. In 1896, Italian inventor Guglielmo Marconi brought a small black box with him to Britain. It was an early prototype for a radio: a makeshift device that transmitted Morse code via electromagnetic waves. As soon as he crossed the border, customs officials smashed the box to pieces, fearing that it would inspire violence and revolution. So Marconi went home and continued to work on his invention, eventually creating a firm designed to develop the radio for commercial use. Before long, though, the government reappeared and declared a security interest in Marconi's device. By the start of the first World War, the Marconi Company had become a full-time contractor for the British government and the British Navy controlled the fledgling technology of radio transmission.

If we apply the patterns of printing and radio to today's information revolution, they imply an outcome strangely at odds with the view from Partenia. They suggest that there is a certain give-and-take along the technological frontier, a dance of regulation that moves power back and forth between firms and governments, between pioneers and bureaucrats. In this view, even the most radical technology will not necessarily force any particular authority to disappear, or even to change its fundamental mission. Instead, technology challenges authority for some period of time, but then, ironically, seems to invite this authority back in: the Church embraced the printing press in the end, the British government absorbed the company it once saw as dangerous. Perhaps the Internet is different. Perhaps it is so revolutionary, so international that it will disrupt the patterns that have prevailed in the past and dispel the myths we have built around them. Perhaps it really will deal governments a fatal blow and usher in new forms of social organization. Perhaps the future of power will be a world of Gaillots and cybercafes, circles of communication or commerce without any central authority. But perhaps not. Certainly history suggests that we have been here before and come out largely where we started. With new rules, to be sure, and new forms of commerce, but still with a basic structure of authority and a recognizable state.

If we view cyberspace from history, therefore, rather than from Partenia, we see a more complex vision. Instead of a one-way scramble to a brave new world, it is a journey of twists and turns, a movement along a frontier whose boundaries shift and stumble and collide. It is a view filled with the normal characters of a frontier town: there are still the pirates and the pioneers, the tinkers and the traveling salesmen. Only in this view, the pirates and the pioneers aren't necessarily the winners. Instead, once the technological frontier has moved beyond a certain point, power—and profits—seem to shift away from those who break the rules and back to those who make them.

Ruling the Waves is essentially a book of frontier stories. It begins with Portuguese explorers of the fifteenth century, follows the development of telegraph and radio in the middle of the nineteenth century, and then turns to the advent of satellite television and the Internet in the twentieth. Part of the book's intent is simply to tell these stories, for they are fascinating in their own right and little known outside of academic enclaves. The broader intent, though, is to use these stories to link current developments along the technological frontier back to their rightful ancestors: to show how Rupert Murdoch and Bill Gates are descended in many ways from Prince Henry and Samuel Morse, and to think about what Murdoch and Gates might be able to learn from these older pioneers.

The stories that are presented here were not chosen with scientific precision. Indeed there are dozens of other technologies and hundreds of other pioneers whose stories are just as intriguing and important: Edison and electricity, Bell and telephones, Watt and the steam engine. There are also dramatic developments outside the western world—particularly in China and the Islamic states—that fell beyond the research scope of this book. All the stories that are here, however, contain some common themes and parallels. All involve, for example, a sharp movement along the technological frontier—a moment in time when innovation leapt suddenly outward, creating new opportunities for commerce and tremendous enthusiasm among aspiring entrepreneurs. In each case, moreover, the technological leap also created a political gap. Innovation, in other words, enabled firms to play in some new sphere of activity, free from the rules or regulations that might bind them in another, more established realm. Finally, all of the stories presented here are about a particular type of technology. They are all about communication, about bringing information from one spot to another. There is nothing unique about these technologies, of course, and the patterns that emerge here might well apply across the technological spectrum. But communications technologies have a certain force to them, and a particular import. For communication is the sinew of both commerce and politics, the channel through which information—and thus power—flows. Ever since God warned Eve to resist the apple, authorities have tried to control information flows. And ever since Eve took that first bite, pioneers have resisted these controls and tried to find ways around them. This is a book about their stories—about life on the technological frontier and the pirates, prophets, and pioneers who struggle to build their empires upon it.

This is also a book about ideas. In particular, it is a book about how markets get established and how firms and governments together shape their creation. Frequently, the worlds of business and politics are described as belonging to wholly different spheres. There is business on one side, following the laws of the market and the dictates of competition, and government on the other, pulled by the demands of power and the desire to maintain it. This book takes a very different tack. Business, I argue, is inherently political and politics is—and has always been—marked by the interests of commerce. This overlap is everywhere: in trade policy, defense policy, and the politics of procurement or privatization. But it is particularly strong along the edges of the technological frontier. For it is here that markets are actually created, where industries spring to life and then settle, eventually, into some kind of ordered existence. As this process unwinds, power is distributed and structures are established. It is a hugely political enterprise, even if governments are not actively calling the shots or regulating commercial activity.

In fact, it is precisely the lack of established regulation that makes the technological frontier so political. In order for commerce to grow in any uncharted territory there need to be rules. Not regulation necessarily, or even governments—just rules. There need to be property rights, for example, and some sense of contracts. In higher technology areas, there need to be rules for intellectual property (who owns the operating system? under what terms?) and provisions for standardization (how do different products work together? which technical platform becomes the norm?). Without these rules, commerce may still emerge, but it will not flourish. There may be bursts of commercial activity and a handful of pioneers who cherish life on the edge, but wide-scale commerce will remain elusive. This is a powerful lesson of history and a tragedy that still affects large swaths of the global economy. Without rules, and particularly without rules of property and exchange, markets simply do not grow. Just look at Russia in the 1990s, or some of Africa's more chaotic regions. A similar dynamic prevails along the technological frontier. New markets need new rules if they are to flourish, and their creation is a distinctly political act.

This connection between politics and business, rules and markets, unwinds slowly along the technological frontier. At any particular point in time, it is difficult to see how politics is shaping markets, or why entrepreneurs might want to do anything other than expand their empires and push technology's edge. Which is why history is so critical. It provides a sense of perspective that can't possibly exist in the present; it displays patterns that can appear only once the dust of the frontier has settled and the pioneers have moved on. And these patterns are strong. Indeed, it appears from the stories presented here that life along the technological frontier moves through four distinct phases: innovation, commercialization, creative anarchy, and rules. Each has its own rhythm and speed, a movement that shifts with the tenor of the times and the nature of technological change. Yet there are clearly patterns and lessons to be drawn from them.

Phase One: Innovation

In the beginning, of course, there is innovation. This is the stage of tinkerers and inventors, a stage marked by laborious exploration and the sudden thrill of discovery. It is the sexiest phase along the technological frontier, a time that sparks the imagination and provides motivation for the next generation of dreamers and planners. It is not a phase in which lots of commerce occurs. Instead, most of the excitement that surrounds a technology during its earliest days is from fellow enthusiasts—people who treasure innovation simply for what it does, rather than for any commercial potential. Others tend to ignore technological breakthroughs or even disdain them. Telegraphy, for example, was derided for years as a worthless game, a newfangled obsession with invisible communication. When Samuel Morse demonstrated his machine before Congress in 1838, people just laughed, comparing the contraption to mesmerism or "animal magnetism." Likewise, when the radio arrived at the turn of the twentieth century, most people saw it as a hopelessly complex machine, good, perhaps, for military functions but without any broader appeal. Early users of radio were almost entirely amateur mechanics, people who were much more interested in the workings of the set than in anything that might be transmitted across it. Even the Internet was distinctly noncommercial at the outset. It was a security tool, a means of communication among a small and specialized group. But a mass market? An instrument of commercial revolution? No one saw it coming.

During this first phase, there are no rules because none are needed: innovation hasn't developed to the point where property rights are critical; there are no questions yet of access or unfair competition; and the societal impact of the new technology is minimal. Indeed, because the technology is still so experimental at this stage and confined to such a small group of users, there simply will not be many people outside this community who either understand the technology or have any concerns about its use. Even if concerns do arise, moreover—if, say, military officials suspect that the new technology has incendiary implications—then governments can generally still rein things in at this stage, imposing or establishing rules before a commercial market has had time to develop. Such was the case with television, which emerged out of radio technologies in the early twentieth century and was instantly besieged by governments wary of its social intent.

So long as governments do not suspect the potential for subversion, though, the innovation phase remains relatively free, open, and uncluttered. The scientists labor in their labs, the tinkerers work in their basements or garages, and innovation occurs. It's the most important phase of the technological frontier but also in many ways the most peaceful. It often ends abruptly.

Phase Two: Commercialization

Once technology is out of the labs and in the public eye, a whole new cast of characters moves onto the frontier. These are the characters usually associated with the frontier: the pioneers, the pirates, the marshals, and the outlaws. They are the ones who define the new territory and bring it to life. In this second phase, the commercial benefits of innovation have become clear. People can now see how the technology will transfer to a mass market and what kinds of profits can be made from it. When the technology is truly revolutionary, they can also see how it carves out new spheres of commerce, spheres that exist beyond the realm of existing markets and beyond the reach of existing authorities. This potential ignites the frontier and draws more people toward it. And thus the familiar scramble ensues. Tempted by the dual visions of anarchy and wealth, entrepreneurs of all sorts rush onto the frontier to stake their claims. Speed is essential during this phase, as is a certain ability to see beyond the confines of established business practice. Not surprisingly, then, most of the pioneers who rush along the technological frontier are young: Marconi was twenty when he brought his black box to London; Marc Andreessen was twenty-three when he founded Netscape. These are the entrepreneurs who can see the opportunity that technology creates and are eager to build their own empires upon it. They also tend to be free spirits, individuals who delight in building their own worlds and operating by their own rules. Their interests during this early phase are largely territorial. Like all pioneers, they want to grab land, to stake a claim and call it their own, even if the property rights to this land are not yet fully secure and the commercial prospects are uncertain.

Pioneers, though, are not the only free spirits who move out to the frontier. Pirates come too, following naturally in the footsteps of the pioneers and often blending in easily with them. In the seventeenth century, pirates mingled indistinguishably with merchants, trailing ships across the Atlantic and strangling the trade that technological advances in navigation had created. In the nineteenth century, pirates plagued the nascent telegraph industry, "borrowing" patented technology to create their own competing systems. And in our own times, pirates wielding satellite dishes and smart cards have invaded the waves of the digital age, stealing television signals from the skies and encryption codes from the Net. Like their predecessors in the Caribbean, these pirates have a certain romance to them. They are rebels who delight in flouting society's rules: when they hang, they hang proudly.

The funny thing about pirates, though, is that they seem to adhere to a certain historical rhythm. When technology is new, it doesn't attract too many rogues. It's simply too technical in the first phase of evolution, too specialized and uncertain. Once technology slips into the commercial realm, however, and begins to generate the extraordinary profits that can occur during this second phase of expansion, the pirates flock. They follow the pioneers along the technological frontier, shadowing their gains and borrowing their technology. Because rules during this period are inherently ill defined, pirates can operate almost without restriction. If there are no rules, after all, no one can break them. This was the case during the heyday of ocean piracy, when states such as Britain and Spain simply were not able to draw any kind of workable distinction between pirates, privateers, merchants, and admirals. It is also the case today. How are we to define the teenager who downloads her favorite party music from an Internet site and sells copies to her friends? Is she a pirate of the digital age, or an aspiring entrepreneur? What about people such as Philip Zimmermann, the mathematician who created one of the world's most sophisticated encryption algorithms and posted it on his web site? Is he a mathematical genius trying to share his knowledge, or a renegade intent on violating the security of the United States? It's hard to tell. During these times of technological flux, the rules are just too flimsy.

What accounts for this flimsiness is the ability of new technologies to slip through the lines of existing law. It's not that governments lack the interest or wherewithal to govern new areas of technology; rather it's just that the old laws are unlikely to cover emerging technologies and new ones take time to create—time that, initially at least, seems to favor the pirates and pioneers. For even as governments begin to understand the implications of telegraphy or hypertext, pioneers such as Morse or Andreessen are already forging ahead, grabbing territory and creating industrial structures. Unless governments manage to nip technology in the bud of innovation (as occurred with television), it's very difficult for them to control this same technology once it has entered the expansionary period. Things are simply moving too quickly, and entrepreneurs are consciously trying to avoid the long arm of regulation. Then, as these entrepreneurs get wealthier, they have more at stake in this new realm and more resources available to protect themselves. So when the marshals show up, the entrepreneurs tend to outrun them, or outgun them, or simply ignore their protestations. When Rupert Murdoch began satellite broadcasts into the British market, for instance, the British government really could not do much to stop him since he was operating, quite literally, above their heads. Likewise, there are many aspects of the Internet economy that, at the turn of this century at least, remain far beyond the reach of any national government, such as content that streams in from foreign sources and information that hides under disguised names and slips across invisible borders. In this phase of technological development, therefore, the politics of the frontier are decidedly libertarian. Markets take over, individuals steer their own fate and governments retreat. It is a period of wild expansion and even wilder expectations, of instant fortunes and dreams of anarchy. It is in many ways the defining moment of the frontier economy, and certainly the most romantic spot along technology's arc.

Phase Three: Creative Anarchy

But this phase doesn't last forever. For before long, problems begin to crop up along the frontier, compromising the commerce that has already emerged and threatening its long-term development. These problems are not the same for every technology; they appear with varying ferocity and after different gaps of time. Almost certainly, though, they will develop. And the pioneers that now people the frontier will demand their resolution.

Consider, for example, the issue of property rights. During the early phases of the technological cycle, ownership is a secondary and occasionally even irrelevant concern. Innovators don't necessarily have much to own at this stage and many of them—though admittedly not all—are untutored newcomers to the world of formal rights. Many of the telegraph's early inventors, for instance, never received either fame or credit for their work; and much of the modern Internet was created by enthusiasts who simply distributed their breakthroughs for free. A lack of property rights during the innovation phase, therefore, is seldom a constraint upon innovation itself. When commercial pioneers join the innovators, of course, this calculus changes: people like Henry O'Rielly, who built one of the first commercial telegraph systems in the United States, and Jim Clark, who

founded Netscape along with Marc Andreessen, have a much
more instrumental view of technology and a more explicit interest in controlling it. Yet even as they yank the technology into
its commercialization phase, many of these entrepreneurs are
not particularly concerned about ownership. Instead they simply
want to attack the frontier as quickly as they can, staking a claim
and creating a market before others have had time to arrive.
Certainly this was the case with O'Rielly, who threw himself
into the nascent U.S. telegraph market in the late 1840s, determined to wire the entire eastern half of the country. A century
and a half later, the infant Netscape adopted essentially the
same strategy, storming into an unformed market and planning,
in less than a year, to dominate it. In both cases, the pioneer's
hold on technology was tenuous—O'Rielly had "borrowed"
freely from the patent held by Samuel Morse, and Netscape's
core technology was owned, at the time of its creation, by the
University of Illinois—but ownership at this stage was less important than speed. The pioneers saw the allure of the new frontier and were determined to be there first.

As the technology matures, however, and markets widen, a
demand for property rights is liable to emerge. Having carved
out positions along the frontier, the more established pioneers
no longer want to work in chaos or cavort with pirates. Instead
they want to own the market they now consider theirs, to control the dominant technology and keep interlopers at bay. They
want property rights, in other words, and some means of enforcing them. This is one of the most common demands voiced
during the phase of creative anarchy, and one of the most important. For if property rights are not established, the pioneers
will tire of their labors before long and hopes of large-scale
commerce will tumble back toward anarchy.

In some cases, the concern for property rights is complicated
by what economists term the "problem of the commons." In
these cases, the creation of a new market rests with the use of a
particular resource, one that, like the oceans or the airwaves, is
large but far from infinite. In the early stages of development, the
resource is plentiful enough to serve all comers. Every radio en-
thusiast in the 1910s, for example, could transmit signals to his
heart's content, just as the first settlers in New England could
pull endless streams of cod from the banks of Newfoundland.

The more people who arrive on these frontiers, however, the harder it becomes for any of them to flourish, for the airwaves get congested over time and the fish run out. In these situations, the more established settlers will again petition for property rights, seeking to regain the solitude that success demands.

Meanwhile, two other problems tend to lurk along the technological frontier. These are problems of coordination and competition, and each can foster anarchy.

Consider first the problem of coordination. When a technology is initially evolving, bursts of innovation will tend to produce multiple devices and systems. Each individual inventor will produce his or her model of a telegraph, or his or her particular software package. And for some period of time, these disparate versions can happily coexist: customers simply use whichever model makes most sense for them, or is easiest or cheapest to employ. Eventually, though, this buffet of options breaks down and customers begin to value standardization over specificity. For what good is a telegraph, after all, if it receives messages only from certain sources? Or a mobile phone that can call only a discrete set of numbers? If these technologies are to develop into full-fledged markets, they need to develop some set of common standards, some means of coordinating their systems and allowing users to migrate freely among them. The problem, though, is that standards do not emerge by themselves. And thus as the pioneers race into a new market, each armed with his or her own technology and vision of the future, they may actually exacerbate the coordination problem that they desperately need to solve.

A final problem of the frontier concerns competition. Suppose that in the midst of anarchy a single pioneer manages—through law, guile, or sheer ingenuity—to control the key technologies for a given market. Then suppose that the same pioneer also manages to impose his own technical standard across this market. At this point, the problems of property rights and coordination will have been solved. The dominant firm can now banish outlaws from its field, spread a single technology across an unfolding market, and reap the financial benefits of scale. But the price of this solution is monopoly—putting the levers of a new and potentially vast market under the control of a single firm. And even if such a solution makes economic sense, it is bound to be a political and commercial disaster. Every pioneer will resent this intrusion; every erstwhile pirate will seek revenge. Even consumers are likely to join the fray, protesting against an undue concentration of power and seeking some formal means of redress. To the aggrieved parties, then, monopoly is not a solution to the problems of anarchy but a problem all its own. It is a problem of dominance and control; a problem of innovation (for monopolists are not supposed to innovate); and a problem, at its heart, of justice. And like all good problems, it creates chaos in its wake and demands an appropriate solution.

If commercialization is the most romantic stage along the technological frontier, then, creative anarchy is the most frustrating. The technology is maturing, the market is widening, but the nascent industry is caught by the inevitable struggles of its own success. Resolving these problems becomes the next and final phase of the frontier.

Phase Four: Rules

This last phase is the most difficult to imagine from the viewpoint of Partenia. When a technology is new, it usually looks so radical, so untamable, that those closest to its creation can't conceive of it being governed. This is particularly true—as with oceanic trade, radio, or cyberspace—when the technology reveals a space that, for practical purposes at least, hadn't been there before. How could anyone in Europe ever hope to impose order on the vast and unruly seas? How could anyone own the air? Or patrol the reaches of cyberspace? During the innovation and commercialization phases, the very idea of governance seems absurd. What occurs during the phase of creative anarchy, though, is critical; for it is here that even the pioneers begin to realize the costs of chaos. And once they realize these costs, once they understand that a lack of rules can diminish their own financial prospects, they begin to lobby for what they once explicitly rejected. Admittedly, it is not always the pioneers who clamor for rules. Sometimes it is the state, and sometimes a coalition of societal groups affected by the new technology and the market it has wrought. In general, though, rules get created because private firms want them.

In many ways, therefore, *Ruling the Waves* is a book about rules. It is a book about why rules get established along the technological frontier, and who plays the greatest role in their creation. The book does not suggest that there is a single path along this frontier. On the contrary, all of the stories that are presented here suggest that rules get made in different ways and by different people, depending on where the technology is developing and what kinds of problems emerge during its evolution. In some cases, the state steps in during the earliest days of anarchy, writing rules to control the new technology before it has time to develop. When the power of radio became clear, for example, state officials in Germany, France, and Russia scrambled at once to channel and constrain it. They sponsored the work of chosen pioneers, linked the newly formed companies directly to the state, and then used a series of international conventions to parse the radio spectrum and establish technical standards. The rules worked, for the most part, to the advantage of Europe's radio firms, but they were set incontrovertibly by the state.

In other cases, firms play a much bigger role in setting the rules that bind them. Frequently, for example, firms resolve the problem of coordination by forming industry associations or standards arrangements. These are private groups, composed of erstwhile pioneers, who get together to pursue what eventually becomes a political agenda. They craft rules and technical standards because their businesses demand them, and because governments, operating at their less-than-breakneck pace, may not yet be able to provide them. This, for instance, is how the U.S. telegraph firms tried to address the mass of incompatible wires that had emerged by 1850, and how music firms today are approaching the threat of MP3 technologies. Note that none of these standard-setting exercises is overtly political. They don't involve governments or voting or lobbying. Yet they are all about politics—about who gets power and sets the rules. By setting standards, firms begin to define what is permissible in a certain industry and what is not. They write rules and draw lines without the intervention of the state.

Eventually, though, governments tend to reenter the scene. There is nothing necessarily malicious in their reentry, nor anything inherently anti-business. It's just that the time and technology are ripe for their return. During the third phase of technological development, firms are content to regulate themselves and generally competent to do so. They can form associations and join standard-setting bodies and create some basic rules for their members to follow. As the industry evolves, however, these governance functions become increasingly unwieldy and unproductive. Firms don't really want to waste time dealing with associations of more and more members, nor do they want to police unruly counterparts or erect cumbersome voting structures. Most critically, firms don't have the ultimate ability to punish misbehavior or create property rights, for these are things that only governments can legitimately do. Firms can establish the rudimentary rules of their game. They can lay out technical standards and norms of business practice; they can consider issues of intellectual property and create common platforms. But they can't enforce these rules, since enforcement—at least in the modern era—is one of the few tasks explicitly reserved for government. It is also something that can be hugely important for business.

Consider, for example, the trials that befell the British East India Company. Chartered by Queen Elizabeth I in 1600, the company was given responsibility for British trade with the spice-laden islands of the Indian coast. It was a tremendous job, and the company profited handsomely from it. As part of its mission, however, the company also encountered the pirates who then plagued the Indian Ocean. Company officials tried for a while to deal with the pirates on their own; like legions of traders before them, they sailed with guns and in convoys and engaged in their own forms of piracy. Yet unlike many of the earlier traders, the East India Company was by this point a significant organization, with a myriad of deals and details to contend with. Its officers simply did not want to add defensive military action to their already large portfolio. So they went back to the British government and demanded protection. And the government, which for centuries had dealt half-heartedly with the pirates, eventually complied. It expanded its naval presence and upheld a firmer definition of illegal behavior, and finally cut the pirates out of the Indian trade. A private firm in this case thus used the state to preserve its own commercial empire.

This is a fairly common pattern along the technological frontier. We see it in the telegraph industry, where Britain's Eastern Telegraph Company relied on the state to protect its overseas interests and discourage new entrants, and in radio, where Telefunken, a leading German firm, grew to global prominence on the back of state protection and support. We see it already in cyberspace, too, as firms that otherwise decry government involvement scurry to secure whatever rules support their own interests. The very same encryption firms that have circumvented U.S. export controls, for example, have also lobbied the U.S. government to protect their intellectual property in foreign markets. Likewise, Netscape and Oracle, prominent pioneers of the information economy, have lobbied fervently for the U.S. government's antitrust suit against Microsoft. One firm's constraints, after all, are its rivals' competitive edge.

This relationship is the central irony of politics along the technological frontier. When technologies first emerge, there is a rush away from governments and a surge of individualism. Pioneers want to live along the cutting edge, forging a path away from the state and building empires in the air. Like Gaillot they are rebels, using technology to escape from authority and create their own versions of Partenia. Over time, however, the rebels tend to return to the state—not because they change their minds or lose their nerve, but simply because the state can secure the empires that they've built. The state can defend firms' property rights; it can regulate their interaction with a demanding consumer market; and it can help to keep the pirates at bay. And in fact, if we look at the cycle that prevails along the technological frontier, we see that even many of the original pirates eventually jump to the side of order, prospering from the same rules they once disdained. Sir Francis Drake, for example, was a privateer long before he became a knight (Queen Elizabeth I fondly referred to him as "her pyratte"), and Rupert Murdoch was commonly referred to as a pirate during his early years of traitorous dealings and legal gymnastics. The trick for pirates, it appears, lies in knowing when to jump, and to whom. Drake and Murdoch both became considerably more conservative once their business empires were well secured. They both also had powerful political patrons: Drake his Queen Elizabeth and Murdoch, Margaret Thatcher. There are lessons here for other pirates.

In the spring of 1997, a lively fight broke out during an otherwise sedate Harvard conference. The conference concerned Internet governance, and the participants ran the spectrum of political and commercial views. Early in the day one of the speakers, an Internet entrepreneur and sometime philosopher, gleefully pronounced the end of governments. In cyberspace, he argued, there would be no way for governments to track illegal activity, no way for them to print the money that defined their control, and no way for them to collect the taxes that permitted them to exist. In cyberspace, therefore, people like him could do whatever they wanted, and there was nothing government could do to prevent it. This speech incited the usual round of academic harumphing and noticeable unease in the audience. Then another of the speakers erupted—a gentleman who worked as one of the U.S. government's top Internet policymakers. "You," he bellowed, "are completely wrong. 'Cause we still have black helicopters!" Matters disintegrated from then on, with neither the bureaucrat nor the entrepreneur willing to concede the other's argument or back away from the fight. It was a good fight, in many ways, and an intriguing one. For essentially both men were right. Governments are more limited in cyberspace than they are in the physical world; in the short term, at least, governments will not be able to regulate, or even track, the information that flows across the Internet's myriad paths. That's why Partenia is such a powerful place. But as the Net matures, and as its technologies march further along their own frontiers, governments are likely to return—not only with black helicopters, but also with standards, and property rights, and the order that even the unruliest of pioneers is eventually bound to desire.

That, at least, is the lesson that history appears to offer. The frontier is a wild place, a land of anarchy and endless dreams. It draws pioneers and emboldens pirates and often showers riches on them both. In the end, though, power doesn't flow necessarily to those who stake their claims or guard their turf: it goes to those who make the rules. History shows us how this process unfolds along technology's edge.

CHAPTER I

The First Wave

Piracy is the first stage of commerce.

HENRI PIRENNE

Prince Henry of Portugal was a patient man. Determined to send Portuguese sailors around the Cape of Bojador, he spent years of his life fiddling with navigational instruments and ship design, trying to find some means to push the sailors farther out into the Atlantic and still bring them back safely. In the early years of the fifteenth century (the years of Henry's reign) this was no easy task. Though innovations in the fourteenth century had already allowed ships to determine their course with decent accuracy and sail during overcast weather, sailing in Henry's time remained a daunting prospect. The ships were slow and bulky, dependent on the weather, and essentially incapable of sailing across the open oceans. Henry wanted to change that. So he toiled for years, eventually using the capital that royalty had bestowed upon him to open a school for navigation in Sagres, the first of its kind.

In the end, his patience paid off. By 1457, Portuguese sea-
men had reached the elusive Cape of Bojador and sailed down
the coast of Africa as far as the Cape Verde Islands. Henry was
amply rewarded with slaves and gold from Africa, possession of
Cape Verde, and the appellation that suits him still: Henry the
Navigator.

It wasn't until after Henry's death in 1460, though, that the
great conquests took place, the conquests that emerged from a
nautical revolution and pushed European sailors all the way
across the Atlantic and eventually around the world. Triangular
sails replaced square ones, hulls changed shape, and inventors
perfected navigational devices that enabled ships to go farther
and farther without losing accuracy. These technologies, combined with a growing determination to find the spices and gold
of the fabled Indies, finally enabled European sailors to leave
the relative security of the Mediterranean and head out to the
open seas.

By the seventeenth century, transoceanic voyages had become commonplace. No longer just the province of the death-
defying explorer, the Atlantic was instead a lucrative trade route,
the passage not to spices, as it turned out, but to gold and slaves
and colonies. The seas were full of ships from Spain, Portugal,
England, and France, jostling each other and racing toward the
riches. It was a classic scramble along the technological frontier,
and a massive one. For between the fifteenth and seventeenth
centuries, advances in navigation had opened a new world of
commerce. Throughout this period, a steady stream of innovation had enabled European explorers to push into what for
them was terra incognita, unknown territory and virgin land
that extended along the tropical vastness of Africa, across the
forests of North America, and into the richness stretching from
the Gulf of Mexico to Tierra del Fuego.[1] In the early days, the
Europeans who sailed to these lands were on a combined mission of adventure and greed. They came to push the boundaries
of scientific knowledge and to see what really lay where dragons
used to lurk. They came to be the first to come, and to find the

---

[1] Clearly these territories were not truly virgin land, inhabited as they were by a number
of indigenous peoples. Since the Americas were *perceived* by the arriving Europeans,
though, as open territory, the chapter relies on prevailing convention and refers to these
lands as part of the "New World."

spices and gold that earlier traders had carried back from the East. And they came in relatively small, greatly heralded, numbers: Bartolomeu Diaz, Amerigo Vespucci, Ferdinand Magellan. These were the true pioneers, the ones who trusted the new technologies before they were tested and who sailed, quite blindly, to uncharted seas.

As the seventeenth century unfolded, the pioneers gradually gave way to those of a more commercial bent. These were the traders and fortune-seekers who plied the seas in ever-increasing numbers, looking to stake a claim in the New World or exploit the lucrative commodities—the gold, the furs, the slaves, the sugar—that the first wave of explorers had revealed. They were still pioneers, this second group, and still sailing into rough and untried seas. But the motive was more distinctly economic now and the scramble more obvious. As the technology of long-distance sailing improved, hundreds—and then thousands—of Europeans set out for sea, hoping to seize the riches of this seductive new world and craft a business (or at least a fortune) from them. The prospects for commerce soared, and the markets of Europe were quickly awash in a flood of consumer delicacies. It was during this period that the great trading companies emerged—groups such as the British East India Company and the Hudson's Bay Company, which carved out wide swaths of commercial territory and became, according to some historians at least, the world's first multinational enterprises.[2]

By the time these merchants set sail, however, they were no longer alone. Instead, the seas were filled by this point with pirates—with rogue sailors, or "freebooters," whose business lay in seizing merchant ships and grabbing whatever might be on board. It was a fairly lucrative business at this time, and a crowded one. It was also devilishly difficult to stop, or even define. For what was the difference, really, between a crew that ripped gold from an Incan temple and one that looted it from the hull of a Spanish ship? Or between pirates, who raided on their own behalf, and privateers, who raided on behalf of the state? What, indeed, was the difference between private looting

[2]For an intriguing argument that multinationals actually existed considerably before the formation of the trading companies, see Karl Moore and David Lewis, *Birth of the Multinational: 2000 Years of Business History* (Copenhagen: Copenhagen Business School Press, 1999).

and national conquest? In the raucous days of the seventeenth century, it was very hard to tell.

In retrospect, then, the advent of piracy was just a natural movement along the technological frontier. Once Henry and his followers made ocean crossings relatively secure, the seas were open to all. There were no property rights in the middle of the Atlantic, and no one to enforce them even if they had existed. There were no sheriffs there either, and no political boundaries. Instead, both the sea and the lands to which it stretched were perceived at the time as a truly open frontier, blessed with incomparable riches and bereft of any sort of rules. And thus the pirates simply took advantage of a classical gap between technology and law. They went to sea, reveled in the chaos, and grabbed whatever they could.

Before long, these roving bands of raiders became nearly as frightening as the dragons and demons that were once thought to reside in the depths. They menaced nearly all commercial traffic on the open seas, often killing their victims on the spot or throwing them overboard. During their heyday in the seventeenth and early eighteenth centuries, colorful and menacing figures such as Blackbeard and Bartholomew Roberts controlled key ports in the Caribbean and an entire island off the African coast. They sank hundreds of trading ships and plundered untold billions in cargoes and gold. For merchants, this harassment was more than merely unpleasant. It also substantially raised the risk, and thus the cost, of all oceangoing voyages. As long as pirates ruled the seas, commerce would be curtailed. And in order for commerce to expand into a sustained and predictable international trade, the pirates had to be stopped.

Eventually, as we know, they were. In a process that unfolded over more than a century, newcomers gradually sailed onto the seas and helped to end the scourge of piracy. Much of the initiative came, ironically, from the great merchant companies—companies that had acted for decades as pirates in their own right, sponsored by state governments to engage in war and plunder as well as commerce. Initially, it was these companies that helped to confuse the distinction between pirates and privateers, and that comfortably mixed business with plunder. It was these companies that were arguably the biggest pirates of them all, the ones that had thrown entire countries under their own iron rule. But as their commercial empires grew, even these behemoth firms became less and less interested in pursuing a military agenda, and less and less willing to pay for one. They didn't want to pursue pirates anymore; they didn't want to fund their own navies; they didn't want to expend their own resources administering vast tracts of land. They simply wanted to trade.

And so, over time, the companies took their concerns directly to the state, lobbying for naval support and international agreements. Had these demands been voiced a century earlier, the states would most likely have been either unsympathetic or impotent. They would have claimed, with justification, that they weren't strong enough to defeat the pirates, or that their limited coffers couldn't afford to wage such a never-ending war. By the middle of the seventeenth century, however, Europe's fledgling nation states were stronger and more confident. Bolstered in part by the riches that seeped out of the New World and onto their shores; enriched in many cases by state-sponsored trading in slaves or gold or fur, they simply had more resources by this point, and a greater sense of their own power. Accordingly, in the 1600s, countries such as Great Britain, France, and Spain began to build permanent navies and pass maritime legislation. They began to fund their own fighting forces, instead of relying on the shady world of privateers, and to create laws that defined piracy more precisely and allowed state governments to maintain some kind of jurisdiction over conduct at sea. Together, this combination of law and enforcement was eventually enough to allow the states to suppress the bulk of the pirates. Some of them remained, of course, hidden in enclaves or cloaked in admiral's garb, but by the mid-eighteenth century, pirates no longer ruled the seas. States did instead, and great merchants such as the British East India Company made vast and legitimate fortunes from what once was pirate loot.

In our own time, the story of high seas piracy is distant enough to seem almost quaint. It is a tale of legendary figures such as Blackbeard and Sir Francis Drake, and of battles now reduced to cartoon memories. It is also, though, a tale that maps out particularly well along the technological frontier, tracking the complex evolution of commerce and the gradual transition from anarchy to rules.

The Technology

Until the fourteenth century, the shipping trade in Europe stayed exceedingly close to the shore.[3] There had long been shipping across the lucrative routes of the Mediterranean and along the North Sea, as merchants plied these waters laden with cargoes of wool and wine and metals. But for centuries, their trade was constrained by the basic dangers that adhered to their craft. The ships of this period were light and crude, scarcely different from those that had sailed the same waters during the days of the Roman Empire and the early Middle Ages. They were good for fighting and relatively good for carrying cargoes, but, for various reasons, they rarely sailed more than several miles from the coast.

Part of this reluctance was due simply to fear and ignorance. Because few people had ventured beyond the familiar waters of the Mediterranean, most sailors presumed that dangers lurked in these waters—dragons and demons that preyed upon the unwary and "whirlpools [that] destroy any adventurer."[4] Maps during this time customarily trailed off beyond the western boundaries of northern Africa, marked from this point with the all-purpose warning *Cave! Hic Dragones!*[5] Because merchants had not in recent memory sailed beyond this point, there was no trade with whatever lands might lay in that direction and

---

[3]Some naval historians have argued that, contrary to popular perception, ancient craft did not actually sail close to the shore, since it is in coastal waters that ships are most likely to encounter dangerous conditions such as sandbars and treacherous tides. Yet it still seems clear that even if these ships did not directly hug the coast, they also did not frequently venture too far from shore or out into the open ocean. See E. G. R. Taylor, *The Haven-Finding Art: A History of Navigation from Odysseus to Captain Cook* (New York: American Elsevier Publishing Company, Inc., 1971), pp. 3–4.

[4]Quoted in C. Raymond Beazley, *Prince Henry the Navigator* (New York: G. P. Putnam's Sons, 1894).

[5]The phrase translates as "Beware! here be dragons!" I have borrowed this reference from Susan Strange, "Cave! Hic Dragones: A Critique of Regime Analysis," in Stephen D. Krasner, ed., *International Regimes* (Ithaca: Cornell University Press, 1983), pp. 337–54. For more on mapmaking during this time, see J. H. Parry, *The Discovery of the Sea* (New York: The Dial Press, 1974), pp. 49–68.

thus no commercial interest in journeying toward them.[6] Even within the Mediterranean, meanwhile, pirates were a constant threat, and merchant ships customarily sailed along heavily traveled routes and in heavily armed convoys.

The greatest reason for sticking so close to shore, however, was simply the technology that prevailed during this period. For centuries, any sailor who turned to the seas was essentially traveling blind: once he left the contours of the coast or the landmarks that might extend several miles into the sea, he was left in a vast and indistinguishable space, with no way of knowing where he was or how to return home. There were stars, of course, and the predictable movements of the sun, but these features were so large and distant that they offered early mariners only the crudest tools of navigation. Nor did they work on cloudy days or during storms, leaving sailors stranded in what the Vikings called *hafvilla*, or the loss of all direction at sea.[7] Until the fourteenth century, therefore, trade was severely constrained by the vagaries of weather and the limits of navigation. Merchants refused to sail during the harsh months of winter; they went only in large groups and along the coast; and they rarely ventured too far from their home port. Venetian ships, for example, swooped across the Adriatic and Ionian seas but didn't cross through the Straits of Gibraltar.[8] And the Hansards, merchants

[6]The great historical exception here were the Vikings, aggressive sailors and marauders from the north of Europe who controlled much of the territory around the North and Baltic seas in the early Middle Ages (780–1070) and briefly established settlements as far west as Newfoundland. Most of these western settlements, though, proved quite short-lived, and the Vikings themselves ceased their maritime adventures sometime around 1200, as the climate of northern Europe grew colder and drifting ice blocked many of their traditional sailing routes. For more on this period, see Samuel Eliot Morison, *The European Discovery of America: The Northern Voyages, A.D. 500–1600* (New York: Oxford University Press, 1971); James R. Enterline, *Viking America* (Garden City, NJ: Doubleday, 1972); Gwyn Jones, *A History of the Vikings* (London: Oxford University Press, 1968); and David M. Wilson and Peter G. Foote, *The Viking Achievement* (London: Sidgwick & Jackson, 1970).

[7]See Joel Mokyr, *The Lever of Riches* (New York: Oxford University Press, 1990), p. 47.

[8]During the medieval period, Venice was the western world's preeminent military power and commercial center. For more on this time and Venice's role in maritime trade, see Frederic C. Lane, *Venice: A Maritime Republic* (Baltimore: The Johns Hopkins University Press, 1973); Robert S. Lopez, *The Commercial Revolution of the Middle Ages, 950–1350* (Englewood Cliffs, NJ: Prentice-Hall, 1971); Robert S. Lopez and Irving W. Raymond, *Medieval Trade in the Mediterranean World* (New York: Columbia University Press, 1955); and Irene Birute Katele, "Captains and Corsairs: Venice and Piracy, 1261–1381," Ph.D. dissertation, Department of History, University of Illinois at Urbana-Champaign, 1986.

who sailed from Baltic towns like Hamburg and Lübeck, ended their southern voyages at France's Bay of Biscay.[9] To cross a boundaryless ocean was simply too dangerous.

Around the turn of the fourteenth century, however, a nautical revolution gradually emerged. It was a slow revolution, and an invisible one for some time—a revolution that might better be termed an evolution, or even just a protracted series of shared technological advances. But it was a significant break with centuries of established technology, and it finally enabled Europeans to leave their coasts and cross the seas.

Much of this revolution had to do with changes in the basic construction of ships. Until this point, most European ships were built using "clinker-planking" technology, in which a series of overlapped wooden planks kept water out of the vessel. In the early fourteenth century, however, clinker-planking gradually gave way to caravel construction, a method that replaced the overlapped planks with a simple skeleton and waterproof caulking. With less wood to support, these ships—epitomized by the Portuguese caravels that Columbus and Magellan later used—were larger and lighter than their predecessors. They were also, as a result, cheaper to sail and generally more seaworthy.[10] Other advances in ship-building around this time included the trireme, a ship outfitted to hold three rowers on each bench, and the cog, a high-sided vessel with a square sail and stern rudder.[11] Both of these developments led to the construction of relatively larger ships that were able to sail with relatively smaller crews.

The major breakthroughs, though, came in navigation. In the middle of the thirteenth century, a series of anonymous scribblers and mathematicians slowly began to improve a technique known

---

[9]Henri Pirenne, *Economic and Social History of Medieval Europe*, trans. by I. E. Clegg (New York: Harcourt, Brace & Company, 1956), p. 90.

[10]See Joel Mokyr, *The Lever of Riches: Technological Creativity and Economic Progress* (New York: Oxford University Press, 1990), p. 46; and Nathan Rosenberg and L. E. Birdzell, *How the West Grew Rich: The Economic Transformation of the Industrial World* (New York: Basic Books, 1986), pp. 80–86.

[11]For more on these developments, see Frederic C. Lane, *Venice: A Maritime Republic* (Baltimore: The Johns Hopkins University Press, 1973), pp. 122–24; and Lane, *Venetian Ships and Shipbuilders of the Renaissance* (Baltimore: The Johns Hopkins University Press, 1934).

as dead reckoning, a mathematical means of estimating distances from one port to the next. As these guesses were tested and then improved, their creators compiled them into "port books," or *portolanos*, which allowed sailors to use the estimated plots in their own travels. Sometime around 1250, the best of the estimates were compiled into a book covering the whole Mediterranean—the first scientifically rigorous map of the seas. This was the beginning of systematic sailing, of applying math and engineering to what had previously been a physical, nearly an intuitive, pursuit.[12]

The next breakthrough started earlier, but wasn't widely applied until about 1410.[13] In 1180, an English monk named Alexander Neckham first described a device that approximates our modern compass. It was a simple device, a magnetized needle set floating in a bowl of water, but, as Neckham noted, it resolutely swung round to the north, helping mariners "when [they] cannot see the sun clearly in murky weather, or at night."[14] However, mariners for several centuries refused to rely very heavily on the compass, since they distrusted its strange magnetic pull and associated its power with necromancy, or black magic. Captains who relied on compasses during this period usually did so only at night or in secret, when crews would be less likely to see the device and accuse the captain of devil worship.

---

[12]In China, a similar rigor appeared much earlier. Cartography was an important pursuit from the time of the Chin dynasty (265–420) and early mapmakers strove to develop a system of coordinates based on the vertical and horizontal lines of a rectangular grid. In ancient Greece, Ptolemy had also championed the science of geography and worked out an early scheme for calculating latitude and longitude. His works disappeared for hundreds of years, however, and were only revived in Europe in the early fifteenth century. See Daniel J. Boorstin, *The Discoverers* (New York: Random House, 1983) pp. 111–13; 149–53; and Joseph Needham, *Science in Traditional China* (Cambridge, Mass.: Harvard University Press, 1981).

[13]All dates here are estimates, since precise records of technological adaptation during this period do not exist. For more information about the use of the compass during medieval times, see Lloyd A. Brown, *The Story of Maps* (Boston: Little, Brown & Company, 1950), pp. 126–34; E. G. R. Taylor, *The Haven-Finding Art* (New York: American Elsevier Publishing Company, Inc., 1971), pp. 92–96; Frederic C. Lane, "The Economic Meaning of the Invention of the Compass," *The American Historical Review*, vol. 68, no. 3, April 1963, pp. 605–617; and Barbara M. Kreutz, "Mediterranean Contributions to the Medieval Mariner's Compass," *Technology and Culture*, vol. 14, no. 3, July 1973, pp. 367–83.

[14]Alexander Neckham, *De Utensilibus*, quoted in Boorstin, *The Discoverers*, p. 221.

By the start of the fifteenth century, however, fear of the compass had yielded to a begrudging respect for its powers. Slowly, mariners saw how it could make their voyages safer and, more important perhaps, open the Mediterranean during those times of the year when weather had previously rendered it impassable. Emboldened, they also began to turn to other technologies—to the astrolabe, the cross-staff, the quadrant—that emerged from the compass, and from a gradually growing belief that navigation could be treated as a science as well as an art.[15] Much of this innovation came from a series of unknown tinkerers and sailors, whose individual contributions were probably rather small and whose names have been lost to history. But much of it culminated in a single great wave, a burst of innovation that rolled out of Portugal in the fifteenth century and then rapidly and unyieldingly traversed the western world.

The Pioneers

Born in 1394, Prince Henry of Portugal did not fit the image of the seabound pioneer. The third son of Portugal's King John I (also known as John the Great or John the Bastard), Henry was ascetic and intense, torn between a desire to crusade for Christendom and a passion for exploration. In 1415, Henry and his two older brothers led a short but spectacular attack on Ceuta, a Muslim trading post on Africa's northernmost tip. In a single day, the brothers and their forces overwhelmed the city, massacring the inhabitants and seizing piles of precious goods: spices, rugs, gold, and jewels. This was Henry's first glimpse of African trade, and his first taste of the riches that stretched southward from Portugal. He stayed behind after the attack, waiting for Muslim reprisals and carefully watching the practice of local trade.

---

[15]For more on these technologies, see J. H. Parry, *The Discovery of the Sea* (New York: The Dial Press, 1974), pp. 155–70; and T. K. Derry, *A Short History of Technology* (Oxford: Clarendon Press, 1960), pp. 205–206. For an argument that the development of guns and naval cannons were also critical aspects of the Europeans' naval expansion, see Carlo M. Cipolla, *Guns and Sails in the Early Phase of European Expansion* (London: Collins, 1965).

Henry returned from Africa eager to crusade again, and to capture the nearby island of Gibraltar. But his father refused permission, and Henry retreated, alone and upset, to the southernmost tip of Portugal. It was here, from a desolate promontory called Sagres, that he launched a concerted expedition into the unknown. His motives, it appears, were a grab bag of theology and commerce, mixed with an overwhelming desire simply to discover what lay across the seas. According to one contemporary chronicler, the prince set off from Sagres "to find out what lay beyond Cape Bojador, for quite practical purposes (in order to find the gold-supplying areas of West Africa); to develop new trade routes with Christian peoples; to determine the limits of Moorish political power; to evangelize among pagans; to ally with any Christian rulers . . . ; and to 'fulfil the predictions of his horoscope.' "[16] The latter is especially interesting, for apparently astrologers had already divined that "this Lord should toil at high and mighty conquests, especially in seeking out things that were hidden from other men and secret."[17]

In search of this fate, Prince Henry began with maps, accumulating the *portolanos* that existed at this time and hiring Jehuda Cresques, a Catalan Jew already renowned for his cartographic skills. The fact that Cresques was Jewish was not incidental to his task, for Christian mapmakers were obliged by their religion to follow the biblically inspired maps that prevailed during the Middle Ages—maps that emphatically depicted a dish-shaped world with Jerusalem at its center; three geometrically arranged continents, Europe, Africa, and Asia; and an ocean (the "Great Outer Sea of Boundless Extent") surrounding the masses of land. In visual terms, these maps represented a theology as well as a geography: the world was centered around Jerusalem and the seas led only to Paradise, a forbidden

---

[16]From Gomes Eanes de Zurara, *The Chronicles of the Discovery and Conquest of Guinea,* trans. by C. Raymond Beazley and Edgar Prestage (London: The Hakluyt Society, 1869), p. 30. For a more complete discussion of Henry's complex mix of motives, see Edgar Prestage, *The Portuguese Pioneers* (London: A. & C. Black Ltd., 1933), pp. 165–72.

[17]Gomes Eanes de Zurara, quoted in Boorstin, *The Discoverers,* p. 161. There are a number of excellent biographies of Prince Henry. See, for example, Beazley, *Prince Henry the Navigator*; and Prestage, *The Portuguese Pioneers.*

land that no living man could ever reach.[18] As a Jew, Cresques was freer to abandon these maps in favor of a more rigorous, realistic depiction. And Henry, the erstwhile crusader, emphatically pushed him along.

At Sagres, Henry oversaw a burgeoning operation of sailing and information-gathering. Sailors were instructed to journey farther and farther along the nearby coast of Africa, meticulously recording their travels on navigation charts, which were then pieced together under Cresques's watchful eye. Explorers or traders who learned of the Portuguese efforts offered their tidbits as well, helping to compile what slowly became one of the world's first organized research efforts. Where his forebears had tended to be conservative and famously superstitious, Henry had a scientist's mind and an innovator's passion. He experimented boldly and continuously, urging his sailors to test a range of emerging technologies, including the compass, the quadrant, and the astrolabe. When the sailors reported back, Henry would integrate their findings into the work at Sagres, refining the technology and constantly generating new pieces of equipment. Over time, Henry and his assistants at Sagres were responsible for many of their era's most important navigational advances, including refinements in the art of mapmaking and early measures of latitude.[19] They also made significant improvements in shipbuilding, using the caravel technique to build small, highly maneuverable boats with a critical asset. Unlike earlier boats, which were primarily designed to sail large cargoes along the prevailing winds, Henry's caravels could also sail quite easily *into* the wind and thus were uniquely able to return home.[20]

Armed with this crucial piece of security, Henry sent his ships farther and farther along the African coast. He had no specific plan or goal in mind—only, as one fifteenth-century chronicler wrote, "to begin and to carry out very great deeds . . . to know the land that lay beyond the isles of Canary and that

---

[18]See Brown, *The Story of Maps*, pp. 81–112; George H. T. Kimble, *Geography in the Middle Ages* (London: Methuen & Co. Ltd., 1938), pp. 19–43, 181–204; Boorstin, *The Discoverers*, pp. 101–04, 162; and Bailey W. Diffie and George D. Winius, *Foundations of the Portuguese Empire, 1415–1580* (Minneapolis: University of Minnesota Press, 1977), pp. 118–121.
[19]Taylor, *The Haven-Finding Art*, pp. 151–62.
[20]See Boorstin, *The Discoverers*, pp. 163–64; and Prestage, *The Portuguese Pioneers*, p. 332.

Cape called Bojador."[21] Geographically, Bojador was not such a tremendous obstacle; there were no straits to navigate or dramatically clashing currents. But psychologically, it loomed huge. For according to ancient rumor:

> [B]eyond this Cape there is no race of men nor place of inhabitants . . . and the sea [is] so shallow that a whole league from land it is only a fathom deep, while the currents are so terrible that no ship having once passed the Cape, will ever be able to return . . .[22]

Bojador was where the sea slipped into terra incognita, where no man was supposed to tread. For years, it had been an impassable end to ocean travel. For Henry it became an obsession.

Between 1424 and 1434, Prince Henry sponsored fifteen expeditions to Bojador, which sat only 180 miles beyond Cape Nun, the traditional limit of navigation at this time. Several got close, but none was able to round it, each expedition citing a different obstacle in its path. And then, in 1434, a squire named Gil Eannes finally disembarked. When he returned, and word was spread, Henry pushed even faster down the African coast and into the unknown sea. In 1445, one of Henry's expeditions sailed south to Cape Verde, the continent's westernmost point; in 1456, another discovered the Cape Verde Islands and journeyed inland along the Senegal and Gambia rivers.[23]

Henry died in 1460—still at Sagres, still alone, and clad (according to contemporary reports) in a hair shirt. None of his expeditions had ventured anywhere close to the southern tip of Africa, and none had strayed farther west than the Azores. Yet Henry's contribution to navigation and maritime trade was inestimable. He was the first to make sailing a rigorous art; the first to research the seas; the first to send ships to the purported end

---

[21]Gomes Eanes de Zurara, *The Chronicles of the Discovery and Conquest of Guinea*, p. 27. See also Beazley, *Prince Henry the Navigator*, pp. 160–78.
[22]Gomes Eanes de Zurara, *The Chronicles of the Discovery and Conquest of Guinea*, p. 31. For more on the obstacles as they were perceived at the time, see Prestage, *The Portuguese Pioneers*, pp. 54–55.
[23]There is some dispute as to the precise year in which the Cape Verde Islands were discovered, but most historians of the period place it between 1457 and 1462. See Prestage, *The Portuguese Pioneers*, pp. 130–45.

of the world. Henry never actually did much sailing himself, but
he was a true pioneer of the seas—a man who ventured, in spirit
and intellect at least, where none had gone before.

After Henry's death, a new generation of explorers scrambled
to continue his work. Portuguese sailors reached the western
horn of Africa in 1471, the southern tip of Africa in 1488, and
the western coast of India in 1498, revealing with each trip that
the edges of the sea were indeed navigable and that great riches
lay beyond them. As word of their exploits spread, others shed
their ancient fears and took to sea as well, eager to match the
Portuguese voyages and discover riches of their own.

The most famous of these voyagers, of course, was Christopher Columbus, the Genoese sailor whose westward journey
across the Atlantic was rejected by King John of Portugal in
1484. Fully convinced that the fastest route to China and the Indies lay to the west, Columbus then approached Spain's King
Ferdinand, who sponsored the grossly miscalculated voyage
in the hope of stemming Portugal's relentless advance. King
John was right, as it turned out, about the miscalculation, but
King Ferdinand was the happy beneficiary of the mistake. For
when Columbus stumbled upon the unfortunately named "New
World," he opened vast territories for exploration and conquest—territories that were even richer than the real Indies and,
oddly enough, easier to reach from Europe. He also opened the
Atlantic itself, transforming it in European minds from the edge
of the world to the beginning of a new one. By the turn of the
sixteenth century, the Atlantic was awash in dueling ships and
voyages, each racing to stake a claim in the New World and
build an empire—commercial, political, or religious—upon it.

The first of these pioneers were the true adventurers, men
such as Columbus, Amerigo Vespucci, Vasco Nùñez de Balboa,
and Ferdinand Magellan. Like Prince Henry, they sailed mostly
to find and to conquer, to determine the shape of the lands that
lay west of Europe and claim them for their king. Gold and
riches were motives as well—Columbus, after all, was searching
for a cheaper and faster route to the East's fabled treasures—but
they weren't primary. Rather, the earliest pioneers were determined to see how the seas stretched out from Europe and what
lay across them. And by the 1520s, just a century after Henry's

experiments at Sagres, they had largely succeeded: in 1502, Vespucci followed the eastern coast of South America; in 1513, Balboa crossed the Isthmus of Panama to reach the Pacific; and in 1521, Magellan sailed around the tip of South America into the Pacific and then, at last, to Asia. The shape of the world was finally mapped, and the sea lanes were open for transit.

There was one other piece of the technical puzzle, though, and one other pioneer who helped transform the New World into a raging commercial frontier. The technology was the map, and the pioneer was a Belgian cartographer named Gerardus Mercator. In the early days of the sixteenth century, explorers such as Balboa and Magellan sailed with a combination of technology, intuition, and luck. The maps they had were crude, and still revealed the ancient lines of Christian dogma. There was no precise measurement of longitude or latitude, and there were few accurate representations of others' voyages. Indeed, with the exception of the freely circulating *portolanos,* sea charts were seen during this time as state secrets—information to be hoarded, guarded, and saved. When Gutenberg's printing press began to spread across Europe, however, maps became a sizzling commodity. Suddenly, an enterprising corps of mapmakers could package information about trading routes and sea lanes and sell it to an eager crop of mariners. There was now a market for maps and a technology that suited it perfectly. Mercator took this market and technology and made them both much better.

Until this point, many European maps still centered the world around Jerusalem and dropped off perilously just west of Portugal. They also made no deference to the sphericity of the earth, depicting it simply as a round but flat surface. Mercator's innovation was to figure out how to portray a round surface on a flat map. His answer was what is still known as a Mercator projection, a map that divides the earth along the vertical lines of longitude and then lays the strips aside each other to form a standard rectangle. Toward the poles of the earth, of course, the Mercator projection becomes increasingly inaccurate: Greenland is too wide, for example, and the Scandinavian countries look larger than they actually are. But the core of a Mercator projection is correct, and the maps represented a quantum leap in precision. Armed with these maps and an expanding body of knowledge, even less skillful or experienced sailors could set out to sea, following the pioneers to the purported riches of the New World.

And so they did. Between 1500 and 1550, hundreds of expeditions set sail across the Atlantic. Some were launched by individual adventurers; some were motivated by evangelical hopes or scientific interest; but most were sponsored by the monarchies of Europe and motivated by riches. The details of these expeditions are well known and need no repetition here: there is Francisco Pizarro's bloody conquest of Peru; Ferdinand Cortés's seizure of Aztec gold and Mayan lands; the early Spanish settlements in Hispaniola and Cuba. What is critical to note is how easy the seafaring part of these journeys had become by the middle of the sixteenth century, and how overwhelming the desire for gold. When Pizarro went to Peru, for example, he did not even speak of spreading Christianity or of discovering new lands. "I have come," he said simply, "to take away from them their gold."[24] Bernal Diaz, a member of Cortés's expedition, was only slightly more politic: "We came here," he later described, "to serve God and also to get rich."[25]

And so they did. By the middle of the sixteenth century, Spanish expeditions were swarming across the heart of South and Central America, and the treasures of the Americas were streaming, brightly and even faster, back toward Spain.

The Pirates

Meanwhile, of course, the other nations of Europe grew increasingly restless about Spain's expanding empire and envious of her gold. Convinced now that the riches of the New World might be as valuable as the spices of the Indies, the nations that had fallen behind in the search for new territories— and particularly France and England—began by the mid-1500s

---

[24]Quoted in H. G. Koenigsberger and George L. Mosse, *Europe in the Sixteenth Century* (New York: Holt, Rinehart & Winston, Inc., 1968), p. 208.
[25]Ibid.

to challenge Spain's monopoly on transatlantic shipping. Some of this challenge took place on the already scarred battlefields of Europe. Some of it took place in formal sea battles, with the growing navies of France and England harassing Spain's formidable armada. But much of it also took place in smaller, private battles and assumed the guise of piracy.

Piracy, of course, was as old as the seas itself. In ancient Greece, *peirates* had plagued early shipping across the Aegean and Ionian seas; in ancient Rome, even Julius Caesar had been captured by the widely reviled *pirata*. Pirates had menaced Venetian shipping during the height of that city's Mediterranean empire and plagued the isolated coasts of England, Scotland, and Ireland. Defined vaguely as rogues or plunderers, pirates lived off the maritime trade of others. They lurked aside all the world's most profitable shipping lanes, chasing the ships that crossed their way and grabbing whatever they could. Though its extent has varied greatly over time, piracy has been a constant companion to maritime trade—so much so, in fact, that one historian of piracy begins his work by lamenting, "To write a complete history of piracy from its earliest days would be an impossible undertaking. It would begin to resemble a maritime history of the world."[26]

It is, therefore, not surprising that when passage of the Atlantic expanded, the pirates followed suit. What is remarkable about this period, though, is just how widespread the piracy was, and how easily it moved between the boundaries of public and private trade. In the middle of the sixteenth century, the typical pirate was a low-born adventurer. Hailing from Portsmouth or Broadhaven, perhaps, the pirate was generally poor by birth, young, uneducated, and nearly always male.[27] He went to sea because other opportunities were few, and because he was lured—like all pirates and many pioneers—by the prospect of wealth; of gold in this case, and jewels, and the legendary loot that was flooding across the Atlantic and into the coffers of

[26]Philip Gosse, *The History of Piracy* (New York: Tudor Publishing Company, 1934, republished by Gale Research Company, Book Tower, 1976), p. vii.
[27]There were, however, some notorious female pirates whose lives have been vividly described in Gosse, pp. 202–206; and Frank Sherry, *Raiders and Rebels: The Golden Age of Piracy* (New York: William Morrow, 1986), pp. 266–77.

Spain. Many of these pirates were devout individualists, loners who had foresworn the nationality of their birth and claimed allegiance only to their ship. They made their own rules at sea, enforced their own brand of discipline, and treated their victims with legendary brutality. Although pirates rarely (if ever) made their captives "walk the plank," they regularly dispensed far harsher modes of punishment: captains were tortured, sailors were forced to join pirate crews, and prisoners were abandoned to die on deserted slips of sand.[28]

To the sailors who bore the brunt of their attacks, such pirates were the scourge of the seas. They were vicious rogues and marauders, the fiends who had turned the seas unsafe again and contributed to the cost of oceangoing trade. They were, as all pirates, the "scum and off-scourings of mankind,"[29] and they needed to be stopped.

Yet, despite repeated pleas from legitimate traders and royally sponsored expeditions and a nearly universal concern about the violence and costs of maritime attacks, piracy continued to flourish and expand in the sixteenth and seventeenth centuries. Pirate ships infested the English Channel in the latter half of the sixteenth century; they swarmed through the Indian Ocean in the seventeenth century; and they dominated Caribbean shipping until around 1730.[30] But the states did little to stop them. On the contrary, one of the most striking features of this age of anarchy is the extent to which national governments either turned a blind eye toward piracy or actively supported it. In En-

---

[28]There are a number of books that describe—often in excruciating detail—how pirates conducted their daily lives and enforced their crude brand of discipline. See, for example, Hugh F. Rankin, *The Golden Age of Piracy* (New York: Holt, Rinehart & Winston, Inc., 1969), pp. 22–42; and David Cordingly, *Under the Black Flag* (New York: Random House, 1995). For a fictional, but historically accurate, account, see Nicholas Griffin, *The Shark Requiem* (London: Little, Brown & Company, 1999).

[29]From the description in Rankin, *The Golden Age of Piracy*, p. 7.

[30]Each of the pirate holdouts had its own specific dynamic and characteristics. Indian Ocean piracy, for example, was dominated by a local tribe, the Joasmees, until late in the seventeenth century, when they were joined by English pirates based out of Madagascar or the American colonies. Early Caribbean piracy was the stronghold of "buccaneers," a motley crew of mostly French and Dutch pirates who targeted Spanish shipping and established their own "pirate republic." For more on the history of piracy and the distinctions between various pirate bands, see Gosse, *The History of Piracy*; Rankin, *The Golden Age of Piracy*; and Sherry, *Raiders and Rebels*.

gland, for example, where piracy had been defined as treason since 1413, only 106 pirates were hanged between 1558 and 1578, a time of rampant piracy and frequent capital punishment. In 1578, when a particularly energetic government brought nine hundred pirates to trial, only three were actually sent to the gallows.[31]

Part of the problem lay, understandably, in a gap between geography and law. When Henry and his associates set out to sea, they were truly sailing into terra incognita. There were no rules that applied across the vast oceans and no sense of who, if anyone, could form them. What monarch could actually claim sovereignty over the Atlantic? What rules applied to, say, a Dutch sailor aboard a French ship skimming through the Azores? It simply wasn't clear. And even if it had been, the rulers of six-teenth-century Europe would have been hard-pressed to im-pose their laws at sea. While most states had navies by this point and Spain had built its fearsome armada, most of these fleets were occupied with European squabbles, and few could match the speed and power of the pirates' own craft.

An even bigger part of the problem, however, lay with the states' own interests. Far from persecuting the pirates or even prosecuting them, the governments of Spain, France, and En-gland all chose instead to back them, using piracy in many cases as a substitute for policy. Throughout this period, for ex-ample, nearly all of the European powers engaged in what was known as privateering, commissioning private ships to redress a maritime wrong and allowing them, as payment, to seize the cargoes of their prey.[32] Although privateering began as a simple form of reprisal, by the sixteenth century it had become a wide-spread and low-cost method of political harassment. Rather than strengthening their own navies or expending public funds, governments would grant open-ended licenses to privateers,

---

[31] See Neville Williams, *The Sea Dogs: Privateers, Plunder and Piracy in the Elizabethan Age* (London: Weidenfeld & Nicolson, 1975), p. 150.

[32] For a description of how privateering operated, see Donald A. Petrie, *The Prize Game: Lawful Looting on the High Seas in the Days of Fighting Sail* (Annapolis: Naval Institute Press, 1999). For a more theoretical examination, see Francis R. Stark, "The Abolition of Privateering and the Declaration of Paris," Ph.D. dissertation, Faculty of Political Science, Columbia University, 1897.

allowing them to prey freely on foreign ships and seize whatever they could. Frequently, these privateers were men who only recently had "retired" from the pirate trade, turning in the sordid garb of one profession for the respectable robes of another. Everyone knew that pirates and privateers were effectively identical; everyone knew that some of the era's most heralded men began their careers as pirates; but in the freewheeling days of the 1500s, such subtleties were often discarded.[33]

Consider, for example, the case of Francis Drake, born of a preacher around 1540 and a pirate most of his life. Drake began his sailing career at the age of thirteen, when he was apprenticed to a small vessel sailing between North Sea ports. At twenty-three, following convention for a young man in his position, he signed on for three trading voyages to the West Indies, sailing under the command of a relative named John Hawkins. Hawkins, who was already well known in political circles, had made his reputation by skirting Spain's commercial monopoly in the Atlantic. Starting in 1562 and backed by some of London's financial elite, he had pioneered what would come to be known as the "triangle trade," capturing slaves along the western coast of Africa, selling them to Spanish settlements in the Caribbean, and taking hides or other cargo back to England in return. Officially, Hawkins's dealings in both Africa and the Caribbean violated the standing trade monopolies of his time, for the Portuguese claimed this part of Africa as their own, and Spain possessed the Caribbean. During his early voyages, though, the authorities turned a blind eye, and merchants in both Guinea and Hispaniola were happy to accept the English seller's business.[34]

In 1568, Drake accompanied Hawkins on the third of his voyages, commanding a fifty-ton ship called the *Judith*. As the ship prepared for its final trip back across the Atlantic to England, Spanish forces launched a surprise attack, burning or de-

---

[33]For an excellent analysis of how states legitimated piracy and used it to their own ends, see Janice E. Thomson, *Mercenaries, Pirates and Sovereigns: State-Building and Extraterritorial Violence in Early Modern Europe* (Princeton, N.J.: Princeton University Press, 1994).

[34]Hawkins was also helped in this regard by England's fragile alliance with Spain at this time. See Williams, *The Sea Dogs*, pp. 32–43.

stroying most of Hawkins's small fleet and killing or enslaving dozens of men. Drake slipped away from the battle in the dead of night and then returned to England—a hero now, and filled with what one account describes as "an inexhaustible lust for revenge."[35]

In 1571, Drake obtained his own privateering commission and set off again for the Americas. He made his first hostile visit in 1572, attacking a Spanish settlement at Nombre de Dios and raiding caravans across the Isthmus of Panama. Bolstered by roughly £30,000 in booty, he returned to the sea in 1577, leading an expedition of five ships through the Straits of Magellan, along the western coast of America, and then westward around the world. Along the way, Drake and his crew plundered whatever they could: wines and silk from Cape Verde, gold from the ports of Chile, and thousands of silver pesos from their most infamous target, the Spanish treasure ship *Nuestra Señora de la Concepción*, known below-decks as the *Cacafuego*. When he returned to England on the *Golden Hind*, Drake carried a cargo worth an estimated £2.5 million, an unprecedented fortune for his time. He repaid his backers with forty-seven times their initial investment, and funneled much of the proceeds directly to the Crown. In 1581, Queen Elizabeth knighted "her pyratte" and Drake returned to the seas.[36]

The Demand for Rules

If we date the age of exploration from the Portuguese exploits of the late 1400s, we see a subsequent age of piracy that stretches well into the early days of the nineteenth century. During this time, vast new territories were discovered and won, tremendous fortunes were wrested from them, and whole new markets—in rum, sugar, and, tobacco—were born.

---

[35]Samuel Eliot Morison, *The Southern Voyages* (New York: Oxford University Press, 1974), p. 634.
[36]For more on Drake's career, see Kenneth R. Andrews, *Drake's Voyages: A Re-Assessment of Their Place in Elizabethan Maritime Expansion* (London: Weidenfeld & Nicolson, 1967); James A. Williamson, *The Age of Drake* (London: Adam & Charles Black, 5th edition, 1965); and Samuel Eliot Morison, *The Southern Voyages*.

It was truly a time of discovery and anarchy, a time when old rules bowed down before new worlds and pirates ruled the waves. The era lasted for an exceedingly long time and must have seemed to be a permanent state of affairs. But in the end, as we know, it wasn't. By the middle of the seventeenth century, anger against the pirates had grown louder and more persistent. Residents of the New World were tired of having their commerce disrupted by pirate attacks; seafaring travelers were aghast at the dangers the pirates posed; and merchants—many themselves descended from piracy—began to petition for redress. As these complaints wound their way through the political core of Europe, they slowly but critically changed political views toward piracy and ultimately led to its demise.

An initial set of demands came from a constant and predictable quarter. Ever since the days of its earliest expeditions, the Spanish government had been urging its European counterparts to help put an end to piracy. In 1560, for example, the Spanish ambassador to London launched a lengthy complaint on behalf of several gentlemen who had been seized for ransom at sea; and in 1618, a later ambassador beseeched King James to punish Sir Walter Raleigh, a favorite English "sea dog," for his attacks on Spanish territory.[37] Such pleas, of course, were largely self-serving. Since Spain already controlled the major shipping routes between Europe and the Americas, and since Spain actually claimed full control over the New World, Spanish ships were the overwhelming target of pirate attack. Indeed, throughout this period, it was the eastward-sailing Spanish galleon, laden with gold or silver ripped from the mines of South America, that ignited other nations' envy and fueled the pirates' dream. What made matters even more infuriating was the infamous Treaty of Tordesillas, a line drawn by the Pope in 1494 to split the New World between competing Spanish and Portuguese claims. According to the Pope's decree, Spain controlled all territories west of the Cape Verde Islands—or, in practice, all of what subsequently became known as the Americas. This arbitrary splitting of the world's lands had caused a tremendous amount of resentment among the other Europeans and was largely responsible for their own attempts at privateering. And

[37]King James complied and ultimately executed Raleigh.

thus, when Spain cried out against piracy, her words fell mostly on deaf ears. Instead, the French, Dutch, and British governments were largely if quietly behind the pirates who preyed upon the Spanish Main, using them as implicit instruments of war against Spanish trade.

Matters changed slightly after 1587, when Sir Francis Drake led twenty-three ships and two thousand men into Cadiz harbor. Catching the Spanish by surprise, Drake and his forces ransacked or burned more than twenty vessels and destroyed roughly £1 million of cargo.[38] In reprisal, Spain sailed for England the following spring, bringing a refurbished armada of 132 ships and igniting the largest battle that had ever before been fought at sea.[39] When it was over, half the Spanish fleet had been sunk or destroyed and the other half, exhausted and dispirited, limped back home. This defeat of the Spanish Armada was a turning point in European history and in the story of maritime trade. For after 1588, Spain's monopoly over Atlantic trade was effectively gone. This meant that ships from other nations could sail to the New World without having to dodge the blasts of Spanish cannon; it meant that traders could do business in the Caribbean or along the isthmus of Central America without having either to hide from local authorities or kill them. It meant, in other words, that states could increasingly rely on merchants, rather than on pirates or privateers, to conduct their trade.

Accordingly, between 1603 and 1609 the newly installed King James I of England issued a series of proclamations that restricted any sort of "Depredations upon the Sea," and declared all privateering contracts null and void[40]; in 1681, France imposed stringent regulations upon its privateers, demanding,

---

[38]The figures here are slightly contentious. Official Spanish reports listed losses of twenty-four vessels and cargo worth 172,000 ducats (about £750,000). The monetary losses, though, were certainly higher, and Drake himself claimed to have destroyed or taken thirty-seven vessels. See Julian S. Corbett, *Drake and the Tudor Navy*, vol. II (London: Longman's Green & Co., 1898), p. 81.

[39]For more on the battle and events leading up to it, see H. G. Koenigsberger and George L. Mosse, *Europe in the Sixteenth Century* (New York: Holt, Rinehart & Winston, Inc., 1968), pp. 266–68; Corbett, *Drake and the Tudor Navy*, pp. 60–107; and James A. Williamson, *The Age of Drake* (London: Adam & Charles Black, 1965), pp. 275–303.

[40]See David Delison Hebb, *Piracy and the English Government, 1616–1642* (Brookfield, Vt.: Ashgate Publishing Company, 1994), p. 162.

for instance, that all privateers post expensive bonds and that they not ransom any prizes above a certain value.[41] There was a certain irony to these proclamations, as well as a distinct air of frivolity to them. Both England and France continued to employ privateers for roughly two centuries after these pronouncements and James, in particular, was able to scorn the privateers only as a result of the riches and power that his predecessor, Queen Elizabeth, had already built as a result of their efforts. Yet these seventeenth-century pronouncements did signal a changing state of mind in Europe. As continental governments grew stronger and more confident, as they developed standing navies and open trading relations, they needed the pirates less and less and resented them more and more. Gradually, like Spain, they began to demand some kind of rules against piracy and some cooperation to this end from neighboring states.

The most important demands, however, came not from Spanish monarchs nor even the broader constellation of European powers. They came instead from the commercial classes and from the economic interests that were increasingly influencing state policy by the middle of the seventeenth century.[42] Until this point in time, piracy was essentially a tool for economic advancement. It was a way for men to make their fortunes and build great wealth, to escape the confines of a class-bound Europe and build new empires on unclaimed lands. It was also a way, surprisingly enough, that did not antagonize many of Europe's established interests, since all the wrangling and all the gains and losses occurred far beyond Europe's shores. As ocean-going trade expanded, though, and particularly once Spain's monopoly was ripped open, commerce on the seas became a much steadier, much more stable pursuit. It became, in other words, a business rather than an adventure, and it began to spawn large and well-funded commercial entities.

The first of these entities, and by far the most important, were the great commercial trading companies: the British East

---

[41]Thomson, *Mercenaries, Pirates, and Sovereigns*, p. 24.

[42]At the end of the seventeenth century, as one historian notes, "most rich men were merchants" and they "formed a wealthy, influential and vocal group which was well organized and to which the government and the [British] Admiralty had to pay careful attention." Patrick Crowhurst, *The Defence of British Trade 1689–1815* (Folkestone, Kent: Wm Dawson & Sons Ltd, 1977), p. 43.

India Company, the Dutch United East India Company, and the Hudson's Bay Company. Established by European governments in the early 1600s, these enterprises were revolutionary in many respects. They pioneered new forms of legal and financial risk-sharing, allowed for an unprecedented accumulation of capital, and created a unique combination of public and private interests.[43] Their stated purpose, though, and their driving force, was trade: trade with the New World, trade with the Indies, trade that would carry the banner and prestige of the state behind it.[44] To this end, states bestowed their trading companies with a rash of privileges not normally ceded to private firms. The Dutch United East India Company, for example, was granted the power "to make war, conclude treaties, acquire territories and build fortresses,"[45] while the Hudson's Bay Company received "the absolute right to administer law and to judge all cases, civil or criminal, on the spot. It was empowered to employ its own armies and navies, erect forts and generally defend its fiefdom in any way it chose."[46] To enhance their competitive position, trading companies were also generally given the exclusive right to trade with a particular region.[47] And because they

[43]There was some variation across countries, with the Dutch companies coming the closest to purely private ventures; the French and Portuguese operating much more like state enterprises; and the English falling somewhere in between. See Thomson, *Mercenaries, Pirates, and Sovereigns*, p. 33. For more on these enterprises and the differences between them, see George Cawston and A. H. Keane, *The Early Chartered Companies, A.D. 1296–1858* (New York: Edward Arnold, 1896); John P. Davis, *Corporations: A Study of the Origin and Development of Great Business Combinations and Their Relation to the Authority of the State* (New York: G. P. Putnam's Sons, 1905), pp. 61–156; and E. L. J. Coornaert, "European Economic Institutions and the New World: The Chartered Companies," in *The Economy of Expanding Europe in the Sixteenth and Seventeenth Centuries*, ed. E. E. Rich and C. H. Wilson, vol. 4 of *The Cambridge Economic History of Europe* (Cambridge: Cambridge University Press, 1967).

[44]Some of the companies had a more distinctly military rationale. According to Thomson, for example, the Dutch West India Company was established not for commerce, but "for preying on Spanish shipping [and] privateering against the enemy's American empire." See Thomson, *Mercenaries, Pirates, and Sovereigns*, p. 36. She is quoting here from Bernard H. M. Vlekke, *Evolution of the Dutch Nation* (New York: Roy Publishers, 1945), p. 177.

[45]Ramakrishnan Mukherjee, *The Rise and Fall of the East India Company: A Sociological Appraisal* (New York: Monthly Review Press, 1974), p. 59; cited in Thomson, *Mercenaries, Pirates, and Sovereigns*, pp. 10–11.

[46]Peter C. Newman, *Company of Adventurers* (Ontario: Penguin Books, 1986), p. 119.

[47]In England, the granting of monopoly privileges often became a contentious issue and, after the Glorious Revolution of 1688, all monopolies had to be confirmed by Parliament. See Thomson, *Mercenaries, Pirates, and Sovereigns*, p. 34; and Cawston and Keane, *The Early Chartered Companies*, p. 66.

often were the only citizens of their country to operate in a given region, they typically assumed some kind of administrative responsibility as well, acting in many cases as the de facto representative of the state.

As the trading companies emerged and expanded, they came to exert a major influence over world affairs. They spread western commerce across the world and became over time a frequent instrument of imperial rule.[48] By many estimates, they were little more than pirates themselves—far-flung concerns whose leaders had no compunction about resorting to violence and thievery in pursuit of their own goals. Clearly, in the early days, such allegations rang true. In 1612, for example, a captain of the British East India Company seized several Indian trading ships and then forced their owners to ransom them back; and in 1621, the Dutch United East India Company took the Banda Islands by force, executing local leaders and taking the citizens as slaves.[49] As the companies settled into the lands they had attacked or persuaded, however, they tended to behave more like rulers than marauders. They established permanent outposts, for instance, appointed governors, and generally settled in for what would become an extended period of colonial rule. In the process, and without any apparent sense of irony, these erstwhile conquerors became major foes of piracy and a central reason for the pirates' decline. For once the trading companies were up and running, once they had their own trade routes to protect and monopolies to preserve, the costs and the risks of piracy were simply too great to bear.

The troubles began in 1692, when a pirate ship out of Rhode Island sailed to the Indian Ocean in search of plunder.[50] After

---

[48]There is a vast literature on the political and economic impact of the trading companies, most of which falls far beyond the scope of this book. For more information, see John Keay, *The Honourable Company: A History of the East English Company* (New York: Macmillan Publishing Company, 1991); Brian Gardner, *The East India Company* (New York: Dorset Press, 1971); Newman, *Company of Adventurers*; and Gary Anderson and Robert D. Tollison, "Apologiae for Chartered Monopolies in Foreign Trade," *History of Political Economy*, vol. 15, no. 4 (1983), pp. 549–66; and the five volumes in Patrick Tuck, ed., *The East India Company: 1600–1858* (London: Routledge, 1998).

[49]Holden Furber, *Rival Empires of Trade in the Orient, 1600–1800*, vol. 2 of *Europe and the World in the Age of Expansion*, ed. Boyd C. Shafer (Minneapolis: University of Minnesota Press, 1976), pp. 44–45.

[50]Officially, the boat was a privateering vessel, commissioned by the governor of Bermuda to attack French shipping. Once at sea, however, the captain convinced his

wandering aimlessly for thousands of miles, the crew of the *Amity* happened upon a Mogul ship and, within minutes, had seized a dazzling cargo of spices, silks, ivory, and more than £100,000 in gold and silver coins. Once the crew returned to New England, news of their plunder spread rapidly, and hundreds of other ships determined to follow in their wake. Before long, the Indian Ocean was awash with pirates, all of whom were looking for easy loot, and nearly all of whom spoke English. Initially, these new neighbors did not cause great concern for the British East India Company, since the pirate ships were more interested in other cargo and many were reluctant to attack English ships. In 1695, however, a pirate ship under English command attacked the *Gang-I-Sawai,* the prized ship of Aurangzeb, leader of the Indian Mogul Empire. Even by pirate standards, it was a heartless attack: nearly the entire Indian crew was tortured and killed, and dozens of women aboard were gang-raped or drowned. When the Great Mogul learned of the attack, he was livid with rage and demanded reprisal from officers of the British East India Company. Although the officers claimed (correctly, in this case) that they had no connection to the pirate attack, nor even to any of the pirates then plying Indian waters, their words had little impact on the infuriated emperor. In his eyes, the pirates were English; the company was English; and the subtleties between them were unimportant.

The emperor's fury put the British East India company in an extremely tenuous position. For even though the company had largely pirated its own way into the Indian subcontinent, it was still on that land as part of a negotiated agreement with the Mogul empire.[51] If the emperor turned against them, the company would lose its commercial advantage. From this point on, therefore, the British East India Company became a major foe

---

crew that it would be far more profitable to head to the Indian Ocean and attack Muslim cargoes instead. For a colorful account of this voyage, see Sherry, *Raiders and Rebels,* pp. 22–34.

[51]No formal treaty of commerce was signed between the company and the Mogul empire, but in 1618 the company procured assurances of "good usage of the English" and an agreement for favorable conditions of trade at Surat. Subsequent agreements and settlements granted the company trading and factory-building privileges in specific areas of the Mogul empire. See Sir William Foster, *England's Quest of Eastern Trade,* vol. I of Patrick Tuck, ed., *The East India Company,* pp. 286–87, 314–23.

of piracy and a strong proponent of military intervention. Repeatedly, the company and its officials would plead with the
British government for some formal operation against the Indian Ocean pirates and some means of attacking their bases. In
the aftermath of the *Gang-I-Sawai* attack, for example, they
asked London to send the Royal Navy to India or, at a minimum, to allow company officials to prosecute local pirates on
their own. "If there be not care taken to suppress the pyrats,"
wrote a company official stationed in Bombay, "your Honours'
trade in India will be wholly lost."[52] When the government declined to send a naval squadron, the company tried another
tack, requesting this time that the government at least take
stronger measures to enforce its own Navigation Acts, which
were designed to stem independent shipping out of the American colonies. Such pleas became louder and more insistent as
the British East India Company begrudgingly agreed to protect
Mogul shipping with its own convoys, a policy that quickly led
to direct hits on the company's merchant fleet.

To be sure, officers of the British East India Company were
not the first to have their lives and livelihoods threatened by pirates. Nor were they the first to lose precious cargoes or face
prohibitively high insurance rates. They were, however, the first
commercial players with the scope and clout to make their concerns heard. When earlier British traders encountered pirates at
sea, the British government had either turned a blind eye or, at
most, enlisted privateers. When Spanish ships were accosted
and plundered during the course of the sixteenth century, Spain
could only hope either to negotiate (a largely futile task at the
time) or to call out the armada. By contrast, when the British or
Dutch East India Company suffered attack at the turn of the
eighteenth century, the British and Dutch governments had
more tools at their disposal and more reason to intervene.[53]

---

[52]Quoted in Keay, *The Honourable Company,* p. 188.
[53]During the same time, ironically, the British East India Company was also undergoing
intense scrutiny in London, where political and commercial forces had rallied against
what had clearly become a highly profitable monopoly. In 1698, Commons granted a
charter for a new trading company, or "General Society." After years of squabbling both
in London and the East, this "New" East India Company merged with and was subsumed
by the old one. See the description in Keay, *The Honourable Company,* pp. 170–217.

Moreover, as transoceanic trade passed out of its tumultuous phase and into a period of calm and consolidation, it wasn't just the major companies that felt comfortable petitioning the state, nor only the most powerful ones that were able to find redress. Instead there were a growing number of concerns—in cotton, tobacco, cloth, and rum—that depended on trade to turn a profit and were therefore quick to seek protection whenever pirates threatened their course. As early as 1608, for example, an English trading company had petitioned King James for relief against the Barbary pirates, a particularly virulent band based along the northern coast of Africa.[54] Another group of London companies made a similar plea in 1617[55]; and in 1626 merchants pushed Parliament to impeach the Lord High Admiral, arguing (among other grievances) that he had failed to protect British trade.[56]

On the other side of the Atlantic, meanwhile, the move to deter piracy took longer. Indeed, for most of the seventeenth century, North American traders (and many colonial governors) were quietly profiting from European piracy. An infamous "pirate republic" in Madagascar was largely supplied by aspiring New England merchants, while Caribbean pirates such as Blackbeard and "Calico Jack" Rackham sought frequent and well-acknowledged refuge in the low-lying coves of North Carolina. Apparently, these were relationships that served both sides well. The pirates used the colonies as a convenient haven and lucrative market, and the colonists—or at least some among them—used the "sweet trade" to evade Britain's commercial rule and build their own fortunes. Once again, though, the birth of widespread commerce led before long to a shift of attitudes. By the end of the seventeenth century, the colonies boasted a growing number of merchants, men who no longer needed the pirate trade or wanted to be associated with it. Between 1689 and 1713, for example, legitimate exports from New

[54]Hebb, *Piracy and the English Government*, p. 15. The Crown responded by working with the merchants to fund an ill-fated naval expedition against the Barbary pirates.
[55]Hebb, *Piracy and the English Government*, p. 7.
[56]Roger Lockyer, *Buckingham: The Life and Political Career of George Villiers* (London: Longman, 1981), pp. 308–325. See also the description in Hebb, *Piracy and the English Government*, pp. 202–203.

England to England grew from £31,254 to £60,000, while New
York's trade rose from £8,763 to £14,428.[57] During this same
time, though, the colonies also harbored a growing and more
rabid strand of piracy, men who scoured the eastern seaboard
from Caribbean bases and seized any cargo they could find.
This combination was not healthy for the pirates. For as piracy
mounted, even its most stalwart allies began to change their
views. In 1698, a group of New York merchants sailed four ships
out of Madagascar, full of black market goods. When pirates
seized three of these vessels, the merchants, themselves only
steps removed from pirates, became passionate opponents of
the pirate trade. A similar shuffle occurred just a year later in
South Carolina, where traders who had historically been quite
tight with the pirates suddenly found their own rice cargoes
under attack.[58] And thus, as one historian writes, "The more
colonial trade expanded, the more likely it was that the pirates
would attack it . . . and the more they turned on local trade, the
more the tension grew between the buccaneers and their old
merchant friends."[59]

In other words, the more prosperous the American colonies
became, the more they experienced the same change of heart
that had already afflicted the British East India Company and
other trading concerns. Once merchants or firms could earn
substantial profits from "honest" trade, they saw the pirates
only as obstacles. It didn't make a difference whether these
"honest" merchants had launched their own careers as pirates,
or if they had turned a blind eye to piracy for most of their lives.
Once their own trade was soundly established, the merchants
turned on the pirates and demanded rules to stop them.

The Supply

By the end of the seventeenth century, then, trade
was becoming too valuable to leave to the pirates. Where once

[57]Crowhurst, *The Defence of British Trade*, p. 111.
[58]Robert C. Ritchie, *Captain Kidd and the War against the Pirates* (Cambridge, Mass.:
Harvard University Press, 1986), p. 233.
[59]Ibid.

the seas had loomed as the final frontier, they now were a more predictable place: a space for commerce more than plunder. And in this space, full of larger interests and a growing social impact, pirates were unsavory. The American colonists didn't want them, the emerging corporations didn't want them, and the European states—more powerful now, and richer—no longer had much need for them.

The problem, though, was how eliminate this ever-present scourge. After hundreds of years of virtually free reign—after hundreds of years, in fact, of being quietly supported by their own governments, pirates were well entrenched in the culture of maritime trade, as well as in their coastal lairs. Getting them out would prove no easy task.

Initially, states responded to private concerns with a time-honored, cost-efficient solution. When private interests complained about piracy, states created private forces—or privateers—to stop them. Even when states realized that privateers weren't all that effective against pirates, and indeed that the constant ebb and flow of privateers was only exacerbating the problem of piracy, they still couldn't easily afford a formal solution. Navies were simply too expensive, and too often needed for other, more pressing concerns.

In 1698, however, the Treaty of Ryswick helped to free naval forces for pirate attack. Negotiated between England and France, the peace put an end to eight years of protracted struggle, with both sides essentially agreeing to halt hostilities, even though their differences remained unresolved. Once the agreement came into effect, both the French and British governments were able to direct their navies out of Europe and toward the pirates cruising in the Indian Ocean and along the Barbary Coast. Accordingly, in 1699 the British government sent four navy vessels to Madagascar, the acknowledged hub of Indian Ocean piracy. Finding fifteen hundred pirates adrift in St. Mary's harbor, the Royal Navy commander resisted a battle of any sort and instead sent an agent to offer all the pirates a formal amnesty if they would renounce their life of crime. Most apparently took the pardon, and took it honestly, for when the navy embarked on a year-long cruise of neighboring waters, it found not a single pirate ship.

The next salvo came in 1701, when the English government publicly hanged William Kidd, one of the era's most notorious pirates and a symbol of its own changing policies. In 1696, the government had replaced Benjamin Fletcher, the openly corrupt governor of New York, with the Earl of Bellomont, a strait-laced Anglo-Irish Protestant who had sworn to break the colony's link to piracy. Realizing that his funds to this end would be limited, Bellomont had agreed to sponsor a private expedition, backed by the English King William, to root out the Indian Ocean pirates. It was basically a privateering contract, supported at the highest levels and given to the "Trusty and well-beloved Captain Kidd."[60] It was also almost completely unfeasible, since the terms of Kidd's commission required him to destroy the entrenched pirates with a single ship; return a considerable share of the proceeds to his well-connected backers; sign his own crew; and then sail back to Boston less than fourteen months after his departure.[61] Once he was at sea, the monumental nature of this task must have become clear. With a mutinous crew, disappearing prey, and an ever-approaching deadline, Kidd turned pirate himself in the Indian Ocean. In the space of a year, he threatened a merchant vessel from the British East India Company's fleet, plundered and seized several other commercial vessels, and murdered a member of his own crew. When Kidd sailed into Boston, somehow convinced that he had merely followed orders, Bellomont promptly arrested him. Two years later, after a trial that often resembled a political and public spectacle, he was hanged by the side of the River Thames.

That Kidd died for his crimes is not surprising. He undeniably committed heinous acts, and had many witnesses to tell their tales. What was so noteworthy about his case was that many pirates, and certainly many privateers, had committed far worse crimes in the past yet received far less fanfare. Kidd, after all, was sailing in the service of the king and at the behest of several of England and America's most powerful men. When Drake returned from his Spanish conquests he was knighted;

---

[60]From King William's commission, cited in Sherry, *Raiders and Rebels*, p. 162.
[61]For these details and more, see the description in Sherry, *Raiders and Rebels*, pp. 149–95. For a more complete account, see Ritchie, *Captain Kidd and the War against the Pirates*.

after Henry Morgan marched through Panama in 1671—looting churches, torturing children, and using nuns as living shields—he was made lieutenant governor of Jamaica. But when Kidd came back, he lost his life.

In retrospect, of course, it is easy to see why. By the time Kidd sailed, the rules of the game had changed. Conduct that had been perfectly acceptable in the past, conduct that had in fact been quietly supported by firms and states alike, was suddenly deemed wholly unacceptable. By the turn of the eighteenth century, piracy had become a much more clearly defined crime, and even states like England—arguably the greatest supporter of piracy in the sixteenth century—were ready and eager to condemn men for it.

In the wake of Kidd's high-profile arrest, Britain and its colonies embraced their piracy statutes with renewed vigor. In 1699, the British Parliament passed the second Act of Piracy, which allowed pirates caught in the colonies to be tried locally and subject to admiralty law. The act also threatened to revoke the charter of any colony that "refused to co-operate in working this system."[62] In this spirit, Bellomont cracked down on pirates and their supporters in New England, seizing ships and dismissing Fletcher's old cronies. He was joined in this resolve by Francis Nicholson, the new governor of Virginia, who organized armed militias to track escaping pirates and even directed a coast guard ship during a successful pirate attack.[63] By 1701, according to best estimates, the transatlantic pirate trade had "all but withered away."[64]

Meanwhile, in the Caribbean, a frustrated group of merchants resolved to get rid of their own local menace, a particularly

---

[62]Neville Williams, *Captains Outrageous: Seven Centuries of Piracy* (New York: Macmillan, 1962), p. 142; cited in Thomson, *Mercenaries, Pirates, and Sovereigns*, p. 50.

[63]See the descriptions in Rankin, *The Golden Age of Piracy*, pp. 64–78; and Sherry, *Raiders and Rebels*, p. 197.

[64]The quote is from Sherry, *Raiders and Rebels*, p. 198, and repeated in Thomson, *Mercenaries, Pirates, and Sovereigns*, p. 51. It is further supported by Stark, "The Abolition of Privateering and the Declaration of Paris," pp. 66–68, and Rankin, *The Golden Age of Piracy*, p. 78. Technically, one of the factors that contributed to a decrease in piracy in the early eighteenth century was the outbreak of war between France and England in 1702. As was the usual case, many of the existing pirates at this time became lawful privateers during the conflict.

vicious cluster of pirates based on the island of New Providence, also known as Nassau. In 1717, they petitioned the British government to appoint a new governor of the island, an upright and well-respected captain named Woodes Rogers. The Crown, eager now to eradicate this last bastion of New World piracy, sent Rogers at once, accompanied by four small warships and one hundred soldiers. When the captain arrived, he faced a formidable task: Nassau was a tangled nest of piracy, full of hopeless alcoholics and some of the era's most notorious thugs: Blackbeard, "Calico Jack" Rackham, and Charles Vane. Against incredible odds, Rogers somehow managed to assert his authority over this motley group, and even to commission a ragtag army of "reformed" pirates. In just three years at Nassau, he rebuilt the island's fort, secured the once pirate-infested harbor, and defeated or killed most of the pirate kings. By 1721, when Rogers left Nassau, the pirates of the Caribbean were all but gone.[65]

The final round of blows came in a natural progression, when many of the pirates who had fled Nassau sought new refuge in Madagascar. Although the British had already demonstrated that they wouldn't permit the island to harbor those who disrupted its East Indian trade, the pirates apparently believed that the British government either had neglected its concern for the island or had lost the ability to police it. They were wrong. In 1721, just as the pirate chiefs were beginning to establish their new fiefdoms, four Royal Navy men-of-war sailed with considerable fanfare back toward Madagascar's waters. Meanwhile, the British government passed a series of laws that made it illegal for merchants to traffic with pirates or for sailors to refuse to resist pirate attack. Apparently, this combination of legal and military pressure was sufficient to cow the pirates of Madagascar. Before a single shot was even fired—indeed, even before the warships reached St. Mary's harbor—the pirates had scattered.

By roughly 1730, then, the golden age of piracy had shuddered to an end. Individual pirates still remained, of course, and a hard-core cluster that sailed from the Barbary Coast would

---

[65]A naval officer killed Blackbeard in 1718; Vane was captured and executed in 1719; and Rogers hanged Calico Jack in 1720. Unfortunately, Rogers never received either the fame or fortune that was rightly his. Instead, the captain (who had also rescued Alexander Selkirk, the model for Defoe's Robinson Crusoe) died bankrupt and unheralded. See, for example, Sherry, *Raiders and Rebels,* pp. 196–279.

not be fully eradicated until 1830.⁶⁶ But the bulk of the pirates—the pirates who had defied and defined Atlantic trade
nearly since its inception—were gone.

How did something that once seemed so permanent end so
resolutely? The answer, as we've seen, is that the rules changed.
From the time of Henry the Navigator through to Columbus
and Drake, the high seas were truly an open frontier. Locked for
centuries by geography and fear, the oceans were finally opened
in the fifteenth century, when technological breakthroughs in
navigation, shipbuilding, and mapmaking allowed the first wave
of pioneers to slip from Europe's shores and venture out to sea.
This wave was followed in rapid succession by a second and
broader wave of pioneers, men who lusted after land and riches,
and then by pirates who, like all good pirates, wanted to follow
others' roads to riches. Because there was no law at sea, and because the states that might have crafted such laws were themselves weak and nearly always at war, the seas remained an
anarchic realm for hundreds of bloody years. Eventually, though,
commerce caught up with anarchy. As the technological frontier
evolved and breakthrough technologies merged into everyday
practice, more and more people were able to take to the seas
and see their future in them: as settlers, as shippers, and as merchants. And in these visions, piracy was increasingly a problem
rather than a conduit. For once this final wave of pioneers had
staked their claim on seaborne commerce, they didn't want the
pirates in their way and decried the extra expense (in losses,
protection, and insurance⁶⁷) they caused. Accordingly, once
commerce at sea became widespread and well established, the

⁶⁶Between 1730 and 1830, individual nations launched a series of attacks against the
Barbary pirates. Thomas Jefferson led a U.S. attack in 1803, for example, and the Dutch
and British participated in a joint campaign in 1816. Yet because neither of these attacks
established any kind of permanent position, the pirates—backed by the relatively independent states that harbored them, continued to thrive. In 1830, however, the French
army invaded and occupied Algiers, putting a permanent end to the Barbary pirates.
For more on these events, see John B. Wolf, *The Barbary Coast: Algiers under the Turks,
1500–1830* (New York: Norton, 1979), pp. 323–38; Peter Earle, *Corsairs of Malta and Barbary* (London: Sidgwick & Jackson, 1970), pp. 265–71; and Christopher Lloyd, *English
Corsairs on the Barbary Coast* (London: Collins, 1981).
⁶⁷For more on the development of maritime insurance during this period, see
Crowhurst, *The Defence of British Trade,* pp. 81–103; and Virginia Haufler, *Dangerous
Commerce: Insurance and the Management of International Risk* (Ithaca: Cornell University Press, 1997).

merchants who stood behind this commerce mustered their resources against the pirates. The problem in this case wasn't congestion or coordination, or even competition; it was instead simply piracy. And the people who most suffered from this problem—the people who griped about it and demanded a solution—were those most anxious to protect their own property rights.

If the merchants were the ones to voice the problem, though, it was states that ultimately resolved it—states that, after centuries of ignoring and encouraging piracy, finally amassed the will and power to enforce their own laws. A key point in this transformation concerned the distinction between piracy and privateering—between pirating on one's own, in other words, and pirating for the state. Inherently, of course, these distinctions are vague, and two centuries of casual substitution had made them at times almost irrelevant. Before states could eradicate piracy, therefore, they first had to define it. This was the process that they launched in the seventeenth century and completed by about 1730. Throughout this period, states were newly vigilant in the pursuit of "real" pirates and much more careful about restraining and confining the privateers they employed: witness, of course, the case of Captain Kidd. These measures were extremely successful, and they put an end to piracy's most raucous phase.

But the formal end of piracy came even later, when states combined their forces and agreed to outlaw privateering. In April of 1856, representatives of seven European nations met in Paris.[68] After a week of debate and consideration, they signed at last the Declaration of Paris, proclaiming (among other laws) that privateering would henceforth be, and remain, abolished. The impact of this agreement was considerable. For within two years, forty-two of the world's nations, including most Western European countries and the United States, had acceded to the Declaration and relinquished the right to employ privateers.[69]

---

[68]The states were France, Britain, Russia, Prussia, Austria, Sardinia, and Turkey.

[69]Of course, not all states had acceded to the Declaration, and not all followed it completely. During the U.S. Civil War, for example, the Confederacy commissioned privateers, as did Chile in the Spanish-Chilean War of 1856–66. The exceptions, though, were relatively minimal. See Thomson, *Mercenaries, Pirates, and Sovereigns*, p. 76.

Over the next several decades, they continued to expand this international legal framework, constructing the principles of free navigation and neutrality at sea.[70] In the process, pirates became the stuff of legend and order came to rule the waves.[71]

And thus the cycle moved full circle. Though they once had been full of demons and gold, by the nineteenth century the seas had become just routes of transport—open to all, dominated by large-scale commerce, and ruled, albeit loosely, by the long arm of the state. Was this the world that Henry foresaw, that Drake admired, or that Blackbeard tried to thwart? We'll never know. But it was the world that the pioneers of navigation had created and one that the pirates, despite their best intentions perhaps, helped to build.

---

[70]See Stark, "The Abolition of Privateering," pp. 75–77. For more elaborate treatments of the subsequent development of an international law of the sea, see J. L. Brierly, *The Law of Nations* (Oxford: Clarendon Press, 6th ed., 1963), pp. 194–211, 223–40; and R. R. Churchill and A. V. Lowe, *The Law of the Sea* (Yonkers, New York: Juris Publishing, 1999).

[71]Some pirates, of course, escaped the rule of law and continue to do so to this day.

CHAPTER 2

The Codemakers

Canst thou send lightnings, that they may go,
and say unto thee, Here we are?

The Book of Job

Until the middle of the nineteenth century, information didn't move all that much faster than man. It moved the
way it had nearly since biblical times, hastened only by sporadic
improvements in transportation: stirrups for horses, for example,
or sails for ships. In 1830, a message sent from London to New
York or Bombay took almost as long to reach its destination as
it had in the days of Vasco da Gama and Magellan. It still had
to be written by hand, taken by coach to the nearest port, and
then shipped across the globe. Along the way, just about anything could happen—either to the message itself or to the circumstances it was supposed to report. As a result, the world
of 1830 was still very much a local one. People focused for the
most part on the news of their town or village, and concentrated
on business that could be conducted within several days' jour-

ney of home. To be sure, the oceangoing revolution of the late Middle Ages had opened vast new territories to Europeans' grasp, and the development of steam technologies presaged rapid gains in transportation by the end of the eighteenth century. But in communication terms, the world of 1830 differed only slightly from the world of 1380: it was unfathomably big, and it took seemingly forever to send information across it.

All of this changed in the course of less than two decades. In 1838, Samuel Morse presented the U.S. Congress with a prototype of his electric telegraph—a machine, he claimed, that could instantly send messages across the nation and potentially around the world. Most members of Congress laughed. Yet by 1852, only fourteen years later, twenty-three thousand miles of telegraph wire had been laid across the United States and the industry was booming. By 1880, almost 100 thousand miles of subterranean cable stretched under the North Atlantic, across the Red Sea, and beneath the Caribbean. The same message from London to New York that had taken weeks or months in 1838 could now be transmitted in minutes. Suddenly, the world had become a great deal smaller.

As telegraph use expanded, it generated the same kind of excitement that surrounds the Internet today. Observers argued that it was the greatest invention of all time, the crowning glory of technological innovation. "Of all the marvellous achievements of modern science," wrote two contemporary authors, "the Electric Telegraph is transcendently the greatest and most serviceable to mankind. It is a perpetual miracle, which no familiarity can render commonplace."[1] When the first transatlantic cable was constructed in 1858, celebrations broke out in England and the United States. "Since the discovery of Columbus," the *Times* of London gushed, "nothing has been done in any degree comparable to the vast enlargement which has thus been given to the sphere of human activity."[2] It was, echoed U.S. President James Buchanan, "a triumph more glorious, because far more useful to mankind, than was ever won by a

[1]Charles F. Briggs and Augustus Maverick, *The Story of the Telegraph* (New York: Rudd & Carleton, 1863), p. 13.
[2]Quoted in Tom Standage, *The Victorian Internet* (New York: Walker & Company, 1998), pp. 82–83.

conqueror on the field of battle."[3] The pioneers of telegraphy, both scientific and commercial, were feted around the world for their achievements and their contribution to global society. Telegraphy, many presumed, would change the face of both business and politics. It would shrink the world, widen commerce, and further the cause of world peace.

And to some extent it did. Telegraphy was indeed a phenomenal innovation—more dramatic, in many ways, than the Internet of today. It made irrelevant distances that had once seemed unbreachable, and widened the perspective of nearly all the world's people. Before the telegraph, there really was no such thing as international news; afterward, there was hardly any way of escaping it. Before the telegraph, both business and government were constrained by the limits of their own geographical reach: it simply was unrealistic for a central office in Boston to relay detailed instructions to an agent in Moscow, or for a central government in Paris to keep a watchful eye over its diplomats in Buenos Aires. After the telegraph, by contrast, information could flow easily and almost constantly between even the most far-flung outposts. The world shrank, and both business and government grew larger.[4]

These shifting dimensions pushed the curve of the technological frontier and caused a rupture within existing norms and rules. Like most technological breakthroughs, telegraphy brought a certain amount of chaos in its wake. It led, first of all, to commercial chaos, with dozens of inventors clamoring for recognition and suing each other over patent rights. It created confusion in the military, since commanders feared that enemy

---

[3]Quoted in Standage, *The Victorian Internet*, p. 81.

[4]There is a considerable literature that traces the impact of telegraphy on business and communications. See for example Glenn Porter, *The Rise of Big Business, 1860–1920* (Arlington Heights, Ill.: Harlan Davidson, 1992); Daniel J. Czitrom, *Media and the American Mind: From Morse to McLuhan* (Chapel Hill: University of North Carolina Press, 1982); Frederick Williams, *The Communications Revolution* (Beverly Hills, Calif.: Sage Publications, 1982); Richard B. DuBoff, "Business Demand and the Development of the Telegraph in the United States, 1844–1860," *Business History Review* LIV, no. 4 (Winter 1980), pp. 459–79; Alexander James Field, "The Magnetic Telegraph, Price and Quantity Data, and the New Management of Capital," *Journal of Economic History* 52 (1992), pp. 401–413; and Richard B. DuBoff, "The Telegraph in Nineteenth-Century America: Technology and Monopoly," *Comparative Studies in Society and History* 26 (1984), pp. 571–86.

spies would instantly transmit their plans abroad, and despair in the news industry, because most observers assumed that telegraphy would make newspapers obsolete. It also created a massive problem of coordination, as entrepreneurs stumbled over one another to erect competing lines, very few of which were even marginally compatible.

For roughly twenty years, these concerns and this confusion held sway. There were no rules in the industry, only a heady sense of potential and the whiff of impending riches. Reveling in their newfound industry, pioneers raced to catch the "lightning" and pirates, armed with competing patents and often indistinguishable from their more legitimate counterparts, followed close behind. Governments watched the industry or supported it, but few—in the beginning, at least—tried to impose any order upon it. It was simply too new and too chaotic.

As the technology emerged, however, and as the most successful pioneers carved the beginnings of an industry upon it, the costs of anarchy became increasingly evident and the progression toward order set in. Rules were established, rates set, and provisions made to connect the expanding web of separate lines and make them speak the same language. Great institutions, both public and private, were created that welded the industry closer together and hammered out procedures to stabilize and regulate it. Few of these institutions attracted the kind of fanfare that had greeted the innovators and pioneers; none evoked the euphoria that surrounded the industry in its earlier, wilder years. But in the final analysis, it was these institutions and rules that allowed telegraphy to slip beyond its initial chaos and become a vital part of modern commerce.

A Brief History of Signals

Since the Middle Ages, dreamers and tinkerers had been trying to sever the link between information and physical movement, between the message and the cumbersome messenger who carried it. They realized that any system would have to consist of two main components: first, a way of replacing the complexity of written language with some kind of shorthand or code; and second, some means of sending these signals through the air, using sound or sight instead of transportation. These basics had been recognized even by the early Greeks, who experimented with complicated systems of fire signaling—lighting fires on adjacent mountaintops in order to warn of enemy advance or herald victory. Yet because none of these systems ever achieved a decent measure of accuracy, important messages were still carried the reliable way: by hand.

The first real breakthrough came in 1791, when two French brothers named Claude and René Chappe sent messages to each other using a rudimentary system of copper pots and synchronized clocks: each set of clangs on the pot corresponded to a number on the clock face, which in turn was translated according to a prearranged code.[5] The system was loud and clunky, but it worked. After centuries of little technological change, the Chappes had at last broken the land barrier, sending messages by sound (which moves at twelve miles a minute) rather than by foot.

Their next innovation was in some ways even simpler, but more dramatic. Realizing that the sound of a clanging copper pot could travel only so far, the brothers turned next to light, replacing the pots with a five-foot-tall wooden tower equipped with adjustable arms. By rotating the arms into various positions, the Chappes could form ninety-eight different combinations or symbols. Sight thus replaced both sound and foot, and symbols could be transmitted almost instantaneously.

From the start, the Chappes' device had political implications. When the brothers first demonstrated their new *télégraphe* outside Paris, a mob rallied and destroyed the machine, worried that it might be used to communicate with jailed royalist sympathizers. Shortly thereafter, France's revolutionary National Assembly formed a committee to investigate the telegraph's military potential. Before long, the convention authorized the construction of three identical telegraph towers, spanning a total distance of about twenty miles. By 1798, France boasted telegraph lines stretching from Strasbourg to Dunkirk, and news of military victories was regularly relayed across the expanding network.

[5]The story of telegraphy's early history follows the excellent account in Tom Standage, *The Victorian Internet,* pp. 6–14.

Once the French system was in place, other European nations rapidly followed suit, constructing their own wooden towers and experimenting with different sets of codes and symbols. By the mid-1830s, nearly a thousand telegraph towers were in operation. Compared to earlier means of communication, this wooden network was extraordinary. It was simple to construct, easy to use, and dramatically faster than any method that had preceded it. Yet it also was plagued by a number of obvious, frustrating problems. First of all, even though the system was so much faster than any of its predecessors, it still remained *relatively* slow, since all messages had to be retransmitted at the end of every line of sight. It was also inherently cumbersome and subject to disruption, relying as it did upon both skilled operators and clear weather.

What promised to solve telegraphy's problems was electricity—a stream of energy that, potentially at least, could instantly relay signals across any length of conducting wire. Since 1729, when Stephen Gray of England first discovered that electricity could be conveyed across wires, inventors had been fiddling with various ways of using electronic impulses to replace the clanging pots and whirling arms of existing telegraphs.[6] Theoretically, the translation was relatively simple: an electronic impulse sent from one end of a wire would activate some kind of receptacle at the other end. Variations in the number or duration of the impulses would act just as the different clangs or arm positions, signaling different letters or numbers that could then be strung together into a recognizable message. In practice, though, the application of electricity proved immensely frustrating. For decades, inventors across the United States and Europe experimented with a wide range of techniques and permutations—with suspended balls, and decomposing water, and multiple conducting wires, one for each letter of the alphabet. Nothing was ever fully developed. And then Samuel Morse came along.

An artist and professor by trade, Morse was an unlikely candidate to pioneer the modern telegraph. Born Samuel Finley Breese Morse in 1791, Morse came late and ill-prepared to the emerging field of telegraphy. He had no technical background,

---

[6]See Alvin F. Harlow, *Old Wires and New Waves* (New York: D. Appleton-Century Company, 1936), pp. 37–38.

and only learned of recent work on electromagnetism in 1832. Bored with painting and his students at New York University, however, Morse was seized almost instantly by the idea of transmitting information via electricity. For the next several years, he toiled to learn the mechanics of this emerging science and then to transform his newfound learning into a practical device. He also worked with chemists and engineers, and found a powerful financial backer in Alfred Vail, a recent graduate of NYU who came from a family of well-to-do manufacturers. After several rounds of tinkering, Morse settled upon a distinctive approach. Words in his system would be coded as numerical digits, which would then be transmitted as electrical pulses along a wire to a receiving station. There, a lever triggered by an electromagnet would record the pulses onto paper. Later, dots and dashes—the familiar Morse code—would take the place of numerical code. By 1835, Morse's idea had taken physical and ingenious shape: he built a telegraph that used electricity to convey and record messages.[7]

Morse was by no means alone in his struggle to construct a telegraph, or even in his conviction that electromagnetism offered the most efficient means for harnessing electricity's power to the service of information. Indeed, as of 1838, no fewer than sixty-two people had already claimed recognition for having invented the first electrical telegraph.[8] Morse, however, was the first to push his invention out of the laboratory and into the realms of business and politics. He spent six years perfecting his rudimentary device, testing and trying it and finally subjecting it to public scrutiny. Then he spent the rest of his life dealing with the implications of his new machine.

Mr. Morse Goes to Washington

From the outset, Morse realized that the telegraph was more than a fancy new way of sending messages. It was, he

[7]For more on the mechanics of his invention, see Lewis Coe, *The Telegraph: A History of Morse's Invention and Its Predecessors in the United States* (Jefferson, N.C.: McFarland & Company, 1993), p. 29.
[8]See the case of *Smith vs. Downing,* cited in George P. Oslin, *The Story of Telecommunications* (Macon, Ga.: Mercer University Press, 1992), p. 19.

understood, a radical new way of transmitting information—
of hastening the reach and speed of global communication, racing faster than any established authorities. It was also a way of
creating vast wealth, since nearly all business enterprises would
presumably want the ability to move and gain information at
the lightning-fast pace that Morse could now offer. Morse saw
all these implications and acted upon them—acted, in fact,
nearly before his invention was even out of the lab. What is even
more remarkable, though, is that Morse responded to his own
foresight in a distinctly noncommercial fashion. Having seen
the future, his initial response was to hand it over to the U.S.
government. As he argued in 1838:

> It is obvious, at the slightest glance, that this mode of in-
> stantaneous communication must inevitably become an
> instrument of immense power, to be wielded for good or
> for evil, as it shall be properly or improperly directed. In
> the hands of a company of speculators, who should mo-
> nopolize it for themselves, it might be the means of en-
> riching the corporation at the expense of the bankruptcy
> of thousands; and even in the hands of Government
> alone it might become the means of working vast mis-
> chief to the Republic.
>
> In considering these prospective evils, I would respect-
> fully suggest a remedy which offers itself to my mind. Let
> the sole right of using the Telegraph belong, in the first
> place, to the Government, who should grant, for a specific
> sum or bonus, to any individual or company of individu-
> als who may apply for it, and under such restrictions and
> regulations as the government may think proper, the right
> to lay down a communication between any two points for
> the purpose of transmitting intelligence, and thus would
> be promoted a general consideration . . .[9]

Morse wasn't being entirely patriotic, of course: he went to
Congress in large part because he wanted Congress to fund

---

[9]Letter from Samuel Morse to Francis O. J. Smith, February 15, 1838, reproduced in
Edward Lind Morse, *Samuel F. B. Morse: His Letters and Journals,* Volume II (Boston:
Houghton Mifflin Co., 1914), pp. 84–86.

commercial development of his telegraph. But he also seems to
have believed seriously in the ills that could surround telegraphy,
and particularly in the ills of monopoly. "The Government," he
later wrote, "will eventually, without doubt, become possessed of
this invention, for it will be necessary from many considerations;
not merely as a direct advantage to the Government and public
at large if regulated by the Government, but as a preventive of
the evil effects which must result if it be a monopoly of a company."[10] Hardly the stuff of your average pioneer.

Congress, meanwhile, was still several decades behind
Morse, convinced that electrical transmission was a tinkerer's
dream and that the future of telegraphy lay with wooden signaling stations. Morse's first demonstration won few supporters,
and so the inventor trudged off to Europe, hoping that governments there would prove more forthcoming. They weren't. In
England, rival inventors persuaded the British attorney general
to not even grant Morse patent rights to his telegraph, and in
France he received patent rights but no financial backing.
Things went even worse in Russia, where Tsar Nicholas I feared
the new machine as an "instrument of subversion" and forbade
even any writings about telegraphy to appear in his country.[11]
Dejected, Morse returned to the United States and his students
at NYU.

In all his spare time, though, Morse continued to toil away at
the telegraph, refining its mechanics and trying to raise support
for its prospects. In 1842 he returned to Congress and proceeded to string wires between two committee rooms. After he
succeeded in relaying a string of messages between the two
rooms, Congress was at last impressed. In February of 1843,
Morse was granted $30,000 and the authorization to build a
U.S. telegraph line from Baltimore to Washington, D.C. Over-
joyed, Morse wrote to his brother Sidney that:

> Such is the feeling in Congress that many tell me they are
> ready to grant anything. Even the most inveterate op-
> posers have changed to admirers, and one of them, Hon.
> Cave Johnson [a Congressman from Tennessee], who

[10]Letter from Samuel Morse to the Honorable W. W. Boardman, August 10, 1842. Reproduced in Morse, *Samuel F. B. Morse: His Letters and Journals*, p. 175.
[11]Harlow, *Old Wires and New Waves*, p. 80.

ridiculed my system last session by associating it with the tricks of animal magnetism, came to me and said: "Sir, I give in. It is an astonishing invention."[12]

By August 1843, Morse had manufactured 160 miles of cable wire and begun construction of the line; on May 24, 1844, he sat at a telegrapher's desk in Washington and transmitted the message "What hath God wrought!" to an associate in Baltimore. One year later, the line became open for business—the world's first commercial telegraph operation. Quickly, however, its operation seemed to confirm Congress's worst fears. In its first three months of public operation, the Baltimore–Washington line had expenditures of $1,859.05 and revenues of only $193.56. The story for the second quarter was equally dismal.[13] Saddled with what appeared to be a losing business, Congress decided to abandon its interest in the Baltimore line, and Morse and his backers were forced to return to private capital. From that point on, telegraphy in the United States remained almost entirely a private affair.

In retrospect, the commercial failure of the Baltimore– Washington line was due to a simple and obvious cause: although the general public was fascinated by the technology of telegraphy, most people saw no practical use for it. Like mesmerism or telepathy, it seemed to be a freak occurrence, something to be marveled at but not utilized. In the first four days of the telegraph's operation, only one person actually paid to use it. Slowly, though, both telegraph wires and telegraph usage began to spread. In 1845, Morse and a group of his supporters (including Amos Kendall, a former postmaster-general, and Ezra Cornell, an entrepreneurial inventor) formed the Magnetic Telegraph Company with a capital subscription of $15,000. Their plans were ambitious and straightforward: to build a network of telegraph lines radiating along the major commercial routes of the United States. Doggedly, Morse remained convinced that if the lines worked, customers would eventually see them as a service rather than a technical oddity. Supply, in other

---

[12]Letter from Samuel Morse to Sidney Morse, May 31, 1843, reproduced in Edward Lind Morse, *Samuel F. B. Morse: His Letters and Journals*, pp. 224–25.
[13]Robert Luther Thompson, *Wiring a Continent: The History of the Telegraph Industry in the United States, 1832–1866* (Princeton, N.J.: Princeton University Press, 1947), p. 32.

words, would generate its own demand. And to a large extent, he was right. As the lines expanded, and as people grew more accustomed to their presence, business began slowly to experiment with telegraphy. The first customers were heavy users of information: newspapers, stockbrokers, and organizers of lotteries. Then came governments and general businesses, and finally private citizens.

In each case, the applicability of the technology was slightly different. For information-intensive industries, the telegraph was truly revolutionary, since its speed created possibilities that simply had never existed. For the first time, news could be reported as it occurred and across a fairly wide region. Similarly, stock prices could immediately reflect the interaction of market forces, and people with information about one market could parlay this information into commercial advantage in another.

For other industries, the telegraph was less revolutionary but almost equally important. It took the same information that these businesses had always employed, such as price lists and word of new products, and markedly increased the speed at which they were communicated. The effect was dramatic.[14] Suddenly, business that used to occur at a leisurely pace was telescoped into minutes or hours of time. Merchants in Boston could place orders in Atlanta overnight; farmers in Ohio could check the daily price of hogs in Chicago. It was the same old business, but conducted at a pace that changed the very tenor of commercial exchange. For once even a segment of the business community began to rely on the telegraph, all of the other businesses that fed into this one were compelled to follow along, demonstrating one of the first examples of a true network effect. By the early 1850s many individual firms, and particularly those clustered on Wall Street or in other financial centers, were sending six to ten messages a day along the "lightning."[15]

Government's relationship with the early telegraph was more complex. When Morse's original hopes for partnership soured

---

[14]For more on the general commercial impact of the telegraph, see DuBoff, "Business Demand and the Development of the Telegraph," pp. 459–79.

[15]For more data regarding early business use of the telegraph, see Alexander Jones, *Historical Sketch of the Electric Telegraph: Including Its Rise and Progress in the United States* (New York: Putnam, 1852).

with the fate of the Baltimore–Washington line, the U.S. Congress retreated from the commercial side of telegraphy. But government remained an active user of the telegraph, and political news was one of the major components of early transmissions. In fact, one of Morse's first successes came in May of 1844, when the Democratic National Convention rejected its leading presidential candidate, Martin Van Buren, in favor of the upstart James Polk. To win Van Buren's supporters, the convention nominated his friend Silas Wright as vice president. Stationed at the two ends of their new telegraph, Vail and Morse managed to convey this news to Wright, and receive word of his decline, before anyone outside the convention building had even heard the news. From this point on, the U.S. government saw the telegraph as a vital technology and an efficient means of communicating across an increasingly far-flung state. They weren't quite sure what to *do* with it (that would come later, during the Civil War of 1861–65), but they realized, at least, that the lightning had power.

A similar awe and confusion characterized the public response to telegraphy. In the 1840s, the cost of sending a telegraph was roughly twenty-five cents a word, far too high for average people or everyday use.[16] People regarded the telegraph, therefore, more as a technical marvel than a useful tool; it was viewed as a feat of magnetism that was destined to be witnessed rather than employed. In the earliest days, in fact, people would line up at telegraph offices just to "say they had seen it," not even caring, as Vail grumpily reported, "whether they understood or not."[17] They often had strange reactions to the machine, such as that of the woman who insisted on sending sauerkraut across it, and even odder notions about how it worked.[18] Over time, however, even private citizens began to see the personal potential of telegraphy. They realized that telegrams could carry vital news— of death, or illness, or change of plans—and that the speed of

[16]The figure is from the original rates charged by the Magnetic Telegraph Company on its New York-to-Philadelphia line, as written in a letter from Amos Kendall to prospective customers, and contained in the O'Rielly Manuscript Collection, II, in the New-York Historical Society Library. Cited in Thompson, *Wiring a Continent*, p. 44.
[17]Letter from Vail to Morse, June 3, 1844. Quoted in Thompson, *Wiring a Continent*, p. 25.
[18]For stories of these reactions, see Standage, *The Victorian Internet*, pp. 66–68.

this information could be worth the cost of sending it. They realized, too, that the value of even personal information was itself dependent on time, and that the real power of the telegraph lay in its ability to shrink time and provide fast, pertinent, and personally critical news. By 1851, according to one contemporary survey, "social" messages (as opposed to business messages) accounted for 9 percent of telegraph traffic in the United States.[19]

Even once this shift of attitude had occurred, however, the cost and constraints of telegraphy restricted its private use to all but the richest customers and most pressing matters. In the 1840s, telegraphy was still on the edge of the technological frontier. It was a plaything for inventors, a growing curiosity for business enterprises, and a development that governments were eager to watch. But it wasn't a mass market, or even the stuff of which such markets are made. For that progression to occur, innovators like Morse had to be succeeded by the next wave of pioneers—by entrepreneurs who extended and standardized the technology and then wove it into the everyday lives of millions of users. Which is, of course, precisely what happened.

Commercial Expansion

In the United States, the commercial telegraph industry began in 1845 and sprang, as one might expect, from a cluster of entrepreneurs who had begun to hover around Morse. The hub of this group was the Magnetic Telegraph Company and its driving force was Amos Kendall, the former postmaster general who had left his position to join forces with Morse.[20]

Having convinced Morse that private telegraphy would succeed where public telegraphy had failed, Kendall and his associates formed the private Magnetic Telegraph Company (or Magnetic, as it was generally called) in 1845 and proposed a most aggressive plan of wiring the entire United States. The obvious problem, though, was that Magnetic by itself couldn't

[19]Marshall Lefferts, "The Electric Telegraph: Its Influence and Geographical Distribution," *Bulletin of the American Geographical and Statistical Society* II (1856), pp. 258–59.
[20]Formally, Kendall became the agent for three-fourths of the Morse patent, leaving one-fourth in the hands of Francis Smith, Morse's long-time business associate.

possibly hope to raise all the capital that its founders' dreams required. And so the founders conceived of an equally ambitious commercial structure, one that would keep control within Magnetic's grasp while pushing the costs and risk of expansion onto a new group of investors. Specifically, Kendall and his associates conceived of the Magnetic Telegraph Company as a holding company of sorts, the backbone for what would become a national web of independent but linked firms. Magnetic would retain the patent rights to Morse's technology and 50 percent of the stock in all linked companies, but individual promoters would raise capital for each of the regional lines and connect them to Magnetic's central trunk. Explicitly, Kendall's notion was that all the telegraph companies would be part of the same "family," joined by compatible technology and sharing in the benefits of an ever-expanding (and thus ever more valuable) network. "Whatever we do," urged Kendall, "we must concert our plans and go on together. It is only by harmony in action that we can do anything, and to preserve it we must make concessions where we cannot agree in opinion."[21]

From the beginning, therefore, Magnetic was an exercise in cooperation, an attempt to grow the telegraph market by pooling funds and patent rights among several independent firms. And in the beginning, it worked. In the very first year of the company's operation, Francis Smith, a longtime Morse supporter and partner in Magnetic, formed his own firm, the New York and Boston Magnetic Telegraph Association, to extend telegraph service between those two cities. Kendall followed suit, launching the Albany and Buffalo Telegraph Company with a capital subscription of $200,000. Then other pioneers joined the game: John Butterfield, for example, a stage-line operator from central New York, purchased the rights to extend service from Massachusetts through to New York; and Henry O'Rielly, an Irish immigrant with a legendary entrepreneurial streak, agreed to extend the network all the way from the eastern seaboard to the Great Lakes region. Over the next five years, O'Rielly would go on to launch seven separate telegraph companies, stringing a web of wires that reached from Philadelphia to Chicago.

[21]Amos Kendall to Francis Smith, August 12, 1845, quoted in Thompson, *Wiring a Continent,* p. 40.

So far, so good. By 1850, Magnetic's network was up and running, and dozens of entrepreneurs were racing to connect their individual lines to the growing web of compatible wires. No sooner had the system established itself, however, than two sets of problems occurred. One was born of external competition, the other of internal rivalry. But both soon pushed the infant market toward chaos.

The first problem stemmed from the many contenders to Morse's supposed throne. Though Morse had beat all other inventors of electrical transmission devices to the commercialization phase, once technology had reached this critical bend, others jumped eagerly into the fray. And the pioneers—and pirates—weren't far behind.

Two variations on Morse's theme proved particularly successful. One, invented by Scotsman Alexander Bain in 1840, recorded electrical signals on a chemically treated paper that changed color when electricity passed through it. The other, designed in 1844 by a New England inventor named Royal House, took Morse's device a step further by converting electrical signals directly into letters of the alphabet. Both systems, to be sure, had their operational drawbacks. While the House system reduced the need for messages to be retranslated, it relied on a cumbersome hand-cranking mechanism that was very difficult to repair. And the Bain system, while faster than those of both House and Morse, was liable to receive false signals if other telegraph lines were nearby. On average, however, both systems worked roughly as well as Morse's. Which meant that entrepreneurs who didn't want to accept Magnetic's terms or its vision of a Morse-dominated empire were able to use the Bain or House technologies to strike out on their own, building rival systems on the basis of proprietary technology. Because each of these systems was explicitly noncompatible with Morse's, though, each new Bain or House line actually reduced the operational size of the U.S. telegraph market, fragmenting it into growing but unconnected chunks. Ironically, then, as the competing systems proliferated, the quality of wire service actually deteriorated and the "lightning" was reduced to snail mail, traveling once again over the nation's postal networks.

Meanwhile, even within the Magnetic "family" relations quickly soured. By the latter half of the 1840s, early investors such as Smith and Ezra Cornell had been joined by a flood of new entrants, the promoters and financiers that Kendall had envisioned as so critical to telegraphy's success. They were joined, as it turned out, by hundreds of enthusiastic investors, many of whom were dizzily innocent of either technical or managerial knowledge. Entranced by the telegraph's commercial prospects, they raced to construct their part of the network, looping lines haphazardly around trees, rowing messages across rivers, and feebly trying to insulate their wires with beeswax. None of these companies had any great capital behind it; many of them had insecure patent rights; and all were moving as fast as they possibly could.

By the mid-1850s, then, just as the "lightning" was making its way across the middle section of the United States, some of the boldest companies began to falter. In the north, two great rivals (the Lake Erie and the Erie & Michigan lines) were both barely able to cover their costs; in the Ohio Valley, O'Rielly's extensive Ohio, Indiana and Illinois Telegraph sank slowly into bankruptcy. And there were dozens, probably even hundreds, more: little firms that never even got off the ground, as well as larger ones that managed to cover their construction costs but never returned a cent of profit to their shareholders. During this time, in fact, nearly every telegraph firm in the United States was teetering on the edge of bankruptcy and a great number of them succumbed, leaving their lines fluttering uselessly across treetops and their promoters in search of other pursuits.

Part of what plagued these early firms was their own enthusiasm. Gripped by the vision of a telegraph empire, many had simply strung their lines too quickly and too sloppily. They had financed their operations as they went, raising capital from local communities and hoping, apparently, that profits would just come pouring in. They didn't. While telegraphy was definitely becoming an accepted mode of communication at this time, it was still too new and too crude for the majority of would-be users. It was also too expensive, especially since many operators compensated for their lack of technical knowledge by trying any method that sounded feasible. Not for a few more years would the technology of telegraphy became truly reliable—and while they waited, patching their wires and searching for insulators, some of the earliest pioneers just ran out of luck.

An even bigger problem than sloppiness, however, was competition. In the minds of Kendall and Morse, the whole purpose of the Magnetic Telegraph Company was to foster a controlled growth of the telegraph—a profitable growth, to be sure, but also a growth that was cautious and connected and technically sophisticated. "If we manage with prudence and act in concert the revenues of a nation are within our reach," Kendall urged. "If we are divided in counsel or raise the public against us, we jeopardize everything and shall live in constant turmoil."[22] Even without public funding, Kendall and Morse still saw the telegraph as a public servant, a tool to foster communication and economic growth. Others, however, including many of their own partners, had sharply different ideas. Telegraphy for them was simply a financial opportunity. They wanted to wire the continent, too—but quickly, and on their own terms.

Before long, these divergent objectives had split the Magnetic "family" into warring kin. Tired of O'Rielly's relentless self-promotion and envious of his extensive network, Kendall and Smith launched a bitter and public feud that would drag on for years, leaving all of their companies burdened in the end with construction debts and shrunken funds. Then the two original partners erupted in a passionate clash, each accusing the other of violating the terms of the Morse patent and both periodically refusing to engage in even the smallest act of cooperation. Other firms complained of messages being lost or garbled as they crossed the system, or even being changed by competing firms. Fights broke out at some of the busiest offices and reports of bribery plagued others. What the network needed, of course, were some basic rules of engagement. But with the rivals biting at each other's heels, no one in the industry was in the mood for engagement, much less cooperation.

Indeed, by 1850, fights over patents and contractual rights had brought several of the telegraph pioneers into pitched courtroom battles. In one of the most stunning defeats, a judge in Kentucky ruled that O'Rielly's entire line was illegal, since it used equipment that "borrowed" far too closely from Morse's

---

[22]Kendall to Smith, October 6, 1848, Smith Papers XI. Quoted in Robert Luther Thompson, *Wiring a Continent* p. 85.

patent. Going even farther than the case demanded, the Kentucky judge declared that Morse had the exclusive right to all electromagnetic telegraphy, implying that even the House and Bain systems were technically in violation of patent law. Although this extension was subsequently reversed by a higher court, O'Rielly took a major hit—especially after the Kentucky judge ordered a marshal to destroy his local line. Meanwhile, the Morse group of patentees brought suit against House, against Bain, indeed against anyone they could find who was creating a competing line of telegraphs. In their eyes, all these second-stage pioneers were simply pirates, scoundrels who had stolen (or at best, copied) Morse's patent and were now trying to compete unfairly against him. To the "pirates," however, it was Morse and his associates who were the thieves, trying to monopolize a vital industry and steal credit for efforts and innovation that rightfully belonged to others. Matters grew so contentious that when Mr. Bain planned to bring his family from Scotland to America in 1849, one newspaper cautioned that "He may be restrained from doing so by the fear that Amos Kendall will claim his wife and children as inventions of Professor Morse."[23]

By this point, the inventors and investors were all at each other's throats, vying for recognition and patent rights and racing to construct whatever might become the dominant system of telegraphy. All of them understood the deleterious effects of their rivalry, but no one was in the mood for cooperation. The fights were far too personal by this point, and the prospective profits appeared almost unfathomably large. In a letter to one of his partners, Francis Smith captures some of the mood of the time and the almost obsessive desire for speed. "I don't want to be humbugged anymore," he begins.

> Out with the plan of our campaign. Show that our Lake lines are to be the great receptacles of the Western intercourse with the Atlantic, and that the connecting lines are open to the people of the West, almost without money and without price, to accomplish this end . . .

[23]Quoted in Harlow, *Old Wires and New Waves,* p. 163.

Whenever you can get money enough raised to get a line up, start it, and Patentees will not hurry for their part, and your share of the benefits shall be made satisfactory. I want no pusillanimous, or doubting movements made—but, dash on with all the battery and thunder and lighting you can command.

Time saved is everything now . . . We will determine whether we or the other party make the best lightning. Again I urge you, don't hesitate—go ahead—and open your fire everywhere—set all the West in a light blaze with your proposals, and keep boldly in view the cheapness of the lines offered and the magnificence of the main arteries.[24]

This was not a man interested in the details of his business strategy or the technical intricacies of his product. Rather Smith, like many of his peers, was a true pioneer, forging into new territory, pushing stakes as fast as his crews could haul them. "Delays," he urged in a postscript, "are ruinous."

Continental Cables

While American entrepreneurs were racing to wire their expanding continent, a similar process was unfolding in Europe. The players, of course, were different and the game was tilted more toward the government's side, but the basic process and problems were largely the same.

In Europe, the dominant innovators were Charles Wheatstone and William Cooke, British inventors who joined forces in 1837 to string an experimental telegraph wire between Euston Square and Camden Town—a distance of about a mile and a quarter. Individually, Wheatstone and Cooke had both been working for several years on the same kinds of problems that had captured Morse and his colleagues: how to send electrical signals across longer wires and how best to record the messages that emerged. When they learned of each other's experimenta-

[24]*Morse v. Smith,* Superior Court of the State of New York for the City and County of New York, Deposition of E. Cornell, O'Rielly Docs., Legal Series, XI. Quoted in Thompson, *Wiring a Continent,* p. 108.

tion, the two men formed a commercial partnership and quickly patented a five-needle telegraph that signaled the letters of the alphabet by moving the needles into various tilted positions. The system was clunky, but it worked.

From the start, the two men made an unlikely pair. Wheatstone, a professor of experimental philosophy at King's College in London, was a true scientist, propelled by a quest for knowledge and largely uninterested in commercial applications. When Cooke first approached him, Wheatstone proclaimed that scientific men should do nothing more than publish the results of their experiments for others to use.[25] (He did get over this stumbling block rather easily, though, demanding a half-share of the partnership's profits.) Cooke, by contrast, was a born entrepreneur who saw telegraphy as the best route to a quick fortune. To Cooke, Wheatstone was a haughty, head-in-the-clouds academic who had no idea how to use—much less sell—the fruits of research. And to Wheatstone, Cooke was a rank amateur, a novice who had somehow managed to stumble into telegraphy without any sense of the science that lay behind it. Yet, somehow, both men were able to swallow their personal distaste and realize the benefits of partnership. Wheatstone made the technological breakthroughs and Cooke figured out how to apply and profit from them.

Perhaps because Cooke was driven by such clear commercial motives, he was one of the first of the telegraph pioneers to grasp the extraordinary importance of railroads to telegraphy, and telegraphy to railroads. While Morse was petitioning Congress and others were working with military officials, Cooke went directly to Britain's expanding network of railroads, urging executives there to see the possibilities: of sending messages as fast as trains, of using the existing right-of-way for a brand new purpose, and of facilitating rail transport by giving conductors and engineers immediate information about what was happening on the tracks.[26] In 1838, Cooke and Wheatstone convinced the Great Western railway company to install a thirteen-mile

---

[25]Standage, *The Victorian Internet*, pp. 34–35.
[26]Eventually, railroads would become a major consumer of telegraph services in both the United States and Great Britain. For more on this development, see Laurence Turnbull, *The Electro-Magnetic Telegraph: With an Historical Account of Its Rise, Progress, and Present Condition* (Philadelphia: A. Hart, 1853), pp. 193–99.

link between Paddington Station and West Drayton; in 1840 they installed another system on London's Blackwall Railway. Soon the system crept northward and west: a new line between Edinburgh and Glasgow was completed in 1841, and an extension of the Great Western line to Slough in 1842. By 1845, Wheatstone and Cooke were so enthusiastic about their possibilities that they formed a new company, the Electric Telegraph Company, to perform contractual work for the railways. A patent suit from a rival inventor soon forced Wheatstone to resign from the company, but Cooke continued and Electric flourished. By 1848 nearly half the country's railroad tracks had telegraph lines strung beside them.[27]

In a comparison that would persist into segments of the digital age, the British system focused on quality rather than price. While Americans such as O'Rielly were stringing haphazard wires in a race to expand, the British entrepreneurs went more slowly. With little competition to worry about, Cooke could afford to build the best technical system and pass the costs along to his consumers. Which was exactly what he did. In the early 1850s, a twenty-word message could be sent five hundred miles across the United States for $1. In the United Kingdom, the same message traveling a shorter distance would cost $7. As a result, few members of the British public actually used the telegraph for anything but the gravest emergencies. It was instead a more confined technology, used primarily by big businesses and the state.

This British picture was echoed—with several important differences—across the European continent. In each of the major countries, telegraphy developed roughly around the same time and along similar lines. A series of local innovators passed their ideas along to local entrepreneurs who built their own version of the telegraph across the dominant paths of communication, such as Brussels to Antwerp, and Moscow to St. Petersburg. In most of these other countries, though, governments moved much more quickly than they had in either the United States or the United Kingdom. In France, the telegraph was controlled by the government right from the start; in Belgium, Russia, and

[27]Standage, *The Victorian Internet*, p. 61.

elsewhere, governments followed rapidly on the heels of the original entrepreneurs, taking over the early lines and then using them as a starting point for much broader state-run efforts. The arguments in each case were largely the same: because the telegraph had such an obvious and crucial potential, it needed to be nurtured by public funds and tied to national concerns; because it was expensive and important, it needed to be overseen by the state. "Like the Post Office," one observer stressed, the telegraph, "cannot be so well conducted by a private Company as by the Government. The latter alone can enforce regularity, and make it accessible to all."[28] It was the same argument, of course, that Morse had tried to apply in the United States; but in Europe, it found a much more receptive audience. By the mid-1850s, every major continental power had its own telegraph service, linked to the state and subservient to its wishes.

Because these entities were public, they also had a host of rules attached to their basic operations. Governments in most European countries reserved the right to inspect messages sent across their wires, censoring content that was deemed to be offensive or injurious to the state. They forbade their citizens, therefore, from transmitting messages in code or using unapproved foreign languages. They also developed elaborate systems of taxation, with tariffs based on the length of words, the number of words, and the distance sent. Ironically, it was the very complexity of the tariff system that often compelled users to write their messages in a shorthand style, which then got them into trouble with the restrictions on codes or foreign language! Not surprisingly, public use of the telegraph remained relatively rare in continental Europe.

Still, in Europe, as in the United States, the 1840s and 1850s were a time of tremendous growth in telegraphy. From a pastime of hobbyists, telegraphy expanded to become a major industrial enterprise, a "servant of commerce" in many respects. Once again, however, the very success of the enterprise revealed its flaws. In the States, these flaws were largely the result of what Europeans might label "excessive competition": there were

[28]See George Sauer, *The Telegraph in Europe* (Paris, 1869), p. 144.

simply too many firms with too many different systems racing haphazardly after the same business. In Europe, the problem was that too many governments with too many differences were establishing separate networks. As of 1852, for example, every major European country had its own relatively sophisticated system, its own set of tariffs and transmission charges, and its own rules and prohibitions. (Not to mention, of course, its own unit of currency and its own language.) None of these systems was connected to, or even compatible with, its neighbor's. And so even when users were located close to one another, as in the various German or Italian states, they often had to rely on a hodgepodge of antiquated connectors, such as runners or horses or boats. Sending a telegram between France and its neighbor, the Grand Duchy of Baden, for example, required a veritable fleet of slow-moving intermediaries, described by one historian as follows:

> [A]n employee of the Baden telegraph administration was posted in the telegraph office at Strasbourg. When a telegram arrived from France destined for Baden, the French clerk handed it to the Baden clerk, who translated it into German, carried it across the river, and retransmitted it on the Baden lines.[29]

The system worked, to be sure, but just barely.

Before long, such obvious inefficiencies led to demands for consolidation and connectivity. Users didn't care so much which system they used, or whose; they just wanted to send their messages as quickly and inexpensively as possible. As these demands made their way through the political systems of Europe, most states had a simple and intuitive response: they sat down with one another and signed agreements that provided, at least, for the free flow of messages across their borders. In 1849, Prussia and Austria took the first steps, providing for wire service between Berlin and Vienna. Under the terms of the agreement, government messages would have top priority on the new line, with telegrams from Austria going first on even days and those from Prussia taking the odd days. The rate due was a simple

[29]Daniel R. Headrick, *The Invisible Weapon: Telecommunications and International Politics, 1851–1945* (Oxford: Oxford University Press, 1941), pp. 12–13.

sum of the rates in both countries. This treaty was soon followed by similar agreements between Prussia and Saxony in 1849 and Austria and Bavaria in 1850. Later in 1850, Austria, Prussia, Saxony, and Bavaria forged the Austro-German Telegraph Union, a consolidation of their separate agreements. Other states, including the Netherlands, quickly joined as well.

Meanwhile, the basic idea behind the Austro-German union began to spread across Europe, revealing the underlying calculus of cooperation. By agreeing to a few common principles, nations (and their businesses) could reduce the emerging chaos of telegraph transmission and thus increase the efficiency of their communication. Like the members of Amos Kendall's proposed Magnetic "family," they could improve the market for everyone by linking their systems and standardizing their technology. Once this logic was made clear, telegraph agreements in Europe multiplied nearly as fast as the telegraphs themselves: France signed with Belgium in 1851, with Switzerland in 1852, and with Spain in 1854. In 1855, these countries formed the West European Telegraph Union, a clear copy of the Austro-German Union and a frequent, though informal, partner with it. France and Belgium also signed a separate convention with Prussia, agreeing to construct uninterrupted lines across international frontiers, to guarantee the secrecy of all information sent over the wires, and to provide efficient and reliable service, including refunds where necessary, to every potential customer. By 1861, eleven additional countries had signed on to this convention.

Together, this series of treaties and conventions helped to solve many of the problems that were already plaguing Europe's early telegraph industry. Governments acknowledged the need to connect their various systems and to provide some base-line assumptions upon which users from across Europe could rely. But still communications did not flow easily or painlessly. Writing in 1869, one observer noted that:

> The great impediment to [the telegraph's] early development was undoubtedly the high rate at which the tariff had been fixed. But, another obstacle, which acted as a restriction in its free use, consisted in the endless confusion in the transmission of international correspondence . . . [E]ach State, it appears, had views of its own in

regard to constructing and managing a line. In every corner of the country improvements were invented, suggested and consequently experimented on. There exists
even to this very day an endless variety of patents for insulators and other appliances. Again, every State had a
separate system of counting the number of words, fixing
a message variously at 15 or 20 or 25 words, while others
counted the address extra and so forth.[30]

Clearly, what was baffling this author and frustrating his
contemporaries was the lack of what we today would call standards. In the mid-nineteenth century, entrepreneurs had used a
breakthrough technology (electric signaling) to create an immediate commercial application. As commercial use of the technology increased, however, business users—indeed all users—
began to suffer from the fact that this technology had sprouted
up so quickly, so randomly, and so variously. Telegraphy, like
the Internet, is inherently a long-distance medium: its value lies
largely in its Superman-like ability to leap great distances in a
single and relatively inexpensive bound. But because each of
the regional systems had its own rules, the value of leaping was
sharply reduced. If messages between France and Prussia still
had to be carried across rivers and translated by hand, they
didn't save their senders as much as they should have. And if
messages between London and Rome incurred high tariffs in
each of the countries along the way, then they didn't save their
senders that much money. What the technology demanded was
a common set of standards—rules that would apply across all
transactions, reducing costs and providing users with a certain
amount of predictability and stability.

In the United States, as shall be seen, standardization was
largely the child of cooperation and consolidation. Firms came
together to form private associations and hammered out their
own sets of common standards. In Europe, by contrast, standardization was largely the work of the state.

In 1865, Napoleon III of France called an international conference on the telegraph, determined to push his European

---

[30]Sauer, *The Telegraph in Europe*, pp. 11–12.

neighbors toward a more efficient international system. It wasn't a particularly difficult goal. Indeed, there was a surprising consensus among European states that cooperation was essential in the world of telegraphy and that, in this arena at least, the individual demands of governments would have to bend before the benefits of coordination. Some states even took the logic a step further, arguing that technical cooperation on telegraphs would lead over time to political cooperation and perhaps even to peace. Accordingly, when the states of Europe sat down to discuss the technical details of telegraphy, they moved with a surprising speed and vigor. By the conference's end, all the participating countries had agreed to adopt the Morse-style apparatus as their technical standard and employ the gold franc as their common currency. They all pledged to use blocks of twenty words as the standard for calculating costs and to formally codify their schedules of tariff and transit costs. Additional agreements covered the hours of operation for telegraph centers, the priority of messages (with government telegrams receiving top priority), and rules for the transmission and delivery of telegrams, for the maximum number of letters that a word could have, for counting words, and for collecting fees. Almost as an afterthought, the conference also created a new organization to oversee the rules it had just laid forth. Remarkably, this organization, the International Telegraph Union, is still alive today.

Like most international documents, those written at Napoleon's conference have a dry and dusty feel. They reek of endless meetings and tactful compromises, of tedious technical provisions and carefully arranged sentences. Yet behind them stands a dramatic turning point in the history of the telegraph. Before 1865, telegraphy in Europe was still a motley collection of individual networks. It was perhaps somewhat tamer than in the United States, somewhat more manicured and more tied to diplomacy than to commerce. But it was still based on webs of independent systems and on competing, and complicating, sets of rules. After 1865, telegraphy in continental Europe settled down. There were still great battles to be fought and territories to be claimed, but these, for the most part, now stretched beyond Europe, across the Atlantic and down into Africa and

Asia. On the continent, telegraphy became more of an established industry, with rules and regulations that quashed much
of the earlier uncertainty and helped to transform the telegraph
from a technical oddity into a mainstream tool of communication and commerce.

Crossing the Atlantic

In 1858, Britain's Queen Victoria sent a cable
across the Atlantic to U.S. President Buchanan. In literary
terms, it was not particularly interesting; diplomatically, it was
irrelevant. But in the history of technology, the message was immense. And at the time, the very fact that it could be sent at all
was seen as perhaps the most amazing thing on earth.

It is difficult, at the start of the twenty-first century, to capture the excitement that greeted this nineteenth-century moment. Until the very instant that Queen Victoria's cable was
received in Washington, the idea of transmitting messages
under the ocean was widely dismissed as preposterous. "Fancy
a shark or a swordfish transfixing his fins upon the insulated
wires, in the middle, perhaps, of the Atlantic, interrupting the
magic communication for months," mused one critic. "What is
to be done against the tides, when they deposit their floating de-
bris of wrecks and human bodies? Even supposing you could
place your wires at the lowest depth ever reached by plumb line,
would your wires, even then, be secure?"[31] While governments
on both sides of the Atlantic had entertained the idea of submarine communication, and even initially funded it, they quickly
withdrew from the plan after a series of expensive and embar-
rassing failures. Meanwhile, many scientists still scoffed at the very
idea of underwater transmission, and most entrepreneurs were
unwilling to invest the time or capital that it would so obviously
entail. And so while land-based telegraphs grew like so many
weeds, underwater technology remained the stuff of dreams.

Then, in 1850, some of these dreamers actually set out to
sea. The first were John and Jacob Brett, brothers from Bristol

---

[31]Quoted in Standage, *The Victorian Internet,* pp. 74–75.

who were determined to string a cable beneath the English Channel. Without any formal scientific training, the brothers really didn't know what they were doing, and their first foray was unpropitious. The thin wire they had dropped out from behind their boat had simply floated along behind them, and when they repeated the experiment with weights attached to the wire, a French fisherman pulled it out of his net and brought it home as seaweed. Yet the Bretts persevered, eventually raising money for a new, more sophisticated cable and enlisting the help of more seasoned engineers. In 1852, the brothers successfully sent a message from London to Paris.

Over the next few years, a slew of pioneers raced to lay cable along some of the world's most strategic crossings: the Irish Channel, the North Sea, and the Mediterranean. These ventures incited even greater interest, and a newfound interest in subterranean lines. The problem, though, was that there clearly was a relationship between distance and transmission: the farther the distance, the more difficult it was to keep the cable whole and the signal distinct. For really long distances, such as across the northern Atlantic, there wasn't even a boat large enough to carry the requisite cable. Many pioneers therefore moved on, especially since routes to Africa seemed more promising at the time. Then Cyrus Field got involved. Field, a thirty-three-year-old self-made millionaire, had just retired from the paper trade. He knew nothing about telegraphy and had never been interested in it. But like Cooke, and O'Rielly, and the Bretts, he saw the commercial prospects and drew the technology along.

In 1854, an English engineer named Frederic Gisborne came to New York looking for money. For several years he had been struggling, alone and without much support, to construct a telegraph line across Newfoundland, a barren and sparsely populated land with fierce winters and a desolate interior. On its own, Newfoundland was hardly an attractive market for telegraphy. But its location—at the eastern tip of America, stretching far out into the northern Atlantic—was strategic. If Gisborne could wire Newfoundland, he reckoned that he could receive information directly from ships sailing across the northern Atlantic, replacing Boston as the first port of call for information and cutting a day or two from the time it took messages to flow between New York and London. And in a world that had already become increasingly addicted to speed, time translated easily into money. With modest backing from Newfoundland's legislature, Gisborne thus set out to construct a new line across Newfoundland and southward to Cape Breton, from whence messages could be transmitted across existing land lines. After nearly three years of intensive labor, he had managed to survey the territory and construct thirty or forty miles of road across it. Then he went bankrupt and, determined not to abandon the project, set sail for New York.

There he stumbled quite fortuitously upon Cyrus Field, who apparently was looking for something new to do. It's not quite clear just who hatched the idea for an Atlantic cable. Gisborne claimed that this had always been his intention, but that fear of public ridicule had forced him quiet. Already, he argued, "I was looked upon as a wild visionary by my friends, and pronounced a fool by my relatives . . . [H]ad I coupled [the Newfoundland connection] with an Atlantic line, all confidence in the prior undertaking would have been destroyed, and my object defeated."[32] Field's brother and biographer, however, asserts that the grandiose scheme belonged to Cyrus: "After [Gisborne] left," he recalls,

> Mr. Field took the globe which was standing in his library and began to turn it over. *It was while thus studying the globe that the idea first occurred to him, that the telegraph might be carried further still, and be made to span the Atlantic Ocean . . .* He cared little about shortening communication with Europe merely by a day or two, by relays of boats and carrier pigeons! But it was the hope of further and grander results that inspired him, and gave him courage to enter on a work of which no man could foresee the end.[33]

Bombast aside, the two men quickly agreed to work together and to stretch Gisborne's plan all the way across the northern Atlantic to Valentia Bay, Ireland, a distance of nearly seventeen

[32]Quoted in Henry M. Field, *History of the Atlantic Telegraph* (New York: Charles Scribner & Co., 1866), pp. 22–23.
[33]Field, *History of the Atlantic Telegraph,* pp. 26–27.

hundred nautical miles. Up until this point, no undersea cable had run successfully for more than three hundred miles, and nothing had ever been laid on a body of water as deep and unpredictable as the North Atlantic. Field, however, was not a man to let practical issues get in the way—especially since a handful of eminent innovators, including Morse, were becoming convinced that long-distance undersea transmission was at least theoretically possible. With a handful of other New York capitalists, Field thus proceeded to get the Atlantic operation underway. He hired electricians, compiled sea charts, and even convinced the British and American governments to back his project with an annual subsidy and to provide the vessels that would be necessary to lay the Atlantic cable. In exchange, Field promised that his new firm, the Atlantic Telegraph Company, would carry all official messages for free. Thus was the transatlantic link an appropriate hybrid of American and European development styles: private entrepreneurship backed—only in part—by public money.

In retrospect, Field's Atlantic cable was a technical disaster. It broke several times during installation, had a close encounter with a large whale, and then, less than a month after its triumphant inaugural, slowly ground to an unintelligible gurgle. Yet, somehow, it didn't really matter. Queen Victoria had, in fact, sent a telegram to President Buchanan. He had telegraphed back. And Cyrus Field had proven that it was both technically and commercially possible to wire the seas. On both sides of the Atlantic, there was profound jubilation and a newfound sense that technology could cure the ills of mankind. As one contemporary history of the telegraph crowed: "The completion of the Atlantic Telegraph, the unapproachable triumph which has just been achieved in the extension of the submarine electric Cable between Europe and America, has been the cause of the most exultant burst of popular enthusiasm that any event in modern times has ever elicited. So universal and joyful an expression of public sympathy betokens a profound emotion that will not immediately pass away. The laying of the Telegraph Cable is regarded, and most justly, as the greatest event in the present century . . ."[34] When word of the completed cable reached Boston,

---

[34]Briggs and Maverick, *The Story of the Telegraph*, p. 11.

one hundred guns were fired on the Common and the city's bells rang out for an hour; in New York, massive celebrations culminated with a torch-lit parade that accidentally set City Hall on fire.

Just as with the telegraph conference of 1865, the sentiment that surrounded the Atlantic cable had a distinctly political and utopian hue. Analysts saw a technological revolution and presumed it would provoke a positive political reaction; they truly believed that the ability to communicate would reshape the political order, erasing national rivalries and causing states to come together in some seamless electronic cocoon. Soon, predicted some of these prophets, "the whole earth will be belted with the electric current, palpitating with human thoughts and emotions."[35] It would become, promised another, "the nerve of international life, transmitting knowledge of events, removing causes of misunderstanding, and promoting peace and harmony throughout the world."[36] And they prophesied: "How potent a power, then, is the telegraphic destined to become in the civilization of the world! This binds together by a vital cord all the nations of the earth. It is impossible that old prejudices and hostilities should longer exist, while such an instrument has been created for an exchange of thought between all the nations of the earth."[37] The parallels to some current strands of cyber-speak are too blatant to ignore.

In many ways, the crossing of the Atlantic (even with faulty technology) brought the technology of telegraphy into mainstream life. Before this point, telegraphy had still been a fairly restricted tool. Some businesses used it extensively; governments relied increasingly upon it; but the vast bulk of citizens, both in the United States and Europe, were still somewhat suspicious of the "lightning" and not entirely sure what it meant for them. Throughout the 1850s, stories still proliferated about people who tried to send food through their local telegraph offices or thought men delivered telegrams by running along the wires. The Atlantic cable changed that. Suddenly, the technology ripped

---

[35]Briggs and Maverick, *The Story of the Telegraph*, p. 12.
[36]Edward Thornton, British Ambassador, quoted in Standage, *The Victorian Internet*, p. 91.
[37]Briggs and Maverick, *The Story of the Telegraph*, p. 22.

into popular consciousness and became too big to ignore. It was technology as symbol, technology as power to change the social structure—even the world. "This," proclaimed one U.S. senator, "is a triumph of science—of American genius, and I for one feel proud of it, and feel desirous of sustaining and promoting it."[38] Once again, the parallels to cyberspace ring clear.

Eventually, of course, the fervor that surrounded the Atlantic cable died down—not because it didn't work at first, but rather because it had, in fact, become mainstream. By the time Cyrus Field strung a second (and much improved) line across the northern Atlantic, businesses on both sides of the ocean simply presumed that they couldn't function without it. On its first day of operation, the new cable generated the princely sum of £1,000. And by 1867, just thirteen years after its launch, the Atlantic Telegraph Company was able to pay off all its debts.

Ruling Britannia

It would be tempting to end the tale of telegraphy at the spanning of the Atlantic, for it is here that the story is at its wealthiest, brawniest peak. But just as Field and his associates were reaping the proceeds from their venture, and as telegrams were becoming a vital component of social and commercial intercourse, political objectives slammed back onto the scene, and governments—in both the United States and Europe—began to play an expanding role in the telegraph industry.

As one might expect, political intervention proceeded along different paths on the two continents and manifested itself in very different forms. But as the nineteenth century drew to a close, nearly all governments were heavily involved with telegraphy.

During the 1850s and 1860s, European governments had primarily been occupied with domestic affairs, struggling to maintain their authority in the face of a revolutionary enthusiasm that swept across the continent. Nationalism became a potent force during this period and, once the radicals had been defeated,

---

[38]Senator Benjamin of Louisiana, quoted in Field, *History of the Atlantic Telegraph*, p. 128.

helped newly emboldened leaders to cement the growing power of the state. This trend culminated, in 1871, in the scattered states of north central Europe joining to form Germany, and creating a sudden power at the very center of Europe. They were followed in due course by the Italian states, which formed a union of their own along Europe's southern flank. Meanwhile, around the periphery of Europe the tussle for colonial possessions in Africa had erupted into a full-scale race, with states vying to control ever-growing slices of the world's southern regions. Suddenly, politics in Europe became a fluid, fast-moving game—the perfect arena for high-speed telegraphy.[39] And European governments, which had scoffed not so long ago at telegraphy's prospects, now wanted wider networks and enhanced control.

Until this point, Britain's relationship with the telegraph industry was marked by a certain degree of schizophrenia. The government was not involved with Cooke and Wheatstone's early projects and had rejected the Bretts' request to subsidize an Atlantic crossing. They resisted supporting the systems that quickly spread across the island and thus were not even present at the 1865 international convention. Yet neither was the government totally aloof from the telegraph industry or unconcerned about its development. As the American government had with Morse, they did eventually subsidize a portion of Field's first foray across the Atlantic and then, on their own, provided backing for an equally ambitious Red Sea cable. When both these ventures ended in failure, the British government vowed never again to allocate public money for the pursuit of undersea telegraphy. But, nearly at the same time, it also moved to extend its control over land-based networks and, somehow, to construct a link to India.

In 1868, the government of Britain formally announced its intention to nationalize the telegraph infrastructure of the British Isles. It was a contentious move that drew the ire of industrialists across the country. The government's perspective, though, was

---

[39]For a study of how telegraphy affected U.S. diplomatic relations with Europe, see David Paull Nickles, "Telegraph Diplomats: The United States' Relations with France in 1848 and 1870," *Technology and Culture* 40 (Jan. 1999), pp. 1–25.

that the telegraph had simply become too important to be left in private hands. It was a public service and a natural monopoly, something akin to the post office and best organized along similar lines. And, as the Chancellor of the Exchequer explained, "The cost of working the telegraph system is greater than it would be in the hands of the State. If telegraphing were made the monopoly of the post office it would be able to work at much lower rates than the companies."[40] To compensate investors, the government spent £8 million—funds that, in many cases, were reinvested in subsequent overseas projects.

It was India, though, that really captured British policy and contributed to its schizophrenia. Since the early years of the nineteenth century, India had been the obvious focus of England's expanding colonial realm—the famous "jewel in the crown" that provided the empire with resources, trade routes, and an inestimable dose of prestige. To maintain relations with this far-flung possession, Britain traditionally had relied on a combination of arm's-length measures: on mastery of the ocean corridors that linked the British Isles with India, on diplomacy with those states (such as Persia) that lay between Britain and India; and on the bureaucratic might of the British East India Company. Together, these measures had proven quite effective, allowing the British government to administer the Indian subcontinent from afar. But in a world of shrinking times and distances, the traditional routes began to feel too sluggish. Britain wanted to wire India, and to ensure that this Indian network was firmly and securely connected to London. As Lord Dalhousie, Governor General of India, complained in 1852:

> Everything, all the world over, moves faster now-a-days than it used to do, except the transaction of Indian business. What with the number of functionaries, bards, references, correspondences, and several Governments in India, what with the distance, the reference for further information made from England, the fresh correspondences arising from that reference, and the consultation of the several authorities in England, the progress of any

[40]Quoted in Frank Parsons, *The Telegraph Monopoly* (Philadelphia: C. F. Taylor, 1899), p. 135.

great public measure, even when all are equally disposed to promote it, is often discouragingly slow.[41]

To hasten matters along, Dalhousie authorized a local inventor, William O'Shaughnessy, to build a five-thousand-kilometer network linking Calcutta, Agra, Bombay, Peshawar, and Madras. Despite a whole series of technical challenges posed by building in the tropics—termites tended to eat wooden poles, for example, and monkeys attacked copper wires—construction continued apace, and by 1856 India had seventy-two hundred kilometers of telegraph line and forty-six telegraph offices.[42] Even before the system was completed, Dalhousie converted it to a government monopoly, making clear his opinion that the telegraph was almost purely an instrument of political control.

Politically, though, the Indian network suffered from a critical flaw: there was no high-speed link to London. Just how serious this gap was became evident in 1857, when rebellion broke out across the north of India. From Lucknow, Sir Henry Lawrence, an official with the British East India Company, sent an urgent telegraph to the governor general in Calcutta. "All is quiet here," he reported, "but affairs are critical; get every European you can from China, Ceylon, and elsewhere; also all the Goorkas from the hills; time is everything."[43] This hasty message arrived in Calcutta on May 10 and reached Bombay on May 27. From there, it was rushed on a steamer bound for Suez and then shipped from Alexandria to Trieste, from whence it finally was telegraphed to London. The journey had taken forty days—time enough for Lawrence to die of cannon wounds and for large parts of India, including Calcutta and Delhi, to spin out of British control. Parliament needed no further convincing. When the Red Sea cable failed the following year, leaving the government with roughly £1,800,000 in losses, policymakers scurried to find some other means of securing a direct line to India. All of these

---

[41]Quoted in Sir William Brooke O'Shaughnessy, *The Electric Telegraph in British India: A Manual of Instructions for the Subordinate Officers, Artificers and Signallers Employed in the Department* (London, 1853), pp. xi–xii, and reproduced in Headrick, *The Invisible Weapon*, p. 51.

[42]Headrick, *The Invisible Weapon*, p. 52.

[43]Quoted in Headrick, *The Invisible Weapon*, p. 19.

efforts concentrated on overland routes, and all became intricately entwined in the complicated politics of late nineteenth-century Europe.

In 1862, the Indian government (which was essentially an overseas department of the British government) formed the Indo-European Telegraph Department and quickly constructed a land line from Karachi to Gwadur, a city on the north coast of the Gulf of Oman. From Gwadur, a second line soon ran to Fao, and then connected to a Turkish line that reached to Baghdad and Constantinople. In a separate agreement with the government of Persia, the Indo-European Telegraph Department also laid a line across Persia, connecting the Indian line with Teheran. This line then connected with a Russian one, which ferried messages across Georgia and into Moscow. By the time this network was complete, India and London were—theoretically speaking—directly connected. The only problem was that politics frequently crowded the lines and clouded the messages. To get from Bombay to say, Manchester, telegrams had to pass through a mishmash of Turkish, Persian, Russian, Italian, French, and Greek hands. Sometimes messages arrived clearly and quickly; sometimes they were garbled, lost, or interminably delayed. Espionage was a frequent game along the European lines, and many government officials suspected that their correspondence was regularly read en route. "In Paris," one reported, "they make four copies of all that they think of importance, and send them round to the different bureaus."[43]

Britain desperately wanted a better, integrated system under its control. In 1866, a committee appointed by the House of Commons reviewed the nature of the problem, concluding that "it is not expedient that the means of intercommunication by telegraph should be dependent upon any single line . . . in the hands of several foreign governments," and recommended government subsidies for a new, fully British line.[44] The British Treasury, however, which was still reeling from the aftermath of several failed telegraph projects, refused to offer its support, arguing instead that high-risk projects such as undersea telegraphy

---

[44]*Parliamentary Papers* 1866 (428) IX, pp. xv–xvi. Also cited in Headrick, *The Invisible Weapon*, p. 22.

were best left to private investors.[45] This, as it turned out, was a very wise move. For in the early 1870s, two private firms rose to the challenge and independently constructed full-scale, integrated lines between the Indian subcontinent and continental Europe. Werner von Siemens, a powerful Prussian industrialist, built an overland line from London to Teheran operated solely by company employees (no prying foreign clerks!) and dedicated to Anglo-Indian communications. Then John Pender, a British investor, built a series of three undersea lines connecting England to Malta, Malta to Suez, and Suez to Bombay.

In the history of telegraphy, John Pender stands as one of the most powerful and influential players.[46] Like Cooke, Field, and O'Rielly, he was a true pioneer, wresting technology out of the laboratory and forcing it into the commercial realm. He built an international empire, one that at its peak stretched nearly as far as Britain's own and was instrumental in keeping this more formal empire afloat. Pender was also a man blessed with a finely tuned sense of politics and of his own political power. In the end, the rules of British telegraphy became his rules, and they worked to advance a finely tuned set of commercial and imperial goals.

Pender started life as a cotton merchant, rising from the middle classes of Scotland to earn a considerable fortune from both trade and speculation. In his thirties he began, like other wealthy industrialists at the time, to dabble in the burgeoning but high-risk business of telegraphy: he was an early investor in the English and Irish Magnetic Telegraph Company and contributed £1,000 to Field's first transatlantic project. Unlike many of these other early investors, though, Pender decided to move all of his assets into the world of telegraphy. When Field's first project failed, Pender became a moving force behind the second attempt, eventually being named a director of the Atlantic Telegraph Company and committing £250,000 of his

---

[45]The biggest fiasco occurred in 1858, when the government financed an ambitious project to connect India and England via a Red Sea cable. The cable never worked, and the government was forced to reimburse the project's stockholders £36,000 a year for the next fifty years. See Headrick, *The Invisible Weapon*, p. 20.

[46]For more on the history of Pender's core firm, the Eastern Telegraph Company, see K. C. Baglehole, *A Century of Service: A Brief History of Cable and Wireless Ltd. 1868–1968* (Welwyn Garden City (Herts): Bournehall Press Ltd., 1970); and Hugh Barty-King, *Girdle Round the Earth: The Story of Cable and Wireless and Its Predecessors to Mark the Group's Jubilee, 1929–1979* (London: Heinemann, 1979).

own money to the next round of experimentation. When the Atlantic cable at last went through, Pender decided to take his experience to the other side of the world and to again pursue Britain's objective of an undersea connection to India. Thus in 1869 he launched a new company, the British-Indian Submarine Telegraph Company, with the ambitious goal of laying a cable from Suez through the Red Sea and then on to Aden and Bombay. Before this project was even underway, Pender launched a second company, the Falmouth, Gibraltar and Malta Telegraph Company. He then added a third (the Marseilles, Algiers and Malta Telegraph Company) and a fourth (the China Submarine Telegraph Company). Clearly the man intended to wire the world.

Over the next twenty years, John Pender built one of the largest commercial enterprises the world has ever seen. His consolidated firm, the Eastern Telegraph Company, controlled at its peak more than a hundred thousand kilometers of cable strung across India, Australia, and Africa. A subsidiary, the Western Telegraph Company, controlled large swaths of the Latin American market. Although there were other immensely successful telegraph firms, and even other immensely successful British ones, Eastern was in many ways the core of Britain's overseas network, the eyes and ears of the empire. Yet it was also throughout this time a completely private, hugely profitable firm. Unlike the national telegraph service or the British–Indian network, the Eastern Company wasn't tied to the British government in any way or hobbled by any particular regulatory restraints. It didn't seem to obey any rules but its own, or to follow any guidelines other than those dictated by its own self-interest. Rather, on the surface at least, Eastern looked like a devoutly independent firm, free from either government intervention or political motive.

If one probes more deeply into Eastern's structure and conduct, however, a different picture begins to emerge. Yes, Eastern was a private firm, and an immensely profitable commercial enterprise. But it also worked very closely—at times almost suspiciously closely—with the British government.

Part of this connection was purely personal. Having been elected to Parliament in 1872, Pender was himself a part of government, and he socialized frequently with Britain's political elite. He also appears to have mixed social and political desires freely, asking commercial favors from high-placed political friends and peopling his corporate boards, as one historian has noted, with a "disproportionate number of aristocrats with connections in the Foreign Office and the Colonial Office."[47] Pender maintained an active correspondence with officials in various arms of the British government, and seems to have had no qualms about securing their aid and support.[48]

The real connections, however, ran even deeper and related to the fundamental mission of Pender's company. At one level, of course, Pender was nothing more than a successful entrepreneur. He built a big business, made lots of money from it, and then used his resources to facilitate an entree to Britain's commercial and political elite. At another level, however, Pender's business was itself a political enterprise, and one that provided the British government with a key set of services. Merely by running his own business empire, Pender was keeping global communications in British hands and in impeccable working order. He was supplying the British government with a crucial communications infrastructure—a system that stretched to all corners of the British Empire and allowed British diplomats to communicate in their own language, at reasonable cost, and with only minimal fear of unfriendly eavesdropping. Using Eastern's backbone, British officials could communicate faster and more securely than any of their rivals across a network whose reach surpassed any other on earth. Indeed, by the end of the nineteenth century, as Tables 2.1 and 2.2 show, two-thirds of the world's cables were British, and nearly 80 percent of these belonged to Eastern and its affiliates.[49]

---

[47]Jorma Ahvenainen, *The Far Eastern Telegraphs: The History of Telegraphic Communications between the Far East, Europe and America before the First World War* (Helsinki, 1981), p. 18.

[48]For evidence of this correspondence, see Barty-King, *Girdle Round the Earth*, pp. 53–56, 80; and Headrick, *The Invisible Weapon*, pp. 36–37.

[49]There were also a number of other telegraph firms that played important roles in their respective markets, including Denmark's Great Northern Telegraph Company, which ran lines across Scandinavia and then in Japan and China; and France's Compagnie Française du Télégraphe Paris à New York. For more on these firms and their operations, see Ahvenainen, *The Far Eastern Telegraphs;* and Store Nordiske Telegraf-Selskab, *The Great Northern Telegraph Company: An Outline of the Company's History, 1869–1969* (Copenhagen, 1969).

TABLE 2.1 Total Undersea Cables in 1892

	Number of cables	Length (km)	Percent of world total
British cables	508	163,619	66.3
American cables	27	38,986	15.8
French cables	74	21,859	8.9
Danish cables	82	13,201	5.3
Other	535	9,206	3.7
TOTAL	1,226	246,871	100.0

Source: Daniel R. Headrick, *The Invisible Weapon* (New York: Oxford University Press, 1991), p. 39.

Presumably, Pender didn't build his global network to wrest favors from the British government; he did it to realize a vision, or reap munificent profits, or some combination of both. But once the network was in place, Pender had a significant amount of political leverage, most of which he appears to have used to keep foreign competitors from encroaching on his markets. An 1897 letter from the Treasury Chambers of the British government to the Under Secretary of State at the British Colonial Office makes this relationship quite explicit. It concerns a proposed telegraph between Canada and Australia and is worth quoting at length:

> Reference has already been made in general terms to the competitive character of the new undertaking. But there is one aspect of this competition to which My Lords wish to draw special attention.
>
> Almost the whole of the traffic which will pass over the Cable will have to be won from the Eastern Extension Telegraph Company and its allies.
>
> Those companies represent a real British interest, and one entitled to great consideration from the Imperial Government. They have been the pioneers of Cable communication with the most distant parts of the Empire, and it is to their enterprise and skill that the establishment of those communications is due.
>
> In times of emergency they have always been ready to render to the Government any services which were in

TABLE 2.2 Private Undersea Cables in 1892

	Number of cables	Length (km)	Percent of world total
Eastern and Associated Companies			
Eastern Telegraph Co.	117	50,843	20.6
Eastern Extension	27	13,597	5.5
Eastern and South African	12	12,586	5.1
Brazilian Submarine Telegraph Co.	6	13,647	5.5
West African Telegraph Co.	12	5,594	2.3
African Direct Telegraph Co.	7	5,086	2.1
Western & Brazilian Telegraph Co.	10	7,341	3.0
West Coast of America Telegraph	7	3,147	1.3
Black Sea Telegraph Co.	1	624	0.3
River Plate Telegraph Co.	3	256	0.1
SUBTOTAL	202	112,721	45.8
Other British Companies			
Direct Spanish Telegraph Co.	4	1,311	0.5
Halifax and Bermudas Cable Co.	1	1,574	0.6
Spanish National Submarine	7	3,998	1.6
Anglo-American Telegraph Co.	14	19,261	7.8
Direct United States Cable Co.	2	5,741	2.3
Cuba Submarine Telegraph Co.	5	2,778	1.1
West India & Panama Telegraph	22	8,440	3.4
SUBTOTAL	55	43,103	17.3
Total of British Companies	**257**	**155,824**	**63.1**
Non-British Companies			
Great Northern Telegraph (Den.)	27	12,838	5.2
Cie. Fr. Du Télégraphe Paris	4	6,475	2.6
Soc. Fr. Des Télégraphes Sous-Marins	14	6,952	2.8
Western Union Telegraph Co.	8	14,340	5.8
Commercial Cable Co. (U.S.)	6	12,849	5.2
Mexican Telegraph Company (U.S.)	3	2,821	1.1
Central & South American Tel. Co.	10	8,997	3.6
Canadian Pacific Railroad Co.	5	78	
Total of Non-British Companies	**77**	**65,350**	**26.3**
Total of All Private Companies	**334**	**221,174**	**89.4**

Source: Adapted from Headrick, *The Invisible Weapon*, p. 38.

their power; and the advantages which have been secured from their cooperation could not in many cases have been obtained from any other body or in any other way.

It is therefore a matter of no small importance that the Government should continue to maintain friendly relations with them.

But this could hardly be expected if the Government were to take an active part in the establishment of a cable in direct competition with the Eastern Company's system...

This competition would not only result in the diversion of traffic from a British Company to an undertaking which would be largely in Colonial hands; but it seems likely that some of the diverted funds would be held to swell the receipts of a foreign company... It should also be borne in mind that any reduction in the gross receipts of the Eastern Telegraph Company might make it more difficult for the latter Company to reduce its rates to India...[50]

So no help should be rendered to any potential competitor to Eastern. Likewise, when British settlers in South Africa began to clamor for a land line from the Cape Colony to Khartoum, Pender instead proposed a submarine cable from Durban (on the east coast of South Africa) up through Zanzibar and Mozambique. After eight hundred British soldiers were killed in eastern South Africa by members of the Zulu tribe in 1879, the Colonial Secretary accepted Pender's offer and even agreed to subsidize the African submarine cable. Several years later, Pender won a similar bid for cable stretching between the Cape Verde Islands and Accra, this time because his lone competitor was linked, according to the Colonial Office, "rather too closely... to a Spanish-owned cable company."[51] Again and again this pattern was repeated. Pender went wherever the British government needed cable to go and once he was there, the government was happy to protect him.

[50]Public Record Office, Kew, England: CO 42/850.
[51]Cited in Robert Jasper Cain, "Telegraph Cables in the British Empire 1850–1900" (unpublished Ph.D. dissertation, Duke University, 1971), p. 182.

It was, to be sure, a rather odd relationship. Eastern was a distinctly private enterprise, and an enormously successful one. Upon his death in 1895, John Pender left a company with a share capital of £6 million and assets that spanned the globe. Pender had never worked for the British government, nor even worked with them in any formal sense. Yet his assets were also Britain's; and it was on the back of these assets, together with her own growing naval power, that Britain maintained control over her far-flung empire. Unlike the situation in the early United States or Europe, there was little confusion in the British cable realm over how to connect the lines or determine rates. There was only one set of rules, and they all belonged to John Pender.

Stateside Rules

In the closing years of the nineteenth century, there was only one other telegraph company in the world that looked anything at all like the Eastern Telegraph Company and that had anything like Eastern's clout or capital or global reach. Its name was Western Union, and it emerged from a wholly different process. While Eastern had grown as an accepted monopoly, Western Union was a scrappy upstart. While Eastern was the product of naval expansion and military needs, Western was the child of landlocked railroads and commercial demands. Eastern's prominence was an accepted thing, supported by the British government and respected by would-be competitors. Western Union's, by contrast, was highly contentious, scorned by competitors and nearly destroyed by government dictate. Yet in the end the two firms were remarkably similar. Each shaped the rules that surrounded their industry and then built an empire upon them.

Western Union wasn't one of the original telegraph firms. It had no great founder or innovator at its helm, no breakthrough technology in its pocket. It arose, rather, from the chaos that had engulfed the U.S. telegraph industry by the early 1850s and the consequent need for some kind of order and stability.

In 1853, when telegraph rivalries and lawsuits were spinning

out of control, Amos Kendall invited all the Morse-related companies to a grandly dubbed American Telegraph Convention.
At this time, competition among the young American firms was
threatening the livelihood of many lines and casting a sinister
shadow across the entire industry. Fights between rival operators and systems had become front-page news, as editors gleefully took sides in commercial races and customers lined up,
like fans, to support "their" telegraph company. Battles were
pitched, and occasionally even described in military terms. "My
men are well armed," one rival reported to Morse, "and I think
they can do their duty." Meanwhile, messages transmitted across
the lines were routinely garbled, lost, or "delayed," and customers were growing frustrated.

Kendall's 1853 Telegraph Convention was designed to ease
these problems, at least among the Morse companies. By coming together, Kendall reasoned, these firms could standardize
their business practices and ease the flow of information along
their lines. In the process, they could also become more competitive. It was the same logic that Kendall had been preaching
since his earliest days with Morse, only couched now in a different kind of language and approached through a different
kind of organization.

Kendall held his convention, appropriately enough, in Washington, D.C. Sixteen firms attended, and after several days they
agreed to lay out some common rules of engagement, including
rights of priority among messages, standardization of signals and
abbreviations, and the mutual responsibilities that the lines had
to each other. They formalized certain language (such as "OK"
rather than "II" to signify a correct message) and dealt with the
complicated issues of tariffs and rate-cutting. The firms also
agreed to come together as the American Telegraph Confederation, and to elect an executive committee that would pursue the
interests of the telegraph industry as a whole. This, as it turned
out, was the first formal movement toward consolidation.

Meanwhile, as Kendall and his colleagues were inching
slowly toward cooperation, a new form of consolidation was
emerging from the West. In 1851, a group of Rochester investors
had formed the New York and Mississippi Valley Printing Telegraph Company. It was one of the dozens of telegraph firms

emerging at that time, and nobody paid it much interest—including potential investors. By 1854, short of funds and running out of time, the company was looking for some other way to enter the telegraph industry. At this point, Hiram Sibley, a former sheriff who had cast his fortune with the company, decided that rather than construct new telegraph wires, he would simply buy up all the telegraph firms that were already dying across the New York and Great Lakes region. To investors, this was further foolishness—proof, indeed, that there was no money to be made from telegraphy. When one friend at last agreed to lend Sibley money, he apparently did so only under a promise of great secrecy. "I'll loan you $5000," he said. "[T]hat means give it to you, for you'll lose it, of course—but you are never to tell that I was such a fool. I believe in you, Sibley, but I don't believe in this telegraphy."[52]

Sibley, though, did believe—not only in telegraphy, but also in the power of consolidation. Like Kendall, he saw that telegraphy would be profitable only once it was organized and orderly. He just chose to go about this organization in a very different way. With only $90,000 in subscribed capital, Sibley began to lease, and then buy, the lines of faltering telegraph firms. First he took the Lake Erie Telegraph Company, weakened by competition with the Erie & Michigan, after which he took the Erie & Michigan itself. He then changed his company's name to the Western Union Telegraph Company and proceeded to buy others' telegraph lines, often for pennies on the original dollar. As Western Union grew, even more prosperous companies quickly came under its sway, joining with Sibley's firm to prevent him from buying a formerly weak competitor. By the time the Civil War broke out in 1861, Western Union's system stretched from the eastern seaboard across to the Mississippi, and from the Great Lakes southward to Ohio.

By this point, Kendall and Morse felt their American Telegraph Confederation to be under direct attack from Sibley and Western Union. And it was. Already, several of the major

---

[52]From Jane Marsh Parker, *How Men of Rochester Saved the Telegraph*, Rochester Historical Society Publishing Fund Series (1926), Vol. 5. Quoted in Thompson, *Wiring a Continent*, p. 252.

Morse firms (including Cornell's) had joined with Western Union, adding to its power and weakening the earlier and looser confederation.

Meanwhile, another pioneer had emerged from along the coast. Flush with the prospects of his Atlantic telegraph, Cyrus Field was beginning to hatch an even bigger dream: a telegraph empire that would link directly to the commercial centers of the eastern United States. In 1855, just as his Atlantic Telegraph Company was beginning to lay cable across the Gulf of St. Lawrence, Field also launched the American Telegraph Company, planning to forge the two systems into an all-embracing transcontinental network. While others in the industry still doubted Field's ability to cross the Atlantic, they didn't doubt the strength of his vision. If Field could cross the ocean, and if he could link his oceanic line to a land-based system, he would control the most powerful communications network in the world. This prospect was simply too much for the others to bear—and too expensive for them to compete against. And thus, nearly two decades after Kendall and Morse first touted the benefits of cooperation, the major U.S. telegraph companies decided to blunt the edges of their brutal competition. It wasn't that they no longer liked the competition, or that anyone forced them to the negotiating table. Instead, they simply decided that the costs of competition had become too steep.

In 1857, representatives of four U.S. telegraph companies— the New York, Albany & Buffalo; Atlantic & Ohio; Western Union; and New Orleans and Ohio—met with American Telegraph to agree on joint procedures and rules for competition. Conspicuously absent from the meeting were some of the older telegraph pioneers such as Morse and Kendall, largely due to the personal bitterness that years of competition had bred. Ironically, it had been Kendall, once again, who had pushed for the meeting and urged his competitors to "devise a plan for harmonizing all interests and protecting existing lines." But when the actual plan was laid out, Kendall—"Amos the pious" to his rivals—was explicitly excluded. The new generation of empire builders, it seemed, had no further patience for the aging pioneers.

Shortly after the initial round of meetings, a sixth company (the Illinois & Mississippi) was drawn into the alliance and a formal "Treaty of the Six Nations" formed. Its terms were stunning. Henceforth, each participating wire company was to be the *sole* provider of telegraph services in a particular area of the country. Western Union, for example, was given the exclusive right to provide telegraph services to Ohio, Indiana, Michigan, most of Wisconsin, and some smaller sections of the east coast. American won the rights to nearly the entire eastern seaboard, and the other firms split key portions of the Midwest and South. All parties to the contract agreed not to compete in each other's territory and to submit any commercial disputes before an impartial arbitration panel. As a final part of the deal, the costs of patents for a new breed of telegraphy, the Hughes system, were to be shared among all members of the group.

The implications of this alliance were profound. It took six of the nation's most prominent firms and brought them into a formal and tightly regulated association. It demanded that these firms renounce competition in favor of cooperation, and bound them to a series of specific, enforceable rules: rules on territory, on expansion, and on patent rights. Most notably, after years of rivalry and fragmented growth, the Treaty of the Six Nations aimed at coordinated, direct control over the U.S. telegraph industry. As Robert Thompson, an insightful historian of the industry, notes: "[T]he new alliance aimed at nothing less than a monopoly of the nation's telegraph business . . ."[53]

Of course, industry insiders such as Kendall and Morse saw these implications as well, and were predictably outraged. In 1857, members of the Magnetic Board (the old conglomerate of Morse companies) decided to renew their alliance and wage an all-out commercial war against the Six Nations. In a remarkable role shift, they planned to compete directly and ferociously with their new rivals, building a second cable line between the major cities of the east coast, as well as a new cable between Europe and North America. The group also intended to lay a series of connecting lines across the South. Their attitude was uncompromising; in an 1857 letter to Morse, Kendall wrote, "I feel a

---

[53]Thompson, *Wiring a Continent*, pp. 316–17.

zeal to punish this perfidy even if my own interests suffer in the process."[54]

While the Six Nations were larger by this point than the Morse group, and considerably more powerful, they still did not cherish the prospect of an expensive and draining competition. So they pursued cooperation with the fervor that had once belonged to Morse and Kendall, floating a series of peace proposals in 1858 and offering to buy patent rights and lease lines directly from Magnetic. Kendall and Morse, though, would not be appeased. Together with the ever-stubborn Smith, they continued to extend their lines and demand exorbitant prices for any possible settlement. Finally, after months of fruitless negotiation, it became clear that the only workable solution was an outright merger between the Magnetic Company and the Six Nations. On October 12, Field's American Telegraph Company, acting on behalf of the North American Telegraph Association, purchased all the patent rights, stocks, and claims of the Morse patent holders. For all practical purposes, there was now only a single organization running the U.S. telegraph industry.

By 1860, then, the situation in the United States was oddly parallel to that of Europe. The anarchy and enthusiasm of earlier days was almost entirely gone, replaced by rules and institutions and a handful of powerful firms. In Europe, most of the key institutions were linked closely to the state: they were the ITU and the state-run companies and the protocols they had hammered out together. In the United States, by contrast, all the firms and nearly all the rules were completely, wholly private. And yet the functions they performed and the ways they behaved were startlingly similar. Across both continents, the leaders of telegraphy had realized—slowly, perhaps, and only after a painful dose of trial and error—that chaos in this business simply didn't work. If there were too many small firms in the industry, or too many competing connections, communication was stymied and commerce halted. To be successful, telegraphy needed to rest on a common set of rules and standards:

[54]Letter from Kendall to Morse, August 27, 1857, quoted in Thompson, *Wiring a Continent*, p. 322.

there had to be easy and reliable connections between various lines, cheap and predictable rates across the entire system, and a shared language. If "OK" meant "yes" on one line, it had to mean "yes" on all connecting lines. Otherwise, the message might easily be garbled to the point of inefficiency.

In a characteristic split, the Americans and Europeans chose to solve these problems in very different ways. The Europeans defined the problem in terms of broader goals—what telegraphy could do for society, what it meant for diplomacy or the military—and generally abandoned market solutions in favor of the state. Where there were private firms in Europe, such as Pender's Eastern Telegraph, they were closely tied to the state and generally responsive to its demands. In the United States, by contrast, the government withdrew from the market in its earliest stages, leaving the field wide open to self-funded pioneers such as O'Rielly, Field, and Sibley. Compared to Europe, there was relatively little concern for public service in the United States, and little interest in telegraphy's public uses. Instead there was an implicit belief that private firms would themselves solve the coordination and regulation problems that were destined to befall their industry. Which they did.

The upshot of this solution, though, in both Europe and the United States, was the emergence of a select group of very powerful companies. In Britain, of course, it was Pender's Eastern Telegraph; in the United States, it became over time Western Union. This was not a wholly predictable outcome, since in 1858, when the Six Nations absorbed the old Morse system, it looked as though Cyrus Field's American Telegraph Company was destined to be the power behind American telegraphy. But the Civil War intervened in 1861 and, as luck would have it, most of Field's assets were strung north to south across the warring states. Western Union, with lines that ran primarily east to west, was in a much safer position. It emerged from the war unscathed and soon became the leading contender for monopoly control. In 1864, Western Union controlled roughly forty-four thousand miles of wire and had a capital stock of $10,066,900. Just one year later, buoyed by an almost insatiable demand for its stock, its capital had risen to $21,063,400. "Nothing else in all business history," one writer proclaimed, "is comparable to the mushroom expansion of this company's capital." Between 1857 and 1867, it grew by roughly 11,000 percent.[55] Meanwhile, the other American companies were groping to find some means of accommodation with this growing giant. But there weren't any. In 1866, Western Union absorbed the United States Telegraph Company (a wartime upstart that had enjoyed a brief meteoric rise) and then merged with the American Telegraph Company. It now possessed a network of nearly one hundred thousand miles of wire and a capital stock of over $40,000,000.

Such size, clearly, had its rewards. Like the Eastern Telegraph Company, Western Union by the 1860s was a company to be reckoned with. It had a virtual empire under its control, and the power to impose its own rules upon customers and competitors alike. Freed from the bloody competition that had marked its earliest years, Western Union was now making a huge profit: its net receipts in both 1867 and 1868 were nearly $3 million.[56] Its founders had reaped tremendous rewards, and early investors built monuments that littered the outskirts of Rochester—Cornell University, for example, was the bequest of Ezra Cornell, who had abandoned the Morse group early on to cast his fortune with Sibley. But size, as the company was about to discover, also has its disadvantages. Once Western Union became a commercial behemoth, it began to draw the scrutiny that thus far had escaped the telegraph industry. It became the topic of opposition and of politics. And it began to be subjected to a whole new range of rules.

Attack on the Union

Ever since the U.S. Congress withdrew from Morse's Baltimore–Washington line in 1845, the telegraph industry in the United States had been resolutely private. While Congress had passed a number of laws regarding telegraphy (prohibiting vandalism on lines, for example, and promoting

[55]Harlow, *Old Wires and New Waves*, p. 259.
[56]Reported in David A. Wells, *The Relation of the Government to the Telegraph* (New York, 1873), p. 6.

the secrecy of messages), its basic position toward the industry had been one of friendly, hands-off support. There was little concern under U.S. law for the social implications of telegraphy, or the military potential, or even the commercial development, of telegraph firms. Instead, Congress's only mission was to protect the *property* that surrounded telegraphy: the patent rights of the technology and the sanctity of physical assets. It was a standard laissez-faire approach, and one that facilitated the frontier spirit of the early telegraph pioneers.

The first crack in this cheerful veneer came during the Civil War, when governments in both the North and South seized telegraph offices and used their facilities for the duration of the war. An even larger crack, though, came after the war, when the undeniable force of Western Union became, for the first time, a subject of popular concern. Suddenly, average citizens were worried that the company had too much power, that it controlled the newspapers and the nation's budding communication industry.[57] They worried about its links to the railroads and the financial prowess of Wall Street. They worried—in a chorus that repeats across the technological frontier—that this firm had simply become too big and too profitable. And this, apparently, was not a good thing. By the late 1860s, segments of popular opinion had moved harshly and aggressively against Western Union. As one historian recalls, the company was now "regarded by liberal and reform opinion and by a considerable segment of the public, with fear, suspicion, even abhorrence. Its enormous capitalization, its merciless elimination of competition toward the close of the Civil War, brought growls of protest, which swelled into a roar of demand that the Government take over the telegraph service and operate it for the benefit of the people."[58]

It is difficult to determine exactly what drove this change in attitude. Part of it was clearly related to a broader shift across the American landscape—a wariness of large corporations and

[57]For more on telegraphy's impact on the shifting news industry, see Menahem Blondheim, *News over the Wires* (Cambridge, Mass.: Harvard University Press, 1994); and Victor Rosewater, *History of Cooperative News-Gathering in the United States* (New York: D. Appleton & Company, 1930).
[58]Harlow, *Old Wires and New Waves,* pp. 330–31.

an urge to beat them back into a more familiar, more manageable size. Only forty years earlier there hadn't been any big businesses in the United States, only local farms and retail shops, and a handful of larger trading companies and financial concerns. Now, in only a few decades, massive enterprises had sprung alive. There were railroads that crisscrossed the country, express companies like Wells Fargo, and the glimmerings of major industrial concerns, such as Procter & Gamble and Carnegie Steel.[59] Telegraph companies were some of the most obvious of these new giants and among the most profitable.

Telegraphs had also slipped by this point into the very mainstream of American life. In 1873, Western Union alone carried more than twelve million messages, a quantum increase over the trickle that had begun only two decades earlier. This growth meant that telegraphs were no longer just the plaything of the rich or an indulgence of Wall Street firms. They were a staple— the backbone, even—of the U.S. communications infrastructure. Businesses used telegraphs to communicate with their suppliers and customers; newspapers relied on "wire services" for an increasingly large share of their news; railroads used the telegraph to conduct operations across their own expanding networks; and average citizens used it to pass along family news. It was this growth, of course, that pioneers such as Morse and Field had predicted and upon which they had bet their careers; it was this growth that had handed Western Union its incredible profits. But such dramatic growth also thrust the telegraph squarely into the public realm. Because once people began to depend upon it, once they saw the telegraph as a necessity rather than a novelty, it suddenly became worthy of policy. The average citizen now cared about Western Union and the fate of the telegraph industry, about telegraph rates and coverage. And so what were once obscure business decisions rapidly became the topic of public discussion and a kaleidoscope of competing demands. Such is the nature of politics along the technological frontier.

---

[59]The classic treatment of this period and its commercial transformation is Alfred Chandler, *The Visible Hand: The Managerial Revolution in American Business* (Cambridge, Mass.: Belknap Press of Harvard University Press, 1977).

The attack on Western Union began quietly. It grew out of a growing murmur of popular discontent, a string of nasty newspaper articles, a cause adopted by groups such as the National Grange and the Prohibitionists, and a belief on the part of numerous government officials that the telegraph should be taken out of the hands of private "monopolists" and converted to a government service. In 1866 these concerns slowly made their way to Washington, where they resulted in a cautious piece of legislation that formally held private telegraph companies to three conditions: first, that they not interfere with travel on rivers and roads; second, that they give precedence, when necessary, to government messages; and third:

> [t]hat the United States may at any time after the expiration of five years from the date of passage of this act, for postal, military, or other purposes, purchase all the telegraph lines, property and effects of any or all of said companies at an appraised value, to be ascertained by five competent, disinterested persons, two of whom shall be selected by the Postmaster General of the United States, two by the company interested, and one by the four so previously selected.[60]

Like many laws, this one was a hedge—echoing public concerns, promising to review them at some distant time, but essentially leaving the industry untouched and unregulated. Once the five-year period drew to a close, however, the same crowd of voices began to clamor for intervention—only this time more loudly, and with the force of growing public sentiment behind them. Criticism during this period focused on several key aspects of Western Union: its size, its profits, and its apparent appetite for devouring would-be rivals. According to the company's opponents, telegraph rates in the United States were considerably higher than they were in Europe: 1½ to 4 times higher, in fact,

---

[60]U.S. Senate, Doc. No. 399, 63rd Congress, 2nd session, *Government Ownership of Electrical Means of Communication: Letter from the Postmaster General, transmitting in response to a Senate Resolution of January 12, 1914, a report entitled "Government Ownership of Electrical Means of Communication," prepared by a committee of the Post Office Department* (Washington, D.C.: Government Printing Office, 1914), p. 20.

and twice as high as in England.[61] This financial gap, they argued, was proof of Western Union's monopoly position and evidence of the benefits that government ownership could bring. If the telegraph were a public concern, rates would be reduced and telegraph operators would focus on the public aspects of their mission: extending facilities to all regions of the country, transmitting weather and other reports, and ensuring that information flowed as freely and easily as possible. As a report by Postmaster General John A. J. Creswell put it:

> The necessity for an efficient and cheap mode of telegraphic communication, which shall be beyond the control of private monopolies, and within the means of all, is daily becoming more apparent. Under the present management the use of the telegraph by the masses of the people is almost prohibited, by reason of arbitrary rates, unnecessarily high charges, and a want of facilities... A Government postal telegraph is the only means by which the full advantage of this great invention can be secured; for, wherever the telegraph is under Government management, it is operated at its minimum cost, and the people receive the benefit in low rates of transmission and greatly extended facilities.[62]

A subsequent report presented before Congress was even blunter:

> It is an outrage upon civilization that one of the greatest inventions of all ages should be permitted to be captured by corporate greed, kept out of reach of the great mass of the people, and reserved for uses in which the business of gamblers forms the chief part... Think of it—50 years

[61]The evidence regarding relative rates was highly contentious at the time, and Western Union argued vigorously against the government estimates. For the government position, see Senate Committee on Postoffices and Postroads, Sen. Report 242, 43rd Congress, 1st session, April 2, 1874. The debate is also well covered in Parsons, *The Telegraph Monopoly* (Philadelphia: C. F. Taylor), pp. 25–38.
[62]Report of Postmaster General Creswell, 1873, in U.S. Senate, Doc. No. 399, 63rd Congress, 2nd session, p. 22.

since the lightning was harnessed to language and literature, and the people cannot even yet avail themselves of
the discovery; 50 years, and gamblers are still the main
beneficiaries; 50 years, and Wall Street is still in possession . . . It is time the telegraph were taken from the gamblers and given to the people.[63]

Similar sentiments appear in a remarkably frank correspondence between William Orton, the president of Western Union
at the time, and Joseph Medill, publisher of the *Chicago Tribune*
and an avid supporter of public telegraphy:

The telegraph is a quicker method of sending the mails; a
method which annihilates time and distance, and, with
the cooperation of the press, "makes all men kin." It is
the noblest of all human inventions, and, while it is a
common carrier, it carries nothing more material than
thought. The lightest of tolls should be charged for its
labors, for it is one of the great educational instrumental-
ities of the nation and world. Its services should be as
nearly free to the whole people as possible.[64]

But Western Union, according to Medill, stood in the way of
this public service. Rather than working for wide access and low
rates, the company objective was to "perform the least service
for the most money." Even worse, it was a company that was
"practically a monopoly," a company with the power to "crush
out, absorb or control all rivals, and exact its own terms from
the public." This frustration with monopoly was a constant
theme among telegraphy's reformers, and a potent one. According to Western Union's critics, the company's sole objective
was domination. It was a conscious monopolist, determined to
destroy any other player in the field and willing, as the postmaster general expressed it, "to use any device which the strong
can employ against the weak."[65]

Another line of criticism was subtler but potentially even

---

[63]Parsons, *The Telegraph Monopoly*, p. 37.
[64]Letter from Joseph Medill to William Orton, December 17, 1872. Reprinted in Wells,
The Relation of the Government to the Telegraph, p. 157.
[65]Quoted in Harlow, *Old Wires and New Waves*, p. 333.

more damaging. According to many of the company's critics, the real danger of Western Union's position lay in its longtime relationship with the Associated Press, a hugely successful news service that had grown up along the backbone of the telegraph industry. For years, everyone in the industry had realized how tightly these two major players cooperated. The Associated Press, which lived from its ability to spread news quickly and reliably across the telegraph network, relied almost exclusively on Western Union.[66] It was a loyal customer, and a mammoth one, accounting by 1872 for nearly 12 percent of Western Union's total receipts.[67] In exchange for this business, Western Union provided the Associated Press with what might now be called a higher level of service—preferential rates, speedier transmission, and a guarantee that its dispatches would get through. When the occasion arose, Western Union also helped to cement its partner's business by denying its lines to competing wire services, or at least tending to lose or garble competing dispatches at a suspiciously coincidental rate.

Such a tight relationship between Western Union and the Associated Press made great commercial sense, but it was also fraught with obvious dangers. Without competition from other wire services or telegraph providers, an alliance between these two giants meant that they could essentially control the flow of news across the United States. Which, apparently, was exactly what they were doing. Any rival to the Associated Press was denied access to Western Union's lines, and any news hostile to Western Union was simply not reported by the Associated Press. According to one congressional report, the Associated Press even cut off its news flow from papers that dared to voice editorial criticism of Western Union.[68] This same report describes a straightforward relationship between the two firms: "The understanding between the telegraph company and the press association secures to the latter low rates and the power of

---

[66]For more on this relationship, see Blondheim, *News over the Wires,* esp. pp. 143–68.
[67]*Ibid,* p. 171.
[68]Senate Committee on Postoffices and Postroads, Sen. Report 242. Reported in Harlow, *Old Wires and New Waves,* p. 332. Similar allegations were made in hearings before the International Typographical Union in 1894, reported in Parsons, *The Telegraph Monopoly,* p. 83.

excluding new papers from the field, and to the former a strong influence upon press dispatches, the support of the papers in such associations, and the right to transmit and sell market quotations."[69] Congressman Sumner of California was more direct: Western Union, he declared, "has a twin connection with another incorporated thief and highway robber known as the Associated Press. They are banded together in the strong bond of mutual plunder and rapacity against the people."[70] It was a powerful combination, indeed, and one that infuriated those who stumbled upon it.

Between 1870 and 1873, two bills were put up for congressional consideration: the Washburn Plan and the Hubbard Plan. Both aimed for governmental control over the telegraph industry, though through very different channels. Under the terms of the Washburn proposal, the U.S. government, like that of the British before it, would simply buy up all existing telegraph lines and place them under the authority of the postmaster general. Under the Hubbard plan, existing telegraph companies would remain in place but would be joined by a new firm, the Postal Telegraph Company, which would be funded partially by the Post Office and dedicated to providing telegraph services at specified rates.

Western Union was outraged. Seeing itself as a beacon of free enterprise, it attacked both plans as political patronage and mounted a full-fledged offensive against Washington. In a series of pamphlets and public debates, officials from the corporation openly mocked the government's efficiency and questioned its motives; they spoke of "despotic" interests in political espionage and the opportunities for political manipulation of information. In 1873, the company even commissioned a special study to chart the progression of the telegraph industry and investigate the proposals before Congress. Not surprisingly, the study's author found that the telegraph was more extensive and more efficient in the United States than in any European country. Telegraph receipts contributed heavily to the government's tax coffers and telegraph rates, contrary to government reports, had

---

[69]Quoted in Harlow, *Old Wires and New Waves*, pp. 332–333.
[70]Quoted in Harlow, *Old Wires and New Waves*, p. 334.

actually declined over time. No, the report concluded, the question of a government telegraph had little to do with either cost or efficiency. What was really at stake was freedom. "Will the people consent to the inauguration of a policy," the author asked, "which revives the old Medieval doctrine of the necessity of State interference with the pursuits and business of the people, and every step in the carrying out of which is a departure from Republicanism and an approach towards despotism and monarchy?"[71] Will they adopt a policy "in direct antagonism with and destructive of the fundamental principles upon which the Government itself has been established"?[72] With the argument put that way, it was rather difficult to argue in favor of the government's schemes.

In the end, Western Union won. It didn't win cleanly, but it won. Both the Washburn plan and the Hubbard plan were defeated, and Western Union continued to run its business precisely as it always had: devouring competitors, aggressively expanding, and building a telegraph network that stretched across the United States. The concerns that had led to the Washburn and Hubbard proposals, however, didn't disappear upon their defeat. Instead, they continued to plague the telegraph industry—and particularly Western Union—for at least another fifty years. By 1900, seventeen separate legislative committees had recommended government ownership of the telegraph industry. None of these bills, however, was ever passed into legislation.

Echoes and Refrains

Compared with the history of oceangoing trade, the story of telegraphy seems small and well defined. It begins with a precise moment—Morse's 1838 presentation to Congress—and ends roughly a century later, when piggyback developments in radio and telephony eroded the telegraph's monopoly over fast and efficient long-distance communication. There were

---

[71]Wells, *The Relation of the Government to the Telegraph,* pp. 55–56.
[72]Wells, *The Relaion of the Government to the Telegraph,* p. 41.

only a handful of true pioneers in the telegraph industry, and a much less colorful crop of pirates.

Within this refined context, however, telegraphy still exhibits the classic attributes of a frontier technology. It emerged from an early phase of innovation, peopled largely by inventors (Morse, Wheatstone, Bain and House) who reveled in the technology for its own sake and figured out how to make it work. Then, once the technology was proven effective, a new cast of pioneers moved in, men such as Cooke, Field, O'Rielly, and Pender, who saw the commercial potential of the "lightning" and fashioned a market upon it. This was a time of empire building and creative anarchy, when there were no rules governing telegraphy and no great demand for them.

As the telegraph market developed, though, it gradually encountered the familiar problems of the technological frontier. Here, the greatest problem was coordination: the problem of too many disparate lines and different codes; of mismatched signals and incompatible languages; of incompatibility that reached such proportions by the 1850s that it threatened to topple the industry. At this point, even the staunchest individualist realized the nature and extent of the quandary. For if the sprouting web of telegraph lines were not made compatible, the value of every strand would be reduced; and if users could not communicate seamlessly and confidently, they might not communicate at all. To ensure the success of their industry, the telegraph pioneers had to create rules of engagement—norms, standards, codes, and prices that enabled them to link their networks and speed the messages across.

In both the United States and Europe, therefore, the problem of coordination created both a demand for and supply of rules, the third and fourth phases of the technological frontier. But the problem manifested itself differently on the two continents and generated a strikingly different kind of response. In Europe, national governments stepped in shortly after the problem of coordination was identified. They realized the issues that incompatibility had already created, foresaw the difficulties that were likely to emerge over time, and recognized that in Europe at least, the core of the problem lay with interstate boundaries and international variation. Accordingly, they crafted a solution that attacked coordination head-on, imposing common rules and technical procedures across nearly the entire continent. Because most of the European telegraph firms at this point were either owned by state governments or closely related to them, this kind of top-level state-sponsored agreement essentially solved the coordination problem in Europe. After the telegraph conference of 1865, the European telegraph industry was stable, orderly, and linked distinctly to the European governments and the newly created ITU.

In the United States, by contrast, the state pulled out of the market well before the problem of coordination even arose. Private firms were left, as a result, to deal with anarchy on their own terms and craft a private solution for it. Between 1845 and 1860, the largest and most entrepreneurial U.S. telegraph firms fashioned themselves into a series of constantly shifting alliances: the Magnetic Telegraph Company, Kendall's American Telegraph Convention, the Treaty of Six Nations. Despite the lofty language that surrounded these efforts at the time, and despite their different membership lists and structures, each of these groups had essentially the same purpose. They were all designed to impose some kind of order on the unruly world of telegraphy and to set some common standards across separate, often rival, firms. They were efforts, in other words, at private governance, and to some extent they worked. By 1860, coordination problems no longer plagued the U.S. telegraph industry and anarchy had given way to order.

As a first cut, then, it appears that the U.S. telegraph firms addressed their industry's growing pains in a novel and efficient way. They joined forces, they created common standards, and they set rules in the absence of the state. A perfect solution, perhaps, for a frontier industry, since it avoided the regulation that characterized the European market, and perhaps constrained it, nearly from the start. Certainly this is the kind of private governance that prevails in many cyber-markets and appeals to these markets' most zealous prophets.

The problem with this type of solution, however, is that it puts governance in the hands of what essentially becomes a cartel. And because cartels are themselves so fragile, it leaves governance vulnerable to a whole new set of concerns. Most

dramatically, any private arrangement, like any cartel, is susceptible over time to the warring interests of its members. In a cartel, after all, members come together to pursue a common goal,
something that will benefit all those who choose to participate
in the arrangement. Yet beneath this cooperative umbrella, members generally continue to compete with one another and contend for each other's markets.[73] If the scope of their cooperation
is narrow enough (focusing, say, on government affairs or response to an external threat), then the members may be able to
sustain this delicate balance between cooperation and competition. The wider and more important cooperation becomes, however, and the closer it moves to the core of an industry, the
harder it will be for even the most devoted participants to sacri-
fice individual interests to the common good.

Certainly this was the case in the U.S. telegraph industry.
Even though the members of each agreement realized how vital
cooperation was, and even though they acknowledged the difficulties of coordination and were determined to resolve them,
they still weren't able to restrain their own competitive tendencies. The Magnetic Company, for example, suffered from the
ruinous rivalry between O'Rielly, Smith, and Kendall; the
American Telegraph Confederation was doomed once its members began to defect to Western Union; and the Treaty of Six
Nations was dogged from the outset by Morse and Kendall's
hostility. In each of these cases, the industry thus destroyed its
own governance mechanism. Private firms could set the rules, it
appears, and they could organize their market for some period
of time, but in the end cooperative governance was killed by
competitive feuds.

The implications of this progression are profound. For if private governance is always subject to the rifts that appeared in
the telegraph industry, then private rules will be, at best, a
short-term solution on the technological frontier. If private
firms can regulate themselves for only some period of time, and
if their regulation is ultimately liable to their own competitive

---

[73]For more on the internal dynamics of cartels, see Debora L. Spar, *The Cooperative Edge: The Internal Politics of International Cartels* (Ithaca, N.Y.: Cornell University Press, 1992).

interests, then private rules must yield at some stage to other types of governance. They must either be supplanted by formal rules—government regulation of some sort or another—or by the dominance of a single private player.

In the case of the U.S. telegraph industry, both of these solutions eventually appeared. After the Treaty of Six Nations dissolved, a single private firm began to dominate the industry—a firm so big and so powerful that it was able, by itself, to complete the piecemeal efforts of the cooperative groups. Through sheer market force, Western Union managed to create the kind of order that in Europe came only from the state. It consolidated a tumultuous industry, set technical and commercial standards, and presided over what quickly became a seamless network of communication. Or, as the president of Western Union proclaimed to his shareholders in 1869: "Practical men saw that there was but one remedy [for the telegraph industry's ills] . . . and that was by a consolidation of all the rival interests into one organization."[74] Theoretically, then, Western Union crafted a nearly perfect solution to the problems of commercial anarchy. It was private; it was effective; and there was no government involved.

The problem with this solution, though, was that it essentially created a monopoly. And Americans, much as they like effective private solutions, have a deep-seated dislike for any firm that resembles a monopoly. This dislike is not necessarily connected to an underlying economic rationale. Indeed, in the telegraph industry—as in telephony or postal services—there are many elements of natural monopoly involved and many reasons why, economically at least, a single firm may be the most efficient provider of universal service.[75] In Europe, where such reasoning predominated, governments created national monopolies from

---

[74]William Orton, *Annual Report of the President of the Western Union Telegraph Company to the Stockholders, July 13, 1869*, p. 6. Quoted in DuBoff, "Business Demand and the Development of the Telegraph," p. 463.

[75]For more on the theory and characteristics of natural monopolies, see Hal Varian, *Intermediate Microeconomics* (New York: W. W. Norton & Company, third edition, 1993), pp. 410–12; and Stephen Breyer, *Regulation and Its Reform* (Cambridge, Mass.: Harvard University Press, 1982), pp. 15–35. For a discussion of naturally monopolistic tendencies in telegraphy, see DuBoff, "Business Demand and the Development of the Telegraph," pp. 459–79.

the start, organizing the telegraph industry like the postal service, and ceding it to a single, national firm. In the United States, by contrast, the government decided not to enter the industry or even to treat it as a natural monopoly. When monopoly emerged, therefore, it was a private monopoly, a creation of the market rather than of the state. And it ignited a fierce storm of demands. Defeated competitors wanted constraints placed on Western Union's power; newspapers lobbied for regulation of rates and business practices; and consumers chafed at being beholden to a monopoly. All of these demands, ironically, involved government's intervention—public rules, in other words, to fix a private problem.

In the end, therefore, public rules found their way back to telegraphy. In 1910, the U.S. Congress passed the Mann-Elkins Act, giving the Interstate Commerce Commission authority to review telegraph rates. Two decades later, it passed the Communications Act of 1934, which formally transferred regulation of the telegraph industry to the newly created Federal Communications Commission.

By this time, though, the technological frontier had moved on. Radio and telephony were now the waves of the future, the technologies that prophets were trumpeting and pioneers were scrambling to commercialize. Telegraphy remained, of course, as did a considerably diminished Western Union and Britain's Cable and Wireless, the nationalized heir to Pender's great Eastern Telegraph Company. But the energy and anarchy that had characterized the industry in its earlier years was gone, replaced by a stable group of firms and a predictable framework of rules.

In the United States and Europe, therefore, rules emerged fairly rapidly in the telegraph industry and governments played a significant role in their creation. To be sure, private rules played a much larger role in the United States than they did in Europe, and private regulation proved, for some time at least, to be a viable alternative to formal governance. Once private governance fell prey to the internal constraints of cartels and external complaints against monopoly, however, even the U.S. government stepped back into the void, providing a basic regulatory structure for what had become a stable and orderly market.

The U.S. government also played a subtle but critical role in the earliest days of telegraphy. In the 1840s, when the telegraph industry was engulfed by anarchy, U.S. courts aggressively defined who held the key telegraph patents and the rights to commercialize them. It was the courts that struck down O'Rielly's line for borrowing too freely from Morse's patent, the courts that upheld Morse's patents against hordes of contenders and yet also protected the validity of House and Bain's rights to separate, albeit competitive, designs. In the United States, therefore, the problem of piracy was curtailed by the clarification of property rights—a clarification that only the state could provide. While rivals in the industry were loose with their terminology and quick to label each other as pirates, the U.S. court system clearly laid out who had patent rights, and thus property rights, in the market. As a result, pirates in the telegraph industry were always more of a nuisance than a threat. They never attacked the fundamental economics of the industry, or questioned its underlying property rights. And without pirates to worry about, pioneers such as Cyrus Field and Hiram Sibley were able to move aggressively into this bold new space, creating an industry that was once considered the most amazing thing on earth.

CHAPTER 3

Radio Days

Transmitter. Picking up something good.
Radiohead—the sound of a brand new world.

TALKING HEADS,
"RadioHead"

On a blustery day in 1901, a young Italian inventor
was flying kites in Newfoundland. They weren't ordinary kites,
to be sure, and he certainly wasn't flying them for pleasure.
Rather, Guglielmo Marconi was trying to receive a tiny message: the letter *s*, three short dots of Morse code, sent from a
radio transmitter in England, nearly 2,200 miles away. It took
four days and a series of mishaps before the message came
clearly through. Gales blew in, and balloons sailed off, and for
hours Marconi could hear only static. But then, on December 12,
he detected three sharp clicks: a lonely "s" transmitted, wireless,
across the sea. This was the start of transoceanic radio transmission and a turning point for the Marconi Wireless Telegraph
Company, one of the most important firms of the early twentieth century.

In Marconi's time, radio was a revolutionary technology. While its impact on time and space was arguably less dramatic than the telegraph's had been, it involved the same function— conveyance of messages across a long distance—and packaged it into a fundamentally new medium. With radio, someone like Marconi could stand alone on a beach (or a cliff, or a plain) and boom an invisible message across the air. There were no messy wires involved in radio, no extensive infrastructure or tapping intermediaries. It was instead a pure signal, sailing unencumbered over tens, and hundreds, and then thousands of miles. With radio, communication promised to leap from earth to the "ether," free of the apparatus that had bound telegraphy and demanded both rules and capital. With radio, communication became an intoxicating combination of fantasy and science: an unseen electromagnetic wave that captured the imagination of its users and confounded even the experts who had created it.

Eventually, what would make radio particularly powerful was its ability—its inherent capacity, really—to send a single message across a wide and unidentified audience. With telegraphs and telephones, communication was a limited, private affair. Messages moved across a fixed wire from one spot to another, with the sender and receiver clearly marked at each end. With radio, the wires came down and the audience expanded. Whoever was tuned into a signal could receive it—easily, unobtrusively, and free of charge. This was broadcasting in its earliest and purest state, designed literally to scatter and disseminate sound. In other eras and with other means, broadcasting was something people did by choice. Orators would choose, for example, to shout their words across a gathered crowd; newspapers chose to sell their stories to the masses, rather than just to a select group of subscribers. With radio, though, broadcasting was technologically determined. Once the signal was released into the air, there was no feasible way for the sender to control who received it. It was just out there in the ether, free to anyone with the means and inclination to listen.

This technical trait had wide-reaching implications. Commercially, it was at first infuriating, since it effectively publicized what had originally been a private function. How could someone make money by conveying messages that all could hear?

How could companies profit from an unidentified, nonpaying customer base? In the early days of radio, no one really knew. As an 1899 article in *Electrical World* commented:

> Telegraphy without wires—how attractive it sounds . . . A little instrument that one can almost carry in the pocket, certainly in a microscopic grip, and if your correspondent be likewise equipped, you may arrest his attention and talk to him almost any time or place, with no intervening medium but the ether . . . But will it pay? . . . NO.[1]

Politically, meanwhile, broadcasting was even worse. Suddenly, messages could be trumpeted across national borders and against national interests. Suddenly, information was free—potentially at least—from any attempt to control it. It was a technological triumph for pioneers such as Marconi, but a nightmare for governments that were just beginning to realize the strategic importance of mass communication. For radio, like the Internet, was an inherently subversive technology.

Not surprisingly, then, governments took an early and intrusive interest in the radio industry. Unlike the cases of navigation and telegraphy, where governments stepped in once the basic patterns of commerce had been developed, regulation of radio occurred in several frantic bursts. Governments took an active role during the formative years of "wireless," exploring how this new technology might be converted into a military tool and funding the men who claimed to be able to do it. They then used radio extensively in World War I and scrambled after the war to translate their control into a peacetime use. During this period, most private firms were either wooed into naval service or scared away from the radio industry. Those that survived, like Marconi's Wireless Telegraphy Company and Germany's Telefunken, grew into commercial giants, igniting the same concerns over monopoly that had entangled Western Union. As peace resumed, however, a wider commercial interest in radio emerged

---

[1] *Electrical World*, June 10, 1899. Quoted in Susan J. Douglas, *Inventing American Broadcasting, 1899–1922* (Baltimore: The Johns Hopkins University Press, 1987), p. 1. For a broader treatment of how industries create business models to align with new technologies, see Henry Chesbrough and Richard S. Rosenbloom, "The Dual-Edged Role of the Business Model in Leveraging Corporate Technology Investments," Harvard Business School, Mimeo, December 1999.

anew, sparked this time by a new round of technological innovation and a glimmering idea that radio might be used to convey music and news rather than just messages.

Yet not even these new pioneers were able to escape government scrutiny or the extending arm of government regulation. Instead, a new wave of formal rule making broke out, driven this time by a curious mix of social and commercial concerns. Socially, the telegraph, like the sextant or telephone, had been constrained by its own finiteness. While it conveyed information quickly and far, its reach was inexorably limited to any two people at a given time. Radio, by contrast, was much closer in temperament to the printing press or the Internet: it had low entry costs, no discernible property rights (initially at least), and an audience that stretched for thousands of miles and encompassed millions of people. This was a mass medium in its purest and most exhilarating form. And it scared many governments to their national-security cores. In the 1920s, therefore, just as broadcast radio was becoming technically feasible, governments across Europe hastened to halt the industry's growth and tie it instead to the service of the state. Rules were created, regulation imposed, and radio in Europe became a stable, government-run enterprise.

In the United States, by contrast, radio remained vividly and energetically private. Pioneers rushed onto the airwaves in the 1920s, fiddling with the emerging technologies and testing all sorts of business models. Radio was the hottest new market of its time, a playground for entrepreneurs and a virtual laboratory of experimentation. The problem, however, soon became evident. With so many pioneers clambering into the ether, the airwaves quickly dissolved into a cacophony of sound. Broadcasters couldn't reach their audiences; listeners couldn't hear their favorite programs; and the entire system ground to a shrill and roaring halt. To solve the problem, private firms turned—reluctantly at first, and then desperately—toward the government, demanding rules and property rights that could break the anarchy of the airwaves into manageable chunks. Which, by the 1930s, was precisely what they got.

The story of radio is thus more nuanced than those of telegraphy and transoceanic trade. There is less of a direct migration from anarchy to order, less of a direct role for private standards and cooperation. Yet the brunt of the story still plays out the same way. Like Icarus, radio's pioneers leaped gleefully into the sky, confident that technology would enable them to escape the political and commercial bonds of earth. They succeeded brilliantly for a while, pushing the edges of the technological frontier and reaping the rewards of a new and unregulated market. But eventually the rules moved in: patents that protected radio's core technologies; antitrust regulation that strove to maintain competition in what many claimed was a naturally monopolistic industry; property rights that gradually defined the radio spectrum and parsed its frequencies out to individual stations. Many of these rules helped to cement claims of those firms that were already dominant in the radio field; some were even the direct creation of the firms themselves. All of them, though, did what observers in Marconi's time would have thought impossible: they took the radio industry out of the skies and brought it crashing back to earth.

A Man and His Machine

In 1896 a young Italian inventor arrived in England. His name was Guglielmo Marconi and he carried with him a small black box that he called a wireless telegraph. According to Marconi, this "wireless" was essentially a telegraph without the wires. It could send messages several miles through the air using only the atmosphere to carry them. Several months earlier, Marconi had introduced his invention to representatives of the Italian government, but they'd shown no interest. So now he was off to Britain, hoping for a more cordial response. The early signs, though, weren't good. As soon as Marconi crossed into England, an overzealous customs official seized his machine and destroyed it on the spot, arguing that any device that could pass information behind the government's back was inherently dangerous. So Marconi retreated and tried again. This was the start of the radio age.

Whether dangerous or not, there is little evidence that Marconi's machine was radical, or even particularly inventive. Indeed Marconi, like Morse before him, was a tinkerer much more than a scientist, an amateur who had stumbled quite accidentally on to other people's work and research. For decades—centuries, even—scientists and philosophers had speculated over the existence of "ether," a sort of otherworldly substance that surrounded matter and explained physical properties such as light and magnetism. The discovery of induction in the late 1820s pushed these speculations further, leading scientists to argue that electrical impulses could be carried invisibly through space: a change in the magnetic force applied to one wire could "induce" a similar change in another, unconnected one. Such hypotheses reached a peak in 1865, when James Clerk Maxwell, a Scottish mathematician, proved logically that both electricity and light moved through the air—the ether—in a series of equally rapid waves. Maxwell didn't identify these waves; he couldn't capture them, but he showed that, theoretically at least, they were out there. For the next two decades, physicists around the world tried to capture Maxwell's waves, producing or detecting the phenomenon that he had described. Finally, in 1887 a German scientist named Heinrich Rudolph Hertz managed to demonstrate physically what Maxwell had mathematically proposed. In his laboratory in Bonn, Hertz connected two ends of a coil of wire to the opposite sides of a small gap. He then sent a high-voltage spark ripping across the gap. This was straightforward stuff by the 1880s, and many scientists had already demonstrated electricity's ability to jump gaps such as these. What Hertz found, however, was that as the spark rushed between the wires, a much smaller spark, only fractions of a millimeter long, flared between two other wires, suspended in a similar configuration on the other side of the room. Somehow, the big spark had radiated energy across to the smaller one. The conduit, Hertz hypothesized, was electromagnetic waves—the same waves that Maxwell had described.

Hertz's discovery was important to both scientists and entrepreneurs. To scientists, it was a major breakthrough—proof of Maxwell's theory and evidence of a world of waves suspended in the ether. To entrepreneurs, meanwhile, it was an obvious link to a new means of communication, a way of sending telegraphic messages without the telegraph's cumbersome and costly network of wires, stretching for miles over the ground. Yet ironically, even though dozens of entrepreneurs were struggling in the late nineteenth century to create a wireless telegraph, none pounced quickly enough upon the practical implications of Hertz's experiments. That honor fell instead to Marconi, an Italian of relatively little education who in 1887 hadn't even heard of wireless prospects or Maxwell waves.[2]

Indeed, in 1887 Marconi was just a boy of thirteen, a lackluster student born to a prosperous Italian father and a well-connected Irish mother. While Marconi had demonstrated an early interest in science and music, he never studied either one formally and even failed to gain admission to the University of Bologna. Angry at his son's inattentiveness, the elder Marconi tried to pull Guglielmo into the family textile business. But Mrs. Marconi had different ideas. Born into a prosperous Irish whiskey family, Anna Marconi (née Jameson) was fascinated by her son's scientific leanings and determined to promote them. Somehow, she managed to persuade Augusto Righi, an eminent professor of physics at Bologna, to allow her son to attend his lectures and use the university's laboratory facilities. Righi was not at all impressed by his student but, to please Anna, he went along.

In 1894, Heinrich Hertz died an early death, leaving behind him a host of scientific admirers and a series of discoveries that were still years or even decades ahead of their practical application. Marconi, who was then twenty, became fascinated with Hertz's work and determined to construct a wireless telegraph based on it. In the attic of his family home, he copied the basic coil and spark gap that Hertz had built, and then added to it a modified spark gap that Righi had suggested. To Hertz's receiver (the smaller spark gap), he also added a device known as a Branly coherer: a tube with electrical contacts at either end and metal dust suspended in the middle. Invented by Edouard Branly in 1891, the coherer served to amplify the energy that

---

[2]As one might expect, there is a fair amount of debate over whether Marconi can truly be called the "father of radio." There were many other scientists who invented key parts of Marconi's apparatus and several who had cobbled together a version of the wireless telegraph before Marconi received his key patent in 1897. In Russia, a scientist named Alexander Popov demonstrated a Hertz-based radio receiver in 1895 and is still revered there as radio's inventor.

was sent from the transmitter (the larger and originating spark) to the receiver. Finally, Marconi attached the coherer to a battery, and then attached the battery to a Morse printer. Here, in primitive form, was a wireless telegraph, set to record messages beamed electronically from the transmitter across the ether to the receiver.

From its first trials, Marconi's wireless worked. The signal could indeed leap to the receiver, and the receiver could record the pulses as the dots and dashes of Morse code. It just couldn't work very far. So Marconi methodically began to tinker in his attic, trying different metals in the coherer and different kinds of electrical contacts. Slowly, and in a completely ad hoc fashion, he discovered that nickel mixed with silver worked best as the coherer and that platinum wires made the best connectors. The distances began to increase. By the summer of 1895 Marconi had moved out of his attic and into the yard below, pushing his range even farther by adding large metal plates to the spark gap and coherer and then suspending one of the plates high above the ground. Soon he was sending messages a full kilometer away.

It is worthwhile to recall at this point that Marconi's "wireless" was not really new. In Russia, an eminent scientist named Alexander Stepanovich Popov had constructed an eerily similar device in 1895, and various Western inventors had already built the key pieces and stages of Marconi's device: the coherer, the receiver, the printer. There was thus no scientific breakthrough inherent in his wireless, no particular bit of genius. Rather, what differentiated Marconi's wireless from the inventions of would-be competitors was that Marconi put it all together, made it go a little farther, and—critically—figured out how to sell it. Even if Marconi didn't really invent radio, then, he invented its market. Which, in the long run, proved equally important.

In Marconi's eyes, there were two great advantages to wireless technology: it got rid of wires, obviously, and it could be used at sea. The logical customers for wireless, he thus reasoned, were national governments, which could use the technology to reduce their reliance on submarine cables and to communicate with their naval fleets. Both of these demands seemed particularly acute in the final decade of the nineteenth century, as Britain was slowly gaining control over most of the world's cables and international tensions in Europe were growing more acute. And so Marconi went to the Italian authorities and urged them to invest in his wireless project. When they demurred, he turned to England, a country with a clear interest in shipping, a dominant international naval presence, and, as it turns out, some profitable connections on his mother's side.

After his unfortunate experience at customs, Marconi retired to the care of Henry Jameson Davis, his mother's cousin and a well-placed member of the Jameson whiskey family. Jameson Davis took the young Marconi under his wing, helping him to repair his equipment, contact a patent office, and arrange for public demonstrations of his work. In a critical move, Jameson Davis also referred Marconi to another research scientist, Campbell Swinton, who was sufficiently impressed to write to William Preece, chief engineer of the British Post Office. This contact would prove to be a turning point for Marconi and for the subsequent development of radio.

Preece was already well versed in the motives and mechanics of wireless telegraphy. Fascinated by the same phenomena as Marconi, he too had tinkered with the practical applications of Hertzian waves and had, by 1892, already sent experimental messages across a three-mile span of the Bristol Channel. In a rare bit of scientific generosity, however, Preece immediately saw the advantages of Marconi's system. As he recalled in an 1897 interview:

> While I cannot say that Marconi has found anything absolutely new it must be remembered that Columbus did not invent the egg. He showed how to make it stand on end.[3]
> Marconi shows how to use the Hertz radiator and Branly coherer. He has produced a new electric eye, more

---

[3]Preece refers to the apocryphal story of Christopher Columbus who, at a dinner in his honor, was challenged by skeptics who claimed that he had achieved nothing spectacular in his voyage across the ocean. "How difficult could it be to sail westward," they asked. Columbus did not reply, but reached for an egg from a nearby dish and challenged them to make it stand on end. When they all failed and said it was impossible, Columbus broke one end of the shell a bit so that the egg could easily stand upright. His lesson to them: anything looks easy once someone else has led the way and shown how it can be done.

delicate than any other known system of telegraphy which will reach hitherto inaccessible places.[4]

Preece suggested that Marconi conduct several tests in his own laboratory. When these proved successful, Preece urged his supervisors to allow Marconi to hold a public demonstration of his invention under the auspices of the General Post Office. This occurred just months after Marconi's arrival in England and only two years after he had first read of Hertz's experiments. On July 27, 1896, Marconi stood before a crowd of spectators and successfully transmitted a signal from the Post Office building in St. Martins-le-Grand to Queen Victoria Street, about a mile away. Two months later he covered a distance of $1\frac{3}{4}$ miles across the Salisbury Plain, and six months after that he publicly transmitted a signal over $4\frac{1}{2}$ miles. In that same month, March 1897, Marconi secured a British patent for "improvements in transmitting electrical impulses and signals."[5]

By this time, the twenty-three-year-old Marconi was a star. He was feted in London as "the wizard of wireless" and received royal treatment back home in Rome. He convinced the Italian navy to use his machines on their ships and then, in July of 1897, formed the Wireless Telegraph and Signal Company with £100,000 in capital (raised mostly by family and friends) and Jameson Davis as managing director. He continued to work closely with the British Post Office, War Office, and Admiralty.

Meanwhile, either by chance or by design, Marconi had also honed his marketing skills. In 1898 he covered the popular Kingstown Regatta for the *Dublin Daily Express,* sailing on a ship called the *Flying Huntress* and sending up-to-the-minute dispatches back to the paper's offices. So intrigued was the public by this invisible coverage that the *Evening Mail,* another Irish paper, published a forty-eight-page supplement on Marconi's reporting that was subsequently translated into Italian.[6] Shortly

[4]Quoted in Orrin E. Dunlap Jr., *Marconi: The Man and His Wireless* (New York: The Macmillan Company, 1937), p. 55.
[5]Hugh G. J. Aitken, *Syntony and Spark: The Origins of Radio* (New York: John Wiley and Sons, 1976), p. 209.
[6]Supplement to *Dublin Evening Mail,* July 21, 1898, recounted in W. P. Jolly, *Marconi* (London: Constable & Company, 1972), p. 54.

thereafter, Marconi scored another major coup, supplying the aging Queen Victoria with a private wireless station so that she could keep in touch with her son, the Prince of Wales, who had injured his knee and was recuperating on the royal yacht. Apparently, the queen was quite impressed. When Marconi sailed to New York in 1899 he was hailed as a hero. His wireless coverage of the America's Cup Yacht Race was described by the *New York Herald* as "a feat unparalleled in the history of journalism. . . . A boon not only to science but to millions of persons who await with eagerness the result of a contest that has excited more interest than any in the history of America's Cup."[7] "We are learning," mused even the staid *New York Times*, "to launch our winged words."[8] All this just five years after Marconi left his parents' attic.

Commercially, though, Marconi was on somewhat more delicate ground. Although his company had contracts by 1899 to equip the Italian Navy and British War Office with wireless equipment, there was a certain lack of trust that surrounded these contracts and tinged Marconi's business relationships. Essentially, Marconi found himself caught by the political implications of his own commercial strategy. He was providing governments with what he described as vital public services: the ability to contact ships at sea, to communicate over long distances, and to break national dependencies on foreign-owned cables. Yet he expected to get paid for these services, and highly: £100 per radio set plus an annual royalty of another £100.[9] This was not a commercial relationship that sat comfortably in the Europe of the late nineteenth century. On the continent, it was seen as just another extension of British power (since Marconi was perceived by this time as an Englishman). And in Britain it

---

[7] *New York Herald*, October 1, 1899, section 4, p. 1. Cited in Douglas, *Inventing American Broadcasting*, p. 10.

[8] Quoted in Douglas, *Inventing American Broadcasting*, p. 23.

[9] These figures refer to Marconi's 1900 deal with the British navy, as recounted in Aitken, *Syntony and Spark*, p. 232. Pocock, describing terms that were eventually agreed to in 1901, gives slightly higher figures: £196 14s 4d per set, plus an annual royalty of £100. And Maclaurin gives even higher figures for the terms that applied by 1903: £20,000 down; £1,600 for each installation; and £5,000 in annual payments. See Rowland F. Pocock, *The Early British Radio Industry* (Manchester and New York: Manchester University Press, 1988), p. 156; and W. Rupert Maclaurin, *Invention and Innovation in the Radio Industry* (New York: The Macmillan Company, 1949), p. 37.

was seen as an affront of sorts, since Marconi had gotten his start from the British Post Office but then quickly—and inexcusably, to many—had favored the creation of his own private firm.

For a while these resentments simply simmered under the surface of Marconi's business dealings. But then, slowly, political interests started to chip away at his commercial position. This was a dynamic that was to haunt the Marconi Company for the remainder of its existence.

The first manifestation came from competing systems—companies set up explicitly to compete with Marconi, often backed by national funding and based on technologies that Marconi claimed as his own. The most potent of these emerged in Germany, where Adolph Slaby, a researcher in electricity, had created a Marconi-type wireless system in 1899. Slaby, who had witnessed Marconi's demonstration at Salisbury Plain, was explicit in his praise of Marconi and his debt to him. As he later recalled:

> In January of 1897, when the news of Marconi's first successes ran through the newspapers, I myself was earnestly occupied with similar problems. I had not been able to telegraph more than 100 metres through the air. It was at once clear to me that Marconi must have added something else—something new—to what was already known, whereby he had been able to attain to lengths measured in kilometres. Quickly making up my mind, I travelled to England, where the Bureau of Telegraphs was undertaking experiments . . . and in truth what I saw there was something quite new.[10]

Yet after Slaby and Marconi failed to establish a proposed partnership, the German had no compunction about starting his own company (backed by government funding) and competing directly with Marconi. In 1898, Slaby joined with Count von Arco, a German aristocrat, and AEG, a leading German electrical firm, to form Slaby-Arco-AEG. Some years later, this firm was joined with Karl Ferdinand Braun's Braun-Siemens-Halske to create the powerful Gesellschaft für drahtlose Telegraphie, known generally as Telefunken. Russia, meanwhile, had

[10]Quoted in Dunlap, *Marconi*, p. 56.

its own system based on Alexander Popov's work, and France a device constructed by Eugene Ducretet, a former maker of scientific apparatus. By 1900, therefore, as naval tensions rose across Europe, there were four competing wireless systems, none of which was fully compatible with the others, and three of which were competing directly with Marconi.

Nationally based competition also affected Marconi's relationships with nonmilitary customers. In Britain, for example, one of Marconi's largest and most obvious targets was Lloyds of London, the giant association of marine insurance underwriters. Lloyds had lots of ships with which it needed to stay in contact; Marconi had a clear-cut interest in outfitting as many ships as possible. It was a perfect fit. And thus, as early as 1897, Lloyds had begun to experiment with Marconi technology. Once a German system became available, though, Lloyds began to investigate that as well, finding it in many ways superior to Marconi's. It pulled back in the end and stayed with Marconi, but only after negotiations that became progressively more hostile and included, at one point, an explicit threat by Lloyds to develop its own wireless network.

Meanwhile, even the British government was already starting to balk at Marconi's price demands and potential levers of control. Within the Admiralty, leading policymakers argued that Marconi's prices were "prohibitive"; in the Post Office, officials considered the idea of a state monopoly of wireless and secretly asked two leading physicists to carve some way around Marconi's patents. Why such hostility toward a man that Britain's elite had so recently celebrated? Part of the reason may have been simple greed: Marconi was perceived to be charging a lot for his equipment, and his associates had an unnerving tendency to raise prices in the middle of negotiations. As even one of the company's officers complained, ". . . as soon as the directors find that their offers are acceptable and people come forward to close the bargain they have invariably imposed fresh conditions and tried to raise the price." This, he noted, "does us harm."[11] Part of the antagonism, however, was also probably due to a basic difficulty in establishing prices and bridging the gap between Marconi's commercial interests and the govern-

[11]Quoted in Jolly, *Marconi*, p. 87.

ment's public purposes. In Britain, at least, Marconi was effectively a monopolist. He had all of the core patents and an embedded relationship with key customers. Yet because his enterprise edged into an area of public concern, Britain's government was loath to grant him the profits of a monopolist. Postal service in Britain, after all, was a *public* service; so was telegraphy. Now Marconi was trying to make money from a new form of communication, and Britain's policymakers weren't pleased.

Yet neither did these policymakers favor a protracted battle over patent rights or a possible nationalization. So they maintained their uneasy relationship with Marconi and continued to buy his equipment. In 1900, the British Navy finally agreed to purchase thirty-two wireless sets from Marconi's company, which was renamed that year as the Marconi Wireless Telegraph Company.

From Machine to Monopoly

By the turn of the century, then, Guglielmo Marconi had made a wildly successful leap from inventor to entrepreneur. He was renowned across the world for his technical prowess and business acumen and had created—in less than a decade—not only a new technology, but a new market as well. By selling telegraph equipment to both governments and private firms, Marconi had demonstrated the commercial promise of Hertz's ethereal waves. And the proof, if he needed any, was in the competition: in the Telefunkens, and Popovs, and others who were racing to carve their own signatures in the emerging wireless market. This, then, was radio's first era of enthusiasm and expansion, a time when one reporter predicted that "[telegraph] cables might now be coiled up and sold for junk"[12] and fledgling American wireless companies were dangling visions of empire before pools of eager investors.

For Marconi, though, it was also a time of commercial uncertainty. He knew there was a technical demand for wireless

[12]Carl Snyder, "Wireless Telegraphy and Signor Marconi's Triumph," *American Monthly Review of Reviews* 25 (February 1902), p. 173.

and that he could meet it, but he was not yet comfortable with the kind of business that was developing, or the customers that he had attracted. Most of these customers, after all, were giant enterprises like the British Navy, enterprises that were tightly linked to their respective governments and increasingly wary of paying the prices that Marconi demanded. Most of them, too, already resented Marconi's power and were actively seeking other ways to get or create Marconi-type systems. Technologically, Marconi was on firm ground. He had a distinct technical advantage in the wireless industry and control over key patents; indeed in 1900 he had cemented this advantage by obtaining a critical patent for the tuning device, an invention that enabled receivers to pinpoint a particular signal or station. But this was still a thin reed on which to base a business empire, especially since many of Marconi's patents were contestable and most were already being widely copied.

So Marconi and his associates began to search for other, more stable and less contentious, ways of organizing their business. They stumbled eventually on a variant of what AT&T was already doing in the U.S. telephone market, and what IBM and Xerox would perfect several decades later. Essentially it was a leasing model, loaning Marconi equipment to customers and then charging them for the services performed with this equipment. The beauty of this model was that it eliminated the customers' need to develop expensive equipment or train their own personnel in its use. Under this system, customers would agree to use Marconi wireless for a fixed (and usually rather long) period of time. They would then pay the Marconi Wireless Telegraph Company a flat royalty fee of £10,000 a year, agree to accept messages from any other ship equipped with Marconi equipment, and promise not to communicate with any ship (except in an emergency) that was not similarly equipped. Marconi, in exchange, would provide each of his customers with a Marconi-trained operator and an ever-widening communications web.[13] Simultaneously, he would use this same ship and land-

[13]Marconi's system also avoided the restrictions of the British Telegraph Acts of 1868 and 1869, which prevented any private company from sending messages over land. See W. J. Baker, *A History of the Marconi Company* (London: Methuen & Co. Ltd., 1970), p. 59.

based infrastructure to compete with the telegraph companies by ferrying messages, for a reduced fee, across the Atlantic. It was a classic model of standardization and scale economies, made even more powerful in this case by the nature of naval communication: if a ship in the Marconi network began to sink, its operator could tap instantly into the broader web, reaching land-based operators around the world and whatever nearby ships were also equipped with the Marconi system. The more ships in the system, the safer everyone felt.

The downside, though, was that Marconi's network led clearly—almost explicitly—to monopoly. To join the network, customers had to use Marconi equipment. They had to stay with Marconi, even if competing firms offered superior technology or more favorable terms. And they had to agree not to communicate with "outside," or non-Marconi, ships, except in emergencies. This rule, according to Marconi, was simply technical necessity, a way of preventing "foreign" signals from interfering with the installed base of equipment. As Marconi explained, "The policy of the Marconi Company has always been that we cannot afford to recognize other systems . . . We cannot be expected to injure our own cause, which we would certainly do if we permitted these stations to communicate with vessels and stations using our system."[14] Perhaps. But "non-interference," as it was called, was also a blunt competitive tool: once Marconi's network reached a critical mass, anyone outside it was effectively penalized. Such is the power of networks and standards. By 1900, the Marconi Company had negotiated contracts with several steamship companies and was supplying equipment to both the Italian and British navies. Shortly thereafter, it signed a fourteen-year exclusive contract with Lloyds of London.

In October of 1900, Marconi moved to cement his technological position and prove that wireless technologies could compete with submarine cables for transoceanic communication. He acquired land in Poldhu, a desolate cove along Britain's Cornish coast, and in Wellfleet, on Cape Cod. On each spot he

---

[14]*New York Times*, October 8, 1901, p. 3. Cited in Douglas, *Inventing American Broadcasting*, p. 71.

built a forest of twenty giant masts, each two hundred feet tall and swathed in a total of four hundred electrical wires. Only months after construction, twin storms destroyed both sets of masts, ripping down £50,000 of investment. Yet Marconi persevered. He rebuilt the station at Poldhu and constructed another in St. Johns, Newfoundland. He also began to experiment with using balloons and kites as more flexible and less expensive receivers. Finally, on December 12, 1901, Marconi heard the letter *s*—three dots of Morse code—transmitted across the Atlantic. It was the first time radio waves had ever traveled so far. And it meant that Marconi could now construct a global network of wireless.

By this point, however, Marconi's technical feats only ignited the political reaction to him. While Canada arranged in 1902 to purchase a new lot of Marconi equipment, the U.S. government remained wary and Germany was positively furious, seeing Marconi's network as a thinly disguised attempt to extend Britain's naval and communications empire. Matters came to a head in 1903, when Kaiser Wilhelm's brother, Prince Henry, sailed from America to Germany on the *Deutschland*, a German ship equipped with Slaby's wireless system. En route, the prince noticed that his ship received virtually no messages and interpreted this silence as evidence that Marconi's affiliates had deliberately jammed its signals. Nothing concrete ever emerged from these allegations, but it was enough for the Kaiser. There arose in Germany what one contemporary writer described as a "malignant Marconiphobia . . . [A] chorus of effervescent indignation, which quickly sped along the cables . . . so that the American hardly less soon than the European public was treated to the unedifying spectacle of a learned professor, a noble Count and various other potent and distinguished personages alike, foaming at the mouth with a species of almost Berserk fury."[15] Shortly after his brother's return, Kaiser Wilhelm called for an international conference on radio. Formally, the purpose of the conference was to replicate in the wireless world the same kinds of rules and regulations that already existed for telegraphy. Informally, it was designed to break Marconi's emerging monopoly.

---

[15]Wilfrid Blaydes, "Mr. Marconi and His Critics," Letter to the Editor, *Electrical World and Engineer,* April 12, 1902, p. 656.

On August 4, 1903, the world's major powers met in Berlin. They denounced noninterference and "Marconism" and in the end signed a very explicit though toothless document. In the Final Protocol of the Preliminary Berlin Conference, seven of the nine attending countries agreed that "Coast stations should be bound to receive and transmit telegrams originating from or destined for ships at sea without distinction as to the system of radio used by the latter."[16] In other words, there could be no policy of noncommunication, no Marconi-like attempts to ignore signals sent by different systems. Since Britain and Italy both refused to sign the document, however, Marconi was under little pressure to comply. Indeed, when the German delegation accused Britain of using the Marconi Company to obtain a particularly British monopoly on wireless communication, the British responded that the Marconi Company had no duty to aid its competitors.[17]

Yet even within Britain, concern over Marconi's power continued to mount. In 1904 a confidential memorandum prepared for the cabinet noted Britain's lonely status at the Berlin conference and cautioned, "[I]f we are to take part in an international Agreement, we must be prepared to impose ... obligations to interchange messages and to avoid interference."[18] Led by the Post Office, which had never forgiven Marconi's original turn toward the private sector, top British officials began to suggest that wireless technologies, like posts and telegraphs, should be licensed and controlled by the state. When Britain's delegation returned in 1906 to a second international radio conference, they thus agreed to an only slightly modified version of the Germans' original position: henceforth, every public radio station would be required "to exchange wireless communication with each and every wirelessly equipped ship, and vice-versa, without regard to the system of wireless telegraphy used by either."[19] Noninterference, in other words, was now formally illegal, meaning that Marconi's business model would have to change.

[16]International Telecommunication Union, *From Semaphore to Satellite* (Geneva: International Telecommunication Union, 1965), p. 143.
[17]Gleason L. Archer, *History of Radio to 1926* (New York: American Historical Society, 1938), p. 80.
[18]Public Record Office, Kew, England: CAB 37/69, No. 39.
[19]*Electrical World* 49, no. 2 (1907), p. 83.

Three years later, Britain's Post Office took the final step and purchased all of the Marconi Company's coastal wireless stations. "The Marconi Company," gloated one representative of the U.S. Navy, "is not dead: but it is mortally wounded."[20]

In retrospect, the story of Marconi's monopoly is a curious one. It extended for no more than ten years and wasn't ever that much of a monopoly. There were always serious competitors to Marconi's technology and companies (most notably Telefunken) that formed around these alternative technologies. It also was never particularly profitable: in 1906, the company's gross earnings were still only $55,170, and its U.S. operations didn't even show a profit until 1910.[21] Why, then, was Marconi so feared? Why did his company attract the combined hostility of so many governments? Perhaps it was an overheated response to Marconi's self-promotion and flair. He was, after all, a genius at promotion and may have managed to make his empire look even more powerful than it actually was. Perhaps it was also just emblematic of the times, a natural reaction to the rise of corporations such as Standard Oil and Western Union. But it seems to have been more personal in Marconi's case, and also more political. Marconi from the outset was a classic pioneer: an aggressive young inventor who brooked no obstacle to his work or dissent to his plans. He created the technology of radio (or at least put others' creations into a workable form) and then went out to create and control the market. It was a classic case of claiming stakes—and hugely valuable ones—along the technological frontier.

The problem, though, was that Marconi's stakes were too big and too valuable, especially in an industry with such obvious military importance. Once the wireless market became as large and important as Marconi alone had imagined it would, governments were unwilling to cede it to his control. And so the rules were written specifically to stop him. By 1909, the British

---

[20]Commander F. M. Barber to Chief, Bureau of Equipment. Quoted in Douglas, *Inventing American Broadcasting*, p. 139. Barber was no impartial observer, but a retired naval officer sent to investigate various European radio techniques.

[21]For historical data on the financial performance of the Marconi Wireless Telegraph Company of America, see Maclaurin, *Invention and Innovation in the Radio Industry*, p. 42.

government purchased all of Marconi's ship-to-shore transmitting stations in Britain, effectively seizing a large chunk of his business for a total of only $75,000. Ironically, Guglielmo Marconi won the Nobel Prize for Physics in that same year, together with his perpetual competitor, Ferdinand Braun.

World War I and the
Intervention of Nationalism

By the time World War I ripped across Europe, wireless technologies had already been almost completely politicized. In Germany, wireless was controlled by Telefunken, which was tightly linked to and supported by the government; in Britain, it fell under the close watch of the Post Office; and in the United States the Navy was attempting to exert its own monopoly over all radio communications. There were still plenty of private firms in the wireless industry, and plenty of ongoing invention, but the wireless market remained closely tied to state interests. This link only tightened, as one might expect, during the war.

Initially, radio rose because telegraphy failed. When war broke out in 1914, telegraphy was still the dominant mode of long-distance communication in Europe. There were over one hundred thousand miles of telegraph wire that connected the continent to the rest of the world and a web of powerful submarine cables. Telegraphy, though, was a vulnerable chain, captive to the very lines that carried it. One snip and the system was down. Quickly, then, the belligerents moved to destroy each other's systems, targeting in particular the submarine cables that were now relatively easy to find and dismantle. On August 4, just hours after Britain issued its ultimatum to Germany, the British cableship *Telconia* yanked Germany's five submarine cables from the bottom of the ocean and destroyed them, cutting in one evening all of Germany's major communication links to the outside world. By 1915 the remaining few links had been severed and Germany, in turn, had cut Britain's Baltic and Black Sea cables to Russia. Once the cables were dismantled or rerouted, radio became a natural substitute.

Technically, radio was an awkward instrument of war. With no way to confine the signal or direct its flow, radio messages were broadcast from one point to another, open to anyone who might be listening in. Such transparency proved disastrous early in the war, when the Russian general Samsonov inexplicably broadcast his plans for the invasion of East Prussia in clear and undisguised language. The Germans heard it all, and were able to take one hundred thousand prisoners during a massive battle at Tannenberg. The hapless Samsonov committed suicide, and other commanders rapidly learned to relay their orders in code. Indeed, code making and code breaking soon became key elements of war, helping to shift momentum to the Allied forces. German naval movements in the North Atlantic, for example, were repeatedly thwarted by the efforts of "Room 40," a hastily assembled team of British linguists and mathematicians who deciphered radio messages passing between the German High Seas Fleet and its ships at sea.[22] A major German offensive in 1918 was similarly compromised by French cryptanalysts who succeeded in cracking the ADFGVX cipher, the most complicated code ever used in battle up to that time.[23]

By the time the war drew to a close in 1918, radio was seen not just as a vital naval technology, but also as a vital security technology—an instrument of power and of war. While most European governments did not directly nationalize their radio industries (that would come later), they did keep a careful and protective eye on what they increasingly saw as "their" radio companies. In Britain, the government continued to work closely with the Marconi Company, even though it rejected some of the firm's most ambitious plans; in France, the newly formed Compagnie Générale de Télégraphie sans Fil, or CSF, established a string of stations across the country's far-flung colonial empire; and in Germany, Telefunken remained both the country's premier radio firm and a closely linked political ally.

Even in the United States, the government moved defini-

[22]On the work of Room 40, see Patrick Beesly, *Room 40: British Naval Intelligence, 1914–18* (London: Hamilton, 1982); David Kahn, *The Codebreakers* (London: Weidenfeld & Nicolson, 1967), pp. 266–97; and Arthur R. Hezlet, *Electronics and Sea Power* (New York: Stein & Day, 1975), pp. 89–94.
[23]See Headrick, *The Invisible Weapon*, pp. 156–57.

tively. Since 1902 the U.S. Navy had been experimenting with radio technologies. It bought sets from Slaby-Arco and a handful of American firms (it adamantly refused to do business with Marconi) and installed radio equipment on several of its ships. In 1912, it won congressional support to increase its radio activities and built a chain of high-powered stations scattered across America's growing list of overseas territories: the Philippines, Puerto Rico, Guam, and Samoa. Even at this point, however, the navy was essentially still just tinkering with the technology. Many of its older officers remained fundamentally opposed to radio, since they saw it as a constraint on their shipboard authority; and many of the most important American inventors, including Lee De Forest, Cyril Elwell, and Reginald Fessenden, had either been pushed into bankruptcy or broken off their earlier relations with the navy.[24]

After the war, however, the navy quickly changed its tune and rushed aggressively into the technology and business of radio. This was a development that was to have major implications for U.S. broadcasting policy, and indeed for the history of radio.

The Birth of RCA

In 1918, the U.S. Navy found itself sitting on a virtual arsenal of radio technologies. There were three foreign stations that it had seized in the early days of the war, and fifty-three American ones taken in 1917.[25] There was the Federal Telegraph

[24]While none of the American inventors achieved the fame or commercial success of Marconi, they each made substantial contributions to the technical development of radio—greater, perhaps, than Marconi himself. For more on these pioneers, see Tom Lewis, *Empire of the Air: The Men Who Made Radio* (New York: HarperCollins Publishers, 1991); Georgette Carneal, *A Conqueror of Space: An Authorized Biography of the Life and Work of Lee De Forest* (New York: H. Liveright, 1930); Lee De Forest, *Father of Radio: The Autobiography of Lee De Forest* (Chicago: Wilcox & Follett, 1950); Helen May Trott Fessenden, *Fessenden, Builder of Tomorrows* (New York: Coward-McCann, 1940); Frederick Seitz, *The Cosmic Inventor: Reginald Aubrey Fessenden, 1866–1932* (Philadelphia: American Philosophical Society, 1999); and Cyril Frank Elwell Papers, 1930–61, Department of Special Collections, Stanford University Libraries, Stanford, California.
[25]Under a clause in the 1912 radio law, the U.S. president was empowered to seize or close any radio station "in time of war or public peril or disaster." See Public Law no. 264, 62d Congress, Section 2.

Company, producer of the powerful arc transmitter, which the navy had purchased outright, and 375 Marconi stations that it had bought as well. There was also a series of patents that the navy had thrust into producers' hands during the war, urging them to forget about liability and concentrate instead on production.

What to do with all these assets? Initially, the navy tried to keep them in its own hands, with the idea of constructing a permanent public monopoly over the radio industry. Convinced that "[c]ommunication is a governmental function and the government must own, control radio, waterways, telegraph and telephone,"[26] Secretary of the Navy Josephus Daniels tried to convince a suspicious Congress to grant the navy a permanent monopoly. They declined. Daniels and his colleagues, though, refused to concede that the radio industry could simply be left to market forces and private hands. They didn't want to lose the technological edge that American firms had finally developed during the war, or the security that radio provided, or control over broadcasting. They also didn't want to leave any room for Marconi to reenter the U.S. market. Thus for the next several months naval officials struggled to concoct a feasible policy—some way to retain control over radio's development without actually placing it under their own formal authority, and to foster an American radio industry without tying it directly to the American government.

In Europe, as shall be seen, policymakers had already solved this dilemma by embracing a public solution: nationally owned or publicly financed stations. Naval officials such as Daniels, however, reasoned that these kinds of solutions wouldn't work in the U.S. context. They needed something subtler, something closer to the market and Americans' prized sense of individual enterprise. The solution they devised was incredibly intricate and secretive. It took years to implement and was in many respects a direct violation of U.S. law. But it was ingenious and, essentially, it worked.

---

[26]Josephus Daniels, *The Wilson Era: Years of War and After* (Chapel Hill, N.C.: University of North Carolina Press, 1946), p. 106. Cited in Hugh G. J. Aitken, *The Continuous Wave: Technology and American Radio, 1900–1932* (Princeton, N.J.: Princeton University Press, 1985), p. 254.

To understand the nuanced logic of this strategy, it is necessary to step back a bit in time, to the years just before the war, when a new crop of inventors had begun to construct the next great wave of radio. During this period, a string of committed pioneers emerged from laboratories, attics, and graduate programs across the United States. They published new theories about electricity and the prospects for "wireless," and fought bruising legal battles when their theories occasionally turned into products. Borrowing a page from Marconi's book, some of these inventors—men like Cyril Elwell and Lee De Forest—rushed eagerly into the commercial market for their invention and sought the support of well-placed financial backers. Emerging industrial giants such as AT&T and Westinghouse had also turned toward the radio industry during this period, nurturing new technologies and grabbing the commercial rights that arose from them.

What most distinguished this next round of invention, though, was its fascination with continuous waves and its concomitant disdain for Marconi's spark-gap technology. Until this point, nearly all radio transmission had been based on some form of Marconi's spark gap, sending single bursts of energy across the airwaves. This was the technology that defined radio and that had created the radio industry. As the technology evolved, however, and stabilized into the commonplace, the next generation of dreamers had begun to yearn for something more. In particular, they wanted some way to send not just sparks across the ether but voice and music as well. They wanted broadcast radio, in other words, even before they knew what it might mean.

By the early years of the twentieth century, the theoretical basis for this transformation had been laid clear. Increasingly, a small band of innovators realized that if radio signals were to transmit anything more differentiated than the dots and dashes of Morse code, they had to become more specific and refined. They had to settle along a finer slice of the available frequency spectrum and ride along "undamped" or continuous waves—waves that fluctuated at a steady and sustained rate rather than in the erratic blasts that characterized spark-gap transmission.

The trouble, though, was that generating such waves was extremely difficult, since no machine at the time was able to generate an electric current that could alternate quickly enough to produce a true continuous wave. Most radio experts, indeed, continued to insist that it simply wasn't possible. In 1906, for example, John Ambrose Fleming, one of the most eminent scientists of the day, had published a comprehensive text on the physical foundations of radio, insisting that Hertzian (that is, radio) waves could not physically be generated without an initial spark.[27]

Stubbornly, though, this next wave of inventors refused to give up. In California, for instance, a young Stanford graduate named Cyril Elwell joined forces with a Danish inventor to launch the Federal Telegraph Company, a firm that used the electrical phenomenon known as arcing to generate an early version of continuous waves.[28] On the east coast, Lee De Forest had been hammering for years at successive versions of his Audion, a pathbreaking device for detecting and amplifying radio signals.[29] And scientists in the burgeoning laboratories of Westinghouse, General Electric, and AT&T were beginning to understand the mechanics of vacuum tubes, the tiny detectors that would soon become a core technology of radio.[30] The most important development around the turn of the century, however, was the two-hundred-kilowatt alternator, the brainchild of a Canadian named Reginald Fessenden who had worked for

[27]See J. A. Fleming, *The Principles of Electric Wave Telegraphy*, 1st ed. (London: Longman, Green & Co., 1906), and the discussion in Aitken, *The Continuous Wave*, pp. 36–40.

[28]For more on the technology behind arc transmitters, see Aitken, *The Continuous Wave*, pp. 87–161.

[29]As the precursor to all later vacuum tube detectors and amplifiers, the Audion is one of the central technologies in the history of radio. For more on its development and De Forest's checkered history, see Aitken, *The Continuous Wave*, pp. 162–249; Erik Barnouw, *A Tower in Babel: A History of Broadcasting in the United States to 1933* (New York: Oxford University Press, 1966), pp. 21–27; Gleason L. Archer, *History of Radio to 1926* (New York: The American Historical Society, Inc., 1938), pp. 92–94; Carneal, *A Conqueror of Space*, and De Forest's own autobiography, *Father of Radio*.

[30]Once again, the best history of this period—and particularly of the role played by larger corporations—is provided in Aitken, *The Continuous Wave*. See also Gerald F. J. Tyne, *Saga of the Vacuum Tube* (Indianapolis: H. W. Sams, 1977); and John W. Stokes, *70 Years of Radio Tubes and Valves* (Vestal, N.Y.: Vestal Press, 1982).

Thomas Edison and then taught electrical engineering at the University of Pittsburgh.[31] Like Elwell and the other pioneers, Fessenden adamantly refused to believe that sparks were the only way to transmit radio signals. He wanted to move away from what he saw as an increasingly obsolete technology and find some way of generating a stable and powerful continuous wave. And thus, through a combination of persistence and persuasion, he had eventually convinced General Electric to attempt to build an alternator with sufficient force to generate a truly continuous wave. Progress was slow, but by 1906 GE had created a one-hundred-thousand-cycle alternator, powerful enough for Fessenden to transmit a Christmas Eve broadcast from his experimental station in Plymouth, Massachusetts. Spurred by this success and by the eerie possibility of transmitting intelligible sound, GE threw itself wholeheartedly into Fessenden's challenge, building alternators that, by the time war broke out in 1914, were capable of transmitting voices across thousands of miles.

Such technical success brought the radio industry to a commercial crossroads. Until this point, radio had been only "wireless"—telegraphy without the awkward infrastructure, a new way of transmitting Morse's dots and dashes. Its principal use lay in ship-to-ship and ship-to-shore communications, and its principal provider was Marconi. After the alternator was developed, however, radio became a broadcast technology—a way of communicating voice and music and sound instead of just the whistles and beeps of code. This transformation meant that the market for radio could theoretically now expand well beyond its traditional base of shipping interests and naval forces. It could become a different kind of market, even more valuable on land—though for very different reasons—than it had already proven at sea.

When news of Fessenden's alternator first spread across the radio community, Marconi was distinctly unimpressed. Firmly committed to spark-gap technology, Marconi and his engineers refused to look seriously at continuous waves, or even to acknowledge that radio waves could feasibly be generated by anything other than a spark-gap system. Moreover, because their

[31]At the time, the University of Pittsburgh was called Western University.

own business was only about messaging, they saw no immediate need to communicate sound rather than code, and no particular demand for Fessenden's clumsy invention. Like many first-wave pioneers, therefore, Marconi and his associates missed the next step along the technological frontier. They presumed that radio technology was *their* technology, and they focused on improving the existing system and business rather than on replacing it with something else. Eventually, though, that something else became too dramatic to ignore. In 1915, Marconi himself belatedly acknowledged the alternator's advantages and approached General Electric with a stunning offer. In exchange for an undisclosed sum, the Marconi Company would purchase twenty-five alternators a year, or essentially all of GE's production.

What made the deal so intriguing, of course, was the strategy that propelled it. Until the alternator came along, radio was linked—both legally and technically—to Marconi and his firm. Marconi was the undisputed pioneer of the radio industry, the holder of its key patents, and the dominant provider of wireless services in the markets that allowed him to compete. From a base in spark-gap technologies, he and his firms had been able to construct a new market and build a business empire upon it. Yet all of this power was vested in a technology that, by 1915, seemed likely to be eclipsed by the next wave of innovation. If Marconi wanted to retain his commercial position, he needed to acquire what he had neglected to create; he had to buy alternators and use them in place of his own spark-gap transmitters. And if the Marconi companies wanted to sustain their distinctive style of commerce, they had to prevent the new technology from falling into upstart hands. What these companies proposed to GE, therefore, was essentially a negotiated approach to a continued monopoly. Marconi would remain the dominant radio firm in a world of continuous waves; GE would become the exclusive provider of this new technology; and both would flourish as wireless turned into sound.

Had Marconi concocted this strategy only a decade earlier, it probably would have worked. But the Marconi companies did not approach GE until 1915, and didn't manage to strike a tentative agreement until 1919—long after the U.S. Navy had realized the strategic import of radio.[32] When the navy learned of Marconi's deal, therefore, they went wild. "[I]mmediately sensing the great advantage that would be placed in the hands of foreigners by the successful conclusion of this transaction," one officer later recalled, "I tried at once to prevent it."[33] For this, after all, was their worst nightmare: the perpetuation of a monopoly in the hands of a private, foreign firm. Stopping the Marconi contract with GE became an integral element of the navy's radio strategy. And also, as it turned out, the key to its radio solution.

On April 8, 1919, two naval officers, Admiral William Bullard and Lieutenant Commander Stanford C. Hooper, met with Owen D. Young, vice-president and acting counsel of General Electric. Appealing to GE's patriotism and referring obliquely to orders from President Wilson, Bullard and Hooper proposed a "policy of wireless doctrine similar to the greater Monroe Doctrine, by which control of radio in this country would remain in American hands."[34] To convince GE to walk away from its lucrative contract with Marconi, Hooper offered a seductive alternative: a new American radio company, bigger and more profitable even than Marconi's, equipped with technology that only Americans had, and that only GE would produce. How would they secure the necessary patents? Well, the navy already had them as a result of war. And how would they convince Marconi to back down? By depriving him of GE's alternators and sending the same equipment instead to a newly formed firm.

There was a certain irony at the core of the navy's strategy, and a delusion that history has never quite managed to explain. The Marconi Company that so irked U.S. officials was not, in fact, a foreign company. It was instead, the *American* Marconi Company—a subsidiary, to be sure, of its British parent, but a largely autonomous firm that had never shown a strong inclination toward any kind of British policy. Nearly all of its officers

[32]Because the negotiations were never completed, there is a fair amount of disagreement over what the final terms would have been. According to Barnouw, however, Marconi's last and most likely offer was to purchase twenty-four complete transmitters for $4,048,000. For more on the negotiations, see Barnouw, *A Tower in Babel*, pp. 57–61; and Aitken, *The Continuous Wave*, pp. 306–326.
[33]Admiral William Bullard, quoted in Archer, *History of Radio to 1926*, p. 164.
[34]Quoted in Headrick, *The Invisible Weapon*, p. 182; and Aitken, *The Continuous Wave*, p. 338.

were American, and most of its business was focused on the
U.S. and South American markets. Still, the navy had determined that American Marconi was Marconi, and that Marconi—in any guise—needed to be stopped.

In a matter of months, GE agreed to the government's proposals. On December 17, 1919, the Radio Corporation of America (RCA) was formed. It was a completely private organization,
though with several unusual twists: only U.S. citizens could serve
as directors or officers of the firm; foreigners could hold no
more than 20 percent of the company's stock; and a government representative would periodically inform the board of the
"views and interests" of the United States. Even more unusual
was that the company had no assets or operations. Rather, all it
contained were shares of American Marconi, purchased one
month earlier from thousands of individual shareholders who
had agreed to exchange their Marconi stock for shares in the
new RCA. The 364,826 shares held by the British Marconi
Company were purchased directly by General Electric.

In retrospect, the final plan for RCA seems to have been as
much Owen Young's creation as the navy's. It was Young who
hatched the details of the new company and presented it to
Washington; Young who first informed the navy of GE's impending Marconi deal. But what Young offered was exactly
what the navy had been searching for: a friendly, private, *American* monopoly. Suddenly, GE was in the radio business.

Within months, any remaining obstacles to RCA fell away,
pushed by Young's adroit maneuvering and the full support of
the U.S. Navy. RCA took over the radio stations that the government had purchased earlier from Marconi and inherited
nearly all of Marconi's employees—including David Sarnoff,
who would quickly rise to become the company's president. The
new company launched an international wireless telegraphy service in February 1920 and swiftly signed exclusive traffic agreements with other countries' national radio systems. By 1923,
RCA was handling 30 percent of the communications traffic
across the Atlantic and 50 percent across the Pacific.[35] Its biggest
coup, though, came in the area of patents, where Young, backed
again by the U.S. Navy, managed to negotiate a critical cross-

[35]From Headrick, *The Invisible Weapon*, p. 182; and Aitken, *The Continuous Wave*, p. 481.

licensing agreement with AT&T, United Fruit, and Westinghouse, each of which controlled the patents for several of radio's most crucial components. Courtesy of the American Marconi Company, for example, RCA had patent rights to produce a two-element vacuum tube, but not the technically superior three-element tube, whose patent rights were held by AT&T. AT&T, meanwhile, could not legally manufacture these tubes for radio purposes unless it acquired rights to the older two-element tube, and until it cleared up matters with General Electric, which owned a complementary patent on the procedure for creating vacuum tubes.[36] Other key patents (for arc transmitters, heterodyne receivers, and the like) were owned by two additional firms, Westinghouse and United Fruit.[37] If these companies were left to themselves, Young reasoned, they might well descend into legal and technical rivalry, fragmenting the industry and stunting its growth. They would also present a serious obstacle to RCA, since they controlled the technologies that RCA would undoubtedly need. To solve this problem Young presided over a series of ambitious negotiations in 1920 and 1921, creating a patent pool that formally brought all the key radio companies into RCA's commercial orbit. Under this rather extraordinary arrangement, RCA secured the right to use any of the radio patents held by the other firms (though not to manufacture radio sets by itself), and they, in turn, received a large ownership interest in RCA: 20.6 percent for Westinghouse, 10.3 percent for AT&T, and 4.1 percent for United Fruit. GE retained 30.1 percent of RCA's stock, leaving just 34.9 percent in outside hands.[38]

---

[36]The total number of overlapping patents was even greater, and considerably more complicated. For a full elaboration, see Report of the Federal Trade Commission on the Radio Industry: in response to House resolution 548, 67th Congress, 4th session, submitted Dec. 1, 1923 (Washington, D.C.: U.S. Government Printing Office, 1924), pp. 39–50.

[37]In 1920, Westinghouse had purchased the rights to Fessenden's radio patents from his original backers and the rights to arc patents from the U.S. Navy. United Fruit, the dominant producer of bananas in Latin America, had purchased several key radio patents through its subsidiary, Tropical Radio, which used radio to communicate across its vast infrastructure of plantations and transportation links.

[38]Figures are from Barnouw, *A Tower in Babel*, p. 73, and describe the pool as of 1921. For more on the patent pool and the negotiations that led to its creation, see Maclaurin, *Invention and Innovation in the Radio Industry*, pp. 105–107; Hiram L. Jome, *Economics of the Radio Industry* (Chicago: A. W. Shaw Company, 1925), pp. 51–57; and Aitken, *The Continuous Wave*, pp. 432–79. Many of the key documents relating to the pool are also reproduced in the U.S. Federal Trade Commission's 1923 Report on the Radio Industry.

Throughout these complicated maneuvers, the U.S. government played the role of benign uncle, looking on affably as some of the country's largest and most sophisticated firms explicitly divided the radio market for their joint control. Property rights—here, in the form of patents—were blithely traded and entrenched, backed by the formal sanction of U.S. law.

Just five years after the war's end, therefore, a new radio company had emerged on the scene—a company conceived, as one account describes it, "in the sin of monopoly,"[39] and destined from the start to wield greater power than Marconi had ever seen. Until the formation of RCA, radio in the United States was still a rough-and-tumble world. There were a handful of firms owned by the technical pioneers (most notably Fessenden, De Forest, and Elwell) and several larger firms, such as GE and Westinghouse, that produced some key components. The only standard in the industry and the only set of rules belonged to Marconi, the foreigner who was determined to control the market he had created. By 1923, though, Marconi and his rules had been brutally pushed aside, replaced by a new game in radio and a new and extremely powerful player. Remarkably, what enabled this new player to arise was an intricate chain of cooperation between the U.S. government and the emerging titans of U.S. industry. Although rarely painted as such, RCA was essentially a joint venture: an investment launched by General Electric, but with the financial and technical participation of Westinghouse, AT&T, Western Electric, and United Fruit. It was a significant concern, capitalized at $40 million and backed by millions of dollars in research. And, in the United States at least, it controlled all the patents and all the processes that were destined to make radio a success.

In retrospect, Marconi hadn't really been mortally wounded by either the International Radio Conference of 1906 or the British nationalization of 1909. He still had his international naval business at that point and the ability—both technical and

---

[39]Kenneth Bilby, *The General: David Sarnoff and the Rise of the Communications Industry* (New York: Harper & Row, 1986), p. 2. As Sarnoff's former executive assistant, Bilby was hardly an impartial observer, which makes his comment even more interesting.

commercial—to run a radio network that spanned the world. By 1923, however, his monopoly was gone.

Radioheads

The creation of RCA had a profound impact on the structure of the international radio industry. It concentrated power and patents in a single firm, and left that firm closely allied with both the American government and the giants of American industry. It did not, however, change the scope of radio at all. Indeed, radio in the early years of RCA was not all that different from radio in the early days of Marconi. Despite the newfound ability to transmit sound, it was still about using radio as wireless telegraphy—about creating the same kind of communications empire that Marconi had envisioned, only transmitted this time with a distinctly American accent.

What changed radio was not the birth of RCA, nor the simultaneous creation of Europe's national radio stations. It wasn't even a technological breakthrough, since the great advances of the early twentieth century—vacuum tubes, high-frequency alternators, and De Forest's Audion—remained firmly in the hands of the major corporate or government players. Rather, what changed radio was the noisy spread of amateur enthusiasts—"radioheads"—who tinkered with the technology and fell in love with its potential. These were men—or boys, as they were customarily called—who had either learned the technology of radio during military service or simply picked it up by reading publications such as *Wireless Age* and *Electrical World*.[40] Using crude devices, these "boys" were able to tap into the ether, picking up whatever signals might be whirring in the air around them. The more sophisticated these sounds became, the more interesting was the game and the harder the boys strained to hear.

[40]Historians generally assert that the great bulk of radio amateurs were male—"young, white, middle class boys and men." (Douglas, *Inventing American Broadcasting*, p. 196.) Some recent scholars have argued, however, that young women were also part of the radio boom. See for example Michele Hilms, *Radio Voices: American Broadcasting, 1922–1952* (Minneapolis: University of Minnesota Press, 1997), pp. 130–50.

Such amateur listening had existed since the earliest days of radio. Spurred by Marconi's youthfulness and do-it-yourself spirit, enthusiasts had built their own crude equipment and formed radio clubs. They communicated with one another in an early form of electronic camaraderie and exchanged tips for various kinds of receiving equipment. Yet in the early days of the twentieth century, radioheads were still a relatively rare breed. The technology was just too difficult to master and the signals too uninteresting—just the dots and dashes of Morse code, sent through the ether with an often-deafening crash and buried in a mass of static.

As Fessenden and others perfected the transmission of continuous waves, however, radio moved rapidly toward a wider and less technical audience. Suddenly, there was a buzz of intelligible language in the ether: words and music that even a casual user could detect. The number of amateur enthusiasts exploded, igniting an instant community and even a gray market of sorts. The radio boys learned, for example, that bedsprings could be used as spark gaps and that a round Quaker Oats box made an excellent core for tuning coils. They figured out how to use crystal-based detectors and stolen telephone headsets and avidly shared these discoveries through a growing network of enthusiasts. Like cloistered birdwatchers, they kept logs of their reception and would listen for hours to the static of their homemade headphones, straining for some signal of a far-off transmission. They weren't hearing much, just weather reports and shipping bulletins and occasionally bits of gossip flung out by other operators. But it was the thrill of discovery that ignited these listeners, the ability to pull a signal out of thin air and hear a voice wafted from hundreds of miles away. In the United States alone, one author estimated, there was an audience of "a hundred thousand boys" a night in 1912, hunched over homemade receivers and waiting for signals.[41]

This, then, was radio's second wave of public excitement. Unlike the first, which Marconi rode, this second wave had little to do with technical glamour or the thrill of discovery. It wasn't about the inventors who had enabled this phase, and it wasn't

[41]Francis A. Collins, *The Wireless Man* (New York: Century, 1912), p. 26.

linked to any commercial objective. It was instead a social breakthrough—a realization that the same technology could be applied to a completely different purpose, offering options that its creators had never even conceived. No longer just a naval tool or messaging device, radio had suddenly morphed into a medium of entertainment—something to be enjoyed, rather than just used.

Implicit in this transformation was a certain sense of destiny—a belief that radio was *technologically* determined to break the bonds of existing authority. Because the ether hovered above the grip of both businesses and governments, the prophets of radio insisted that there was no way to regulate it or profit from it, no way to own a radio wave or parse the air into manageable chunks. Radio, they cried, was destined to be free, open like space itself to all comers and unencumbered by traditional notions of property rights. It was democracy's tool, bound to broadcast without constraints from either big business or big government. Which wasn't at all how matters turned out.

The Business of Broadcasting

For roughly a decade, the radio boys lived comfortably with the wireless industry. The stations sent their messages; the boys listened in; and because technical experts had claimed that the ether was effectively infinite, no one really worried about the physical impact of their presence. As the number of users multiplied, however, and the boys began to talk as well as listen, it quickly became clear that the ether was not in fact infinite, and that congestion of the airwaves was a very real problem. By 1910, transmissions were increasingly tumbling into one another, canceling each other out, and raising a cacophony of voices around and on top of the signals. While the radio boys didn't mind the chaos, the U.S. Navy and established wire companies did, and they blamed the young enthusiasts for creating this congestion. Naval fleets, critics reported, were unable to communicate with their headquarters due to "amateur clamor," and amateurs had even impersonated naval officers in some cases, sending false orders over the airwaves and creating chaos at sea.[42] One article published in 1910 warned, "The efficiency of a number of the coastal stations of the Navy has been cut in half because of the presence of dozens of small amateur stations."[43] Another, less dramatically, complained, "They gossip about everything under the sun . . . They ask each other for the baseball or football scores, make appointments to meet the next day, compare their lessons. And they quarrel and talk back and forth by wireless in regular boy-fashion."[44] The radio boys derived a kind of gleeful pleasure from such complaints, seeing them as evidence of technology's great potential to shift the balance of power closer to the side of the individual. But to naval officers and other government officials, the amateurs' intervention was infuriating and potentially dangerous.

For several years these arguments played out, pitting a large but disorganized group of amateurs against a small but powerful group of opponents. Congress, it appears, was relatively confused by the conflict, still unsure about the technology of radio or the possibility of carving property rights upon it. As Massachusetts Senator Henry Cabot Lodge admitted, "Personally I confess I do not understand the questions involved and I certainly should not be willing to vote until I am fully informed."[45] Representative Ernest W. Roberts of Massachusetts was even blunter, identifying the basic problem that confronted radio regulation and supporting, most likely by accident, the ideological position of the radio boys: "We have been brought up with the idea that the air was absolutely free to everyone."[46] How, then, could interference be prevented?

[42]Reported in Harlow, *Old Wires and New Waves*, p. 469.

[43]*New York Times,* January 28, 1910, p. 8; cited in Douglas, *Inventing American Broadcasting*, p. 209.

[44]Francis A. Collins, "An Evening at the Wireless Station," *St. Nicholas* 39 (October 1912), p. 113. Quoted in Douglas, *Inventing American Broadcasting*, p. 208.

[45]Letter from Henry Cabot Lodge to Elihu Thomson, February 3, 1908, Clark Collection. Quoted in Douglas, *Inventing American Broadcasting*, p. 217.

[46]Roberts, however, who chaired a congressional hearing on the regulation of wireless telegraphy in 1910, went on to argue that some kind of regulation was indeed necessary, since "the march of civilization" now demanded "some change of the old common law with regard to rights in the air." See "Wireless Telegraphy and Wireless Telephony," Hearings Before a Subcommittee of the Committee on Naval Affairs of the House of Representatives on H. J. Resolution 95 (Washington, D.C.: Government Printing Office, 1910), p. 4.

Just as the issue appeared to be grinding to a standstill, however, a major event shook the wireless world—and indeed the world in general. On April 14, 1912, the luxury liner *Titanic* hit an iceberg in the North Atlantic and sank. Of the 2,227 people on board, 1,522 died. Although the *Titanic* was fully outfitted with radio equipment, and although help eventually came from a ship, the *Carpathia*, that heard its distress call, radio had nevertheless played a role in the liner's disastrous plight. When the *Titanic* went down, other ships much closer than the *Carpathia* had failed to hear its call: one nearby wireless operator was asleep when the call came in; another passing ship was not equipped with radio equipment. Even the *Carpathia* received the emergency signal sheerly by accident, when its off-duty operator returned to his headset to check the time. After the *Titanic* sank, moreover, radio waves along the eastern seaboard were clogged with the calls of desperate friends and family, each struggling to determine what had happened in the frozen North Atlantic and whether their relatives had managed to survive. Wireless operators were overwhelmed by the frenzy, and congestion was so bad that false information was passed along. In the early hours after the disaster, two messages apparently crossed in the air: the question "Are all *Titanic* passengers safe?" and a completely unrelated message that stated "towing oil tank to Halifax." The crossed message—"all Titanic passengers safe. Towing to Halifax"—was subsequently reported in the press and then revealed to be false, leading to outrage by those who believed that their friends and family had been saved.

Suddenly, arcane disputes over radio congestion became a matter of international concern—a matter, indeed, of life and death. Bills that had languished for years before Congress were rushed swiftly back into session, where shaken representatives quickly signed them into law. In 1912, the U.S. government passed its first major piece of legislation regarding the radio industry.[47] Aimed directly at the growing ranks of radio boys, the Radio Act of 1912 required that all wireless operators be licensed and that distress calls take priority over any other transmission.

---

[47]Earlier bills in 1910 and 1912 simply required ships of certain types to carry wireless equipment and operating personnel.

It also broke the radio spectrum into discrete chunks and required amateur operators to stay in a small and technically inferior spot. While they could listen to any frequencies, the amateurs could now transmit only along the "short wave" portion of the spectrum, using waves of two hundred meters or less.

On technical grounds, the Radio Act was not particularly harsh. It was a continuation in many ways of a legislative process unfurled as early as 1906, when the countries that met at the Second International Radio Conference split the radio spectrum into "commercial" and "government" spheres and required communication between ships at sea. It was also considerably more liberal than other radio legislation being passed at the time. Britain had already nationalized the Marconi ship-to-shore stations, and governments elsewhere had laid an early and exclusive claim to the radio spectrum.[48] To the radio enthusiasts, though, it was a painful blow—the first step toward parsing out the ether and imposing regulation upon it. After 1912, the frontier world of radio was effectively foreclosed. Boys might still wander along its outer limits, but they couldn't lay claim to it anymore, or even consider it a wide-open space. Instead, radio in the United States became a space like any other, marked by lines of property and ruled, albeit lightly, by the federal government. This was the first phase of radio's regulation.

The second phase was grander and far more sweeping in its implications. During World War I, as we have seen, radio fell promptly into the military sphere and amateur experiments crashed to a halt. Many radio boys went into the service, where they honed their skills as operators and dreamed of returning to the private world of transmission. When they came home, these enthusiasts filled the airwaves once again, sticking to their prescribed frequencies but experimenting within these bounds. Taking advantage of the dramatic improvements in broadcast technology, radio stations cropped up across the country—home-grown affairs for the most part, transmitting weather reports and phonograph music.

And then, early in 1920, something remarkable hit the air-

---

[48]For a discussion of how the U.S. approach to radio regulation differed from that of other countries, see Hugh G. J. Aitken, "Allocating the Spectrum: The Origins of Radio Regulation," *Technology and Culture* 35, no. 4, October 1994, pp. 686–716.

waves. Frank Conrad, an engineer at Westinghouse Electric's Pittsburgh plant, began to transmit music concerts. They weren't anything really new, just basic phonograph records sent across the airwaves. Because Conrad was a particularly well-qualified amateur, though, the quality of his broadcasts was better than most, and a larger and larger crowd began to listen to his program. Suddenly, people were calling Conrad, requesting that he play specific songs and even, in some cases, mailing their own phonograph records to him. The concerts became regular events, airing every Wednesday and Sunday night, and a local music store provided records in return for being mentioned on the air. Then a larger store began to link its own advertisements to the concerts, offering "approved radio receiving sets for listening to Dr. Conrad."[49] Conrad himself was not in business. He was an amateur like all the others, tinkering with radio purely for his own enjoyment. But slowly, and without even his participation, business was building around him.

Soon, Harry P. Davis, a vice-president at Westinghouse, realized that Conrad might be on to something. If radio broadcasts could increase popular demand for radio sets, then Westinghouse, which had just begun to produce small crystal receivers, might be able to grow its sales on Dr. Conrad's back. Westinghouse thus built Conrad a dedicated studio and applied to the Department of Commerce for a special broadcasting license. On November 2, 1920, the new KDKA transmitted its first broadcast, revealing the results of that day's presidential election. It continued with daily broadcasts, of live music, news reports, and speeches. The impact was electrifying. Suddenly, radio's band of listeners swelled, taking in not just the mechanically inclined radioheads but people from all walks of life who were stunned by the ability to hear this sound from an invisible source. In a matter of months, radio flew from the arena of messages and communication to the world of song and entertainment. This at last was broadcasting—not because the technology was too flawed to prevent it, but because the technology was eminently suited to it.

[49]Cited in Paul Schubert, *The Electric Word* (New York: The Macmillan Company, 1928), p. 199.

In 1921 and 1922, radio mania hit the United States. People flocked to department stores to purchase sets, and applications for broadcasting licenses flooded the Department of Commerce. Whole sectors of the economy began to suspect that radio would change their world forever: newspapers rushed to create broadcast versions of their news; universities scrambled to offer college courses "on the air"; and churches built radio towers to proclaim their word from on high. Westinghouse gleefully rode the tide, joining the RCA patent group (it had not been a member until this point) and launching two new stations, WJZ in Newark, New Jersey and WBZ in Springfield, Massachusetts. From its perch in Pittsburgh, meanwhile, KDKA continued to broadcast an expanded schedule of news, music, and just plain talk. Now, though, it had a lot of company. By the end of 1922, there were 576 licensed radio broadcasters in the United States.[50] Some of them were good, and many were skilled, but in the boom of broadcasting that was underway, it didn't really matter. Anything that went on air appeared to have an audience.

Once again, the language used to describe this boom was utopian and reverential. In the seven years between 1921 and 1928, one contemporary historian noted, "the popular use of radio spread as nothing before has ever spread, not only into every nook and cranny of the United States, but in growing waves all over the earth." In this new era of intercommunication, he continued, "an entire nation has come to the point of absorbing some new thing into its life, a thing that will henceforth play a profounder part in its environment than it can guess."[51] *Radio Broadcast*, a publication launched in 1922, predicted that in the age of radio, "the government will be a living thing to its citizens instead of an abstract and unseen force . . . elected representatives will not be able to evade their responsibility to those who put them in office."[52] Henry Luce, who

---

[50]This figure is from Barnouw, *A Tower in Babel,* p. 104. Schubert (p. 214) gives the slightly lower figure of 508. For more on this growth and the chaos that ensued, see Jora Minasian, "The Political Economy of Broadcasting in the 1920's," *Journal of Law and Economics* 12 (October 1969), pp. 391–403.

[51]Schubert, *The Electric Word,* pp. 212–13.

[52]Cited in Barnouw, *A Tower in Babel,* pp. 102–03.

would go on to found *Time* magazine the following year, was particularly philosophical, proclaiming in a 1922 address that:

> In radio broadcasting we have a force, an instrumentality greater than any that has yet come to mankind . . . When you transmit the human voice into the home, when you can make the home attuned to what is going on in the rest of the world, you have tapped a new source of influence, a new source of pleasure and entertainment and culture that the world thus far has not been able to provide with any known means of communications . . . I regard radio broadcasting as a sort of cleansing instrument for the mind, just as the bathtub is for the body. Now the broadcasting station makes possible, for the first time in the history of civilization, communications with hundreds and thousands and, perhaps, millions of people, simultaneously.[53]

On the surface, all of this activity should have been a boon for the radio boys. It was a validation of their belief in popular radio and a tremendous opportunity to jump into a new market that they already knew so well. By the same token, the emergence of popular radio should have been a slap at RCA, since the company's business model was based entirely on radio-as-communication rather than radio-as-entertainment. Indeed, RCA seemed to have missed the radio boom entirely: it had no radio station of its own and no apparent interest in broadcasting. "We had everything," one GE executive recalled sheepishly, "but the idea."[54] Yet as the market developed, it was RCA that rose to power and became the beacon of broadcasting in the United States. By 1930, amateur radio had all but disappeared from the world of broadcasting and the radio boys were gone.[55]

---

[53]Quoted in Bilby, *The General,* pp. 64–65.
[54]William C. White, *Reminiscences,* p. 19. Unpublished manuscript quoted in Barnouw, *A Tower in Babel,* pp. 73–74.
[55]Amateur radio itself remained, but the enthusiasts moved out of the broadcast field and back into communications, where they focused on "short waves," or waves below two hundred meters. See Clinton B. DeSoto, *200 Meters and Down: The Story of Amateur Radio* (West Hartford, Conn.: The American Radio Relay League, Inc., 1936).

Chaos, Sarnoff, and NBC

Under the law of 1912, the U.S. Secretary of Commerce had no power to deny a radio license or enforce the use of a particular wavelength. He simply assigned wavelengths to those who applied for them, trying only to minimize interference and keep all tiers of users in their assigned bands. When the new crop of broadcasters applied to Washington, Herbert Hoover, the Secretary at the time, had just handed them out, assigning the applicants to the somewhat arbitrary wavelength of 360 meters. As the number increased, he continued to arrange the smaller stations across the 360-meter spectrum, and pushed the larger ones to four hundred. Emphatically, the pro-business Hoover refused to limit the number of licenses, arguing that the spectrum was a national resource that could not be garnered by a select few. "You will recognize," he stated in a 1924 interview, "that if anybody should be able to have the exclusive use of a certain wavelength, he would have a monopoly on that part of the ether. That cannot be permitted."[56] And thus, even when the number of applicants exceeded the available spectrum and congestion began to plague the ether, Hoover refused to restrict the number of licenses. Instead, he simply reapportioned the spectrum once again and clustered the growing broadcasters more tightly across it.

Eventually, though, the sheer mechanics of transmission caught up with Hoover's system and brought matters to a technical and political impasse. The first spark occurred in 1926, when a desperate Chicago station jumped over to a Canadian frequency in order to expand its broadcasting space. Outraged, Hoover took the station to court and sued to have its license revoked. But the courts only reinforced what he and his associates had long suspected: because there was no formal regulatory structure in place, the Secretary of Commerce had no power to revoke licenses.[57] In practice, it appeared, the airwaves were still free.

---

[56]"The Government's Duty to Keep the Ether Open and Free for All," Interview with the *New York World*, March 16, 1924. Quoted in Aitken, "Allocating the Spectrum," p. 699.

[57]For more on this case, see Aitken, "Allocating the Spectrum"; and Marvin R. Bensman, "The Zenith-WJAZ Case and the Chaos of 1926–27," *Journal of Broadcasting* 14, no. 4 (Fall 1970), pp. 423–40.

Chaos ensued. One hundred fifty-five new stations jumped into the ether, pushing the total number of broadcasters above seven hundred. Existing stations abandoned their assigned frequencies for better ones and everyone started to broadcast louder and louder, hoping to drown out their competitors in the din. It was a time of utter anarchy and no one expected it would last for long. But without a secure regulatory framework—without any *rules* for radio—it wasn't clear just how the anarchy could be contained.

Meanwhile, RCA was struggling to define its own role in broadcasting. After the original merger, control of the company had gradually passed to David Sarnoff, a Russian immigrant who had risen steadily through the ranks of the American Marconi Company. Sarnoff was an unlikely corporate executive. He was poor, lacked a college degree, and—quite unusually for a prominent executive in his era—Jewish. But he was also brilliant, blessed with both a magnetic mind (honed, he would argue, by studying the Talmud from dawn to dusk as a child) and a keen intuition for the politics of business. Sarnoff had met Marconi in 1906 and had rapidly convinced the older man to take a serious interest in his career. He then devoted the same art of persuasion to Edward J. Nally, general manager of the American Marconi Company, who would go on to become the first president of RCA. Sarnoff also nurtured a legend that, as a young man of twenty-one, he had been the telegrapher who first caught word of the *Titanic*'s sinking and relayed it, tirelessly and accurately, across the eastern seaboard. As subsequent biographers have pointed out, the story is almost certainly inaccurate. Why would Wanamaker's, the department store in New York that employed Sarnoff, have been a control point for telegraphs arriving across the North Atlantic? And why would Sarnoff have stayed at his post for the seventy-two hours that reports assign to him when there were plenty of other telegraph operators in New York? Yet Sarnoff clearly enjoyed the tale, and he never took pains to correct it.[58]

Within RCA, Sarnoff became known for his tenacious intelligence and for an early commitment to broadcast radio. In 1915,

[58]For more on this controversy, see Bilby, *The General*, pp. 30–35; and Lewis, *Empire of the Air*, pp. 105–109.

before RCA had even been created, he had written a memo to Nally, urging him to consider "a plan of development that would make radio a household utility in the same sense as the piano or phonograph."[59] At RCA, Sarnoff expanded upon his concept for a "Radio Music Box," considering sports broadcasts and radio sales and pressing his ideas upon all who would listen. Once the launch of KDKA proved him right, Sarnoff was quick to exploit his vision and his enhanced position within the company. In 1921, he jolted to the top of RCA, determined to push the giant company into the emerging world of broadcasting.[60]

It wasn't entirely easy. On the one hand, RCA was perfectly positioned to enter broadcasting. It had, after all, nearly all of the patents that surrounded radio and, through its links with General Electric and Westinghouse, a virtual monopoly on the sale of radio equipment.[61] On the other hand, though, radio mania had pushed thousands of amateurs into the manufacturing business. Radio stations were popping up like mushrooms, as were entrepreneurs who could turn out workable radio sets using commonly available components and a few key pieces that, under the patent arrangements, could still be produced by their original inventors for "amateur" use. To the radio boys, this was ecstasy, an opportunity at last to take their homespun devices directly to the air and to the market. But to Sarnoff, it was hell. As one writer described it: "A raggle-taggle mob of free enterprisers was running away with the business."[62] If RCA were to thrive in this new world of popular radio, Sarnoff had to find some way to eliminate the amateurs. At the same time, though, he also had to prevent his own partners—GE, Westinghouse, and AT&T—from storming into the radio market on their own and capturing either the production or broadcast side of the industry.

---

[59]See Bilby, *The General,* p. 39. This memo has subsequently been lionized by other historians of radio; see for instance the description in Archer, *History of Radio to 1926,* pp. 111–13.

[60]Sarnoff was made general manager of RCA on April 29, 1921, just two months after his thirtieth birthday. For more on his impact on broadcasting in both radio and television see Thomas K. McCraw, *American Business, 1920–2000: How it Worked* (Wheeling, Ill.: Harlan Davidson, 2000), pp. 115–36.

[61]Under an agreement signed in 1921, RCA sold equipment manufactured by GE and Westinghouse, two of its principal owners.

[62]Lawrence Lessing, *Man of High Fidelity: Edwin Howard Armstrong* (Philadelphia and New York: Lippincott, 1956), p. 134.

In 1923, RCA began to apply pressure on its distributors, urging them not to sell vacuum tubes as independent components. The company also brought suit against several small set makers, arguing that their sales constituted a violation of RCA's patent rights. AT&T joined the fray as well, warning radio stations that any transmitters other than those made by the Western Electric Company (an AT&T subsidiary) constituted a violation of AT&T's patent rights.[63] At the same time, representatives of the music industry were also growing restless. While the rise of popular radio had vastly increased the reach of popular music, it had also, according to the American Society of Composers, Authors and Publishers, deprived musicians and composers of their fair reward. If radio stations were going to play music on the air, they should compensate the creators of this music. Eventually, with the weight of several court decisions behind them, ASCAP forced the radio stations to pay an annual license fee to broadcast any ASCAP-controlled music. While this practice applied across all stations, it obviously hit hardest at the amateurs, many of whom had launched their stations with virtually no cash behind them. It was to cover these costs that many radio stations turned to corporate sponsorship, beginning a trend that would eventually define the commercial model of radio.[64]

All of these developments seemed poised to advance RCA's position. But then, in 1923 and 1924, events moved sharply in the opposite direction. First AT&T, a powerful but restless member of the RCA alliance, gave evidence that it was about to put its own radio receiver on the market, competing directly with the GE and Westinghouse brands that RCA sold. This, according to RCA, would be a violation of the original patent agreement. It would also be financially devastating. Even worse news came just months later, when the Federal Trade Commission launched a formal investigation of the "radio trust," claiming

[63]William Peck Banning, *Commercial Broadcasting Pioneer: The WEAF Experiment, 1922–1926* (Cambridge, Mass.: Harvard University Press, 1946), pp. 74–75.

[64]For more on the rise of radio advertising, see Barnouw, *A Tower in Babel*, pp. 199–211; Barnouw, *The Golden Web: A History of Broadcasting in the United States*, vol. II, 1933–1953 (New York: Oxford University Press, 1968), pp. 5–18; and John W. Spalding, "1928: Radio Becomes a Mass Advertising Medium" in Lawrence W. Lichty and Malachi C. Topping, eds., *American Broadcasting* (New York: Hastings House Publishers, 1975), pp. 219–28.

that RCA and its allies had "combined and conspired for the purpose of . . . restraining competition and creating a monopoly" in the radio industry. Which, of course, they had.

It looked as if RCA's power was about to crumble. But then, in a series of moves that would later win him volumes of adulation, Sarnoff managed to wrap the two crises together and turn them to RCA's advantage. First, he and the other top managers at RCA launched a full-scale assault on Washington. They went before Congress and met with the trade commissioners, arguing that RCA's patents had to be protected and that the company was in fact playing a vital role in keeping the budding radio industry organized and orderly. According to Sarnoff, radio was again dancing with anarchy. There were too many little stations; there was too much interference in ever-narrower bandwidths and too great a chance that radio might "smother in its own dissonance." Both publicly and in private, Sarnoff argued that there had to be some "disciplined superstructure" imposed upon the public airwaves, some way of rationalizing and coordinating what was increasingly seen as a public service. And RCA, he intimated, was in the best position to do this.

Meanwhile, the RCA allies flung themselves into a heated and intensely secretive round of negotiations, with AT&T pitted increasingly against the other members. On the table were a host of contentious issues, including whether AT&T could make radios, whether RCA could use telephone lines for radio transmission, and whether any of the partners was already violating their common patent agreement. Ultimately, matters grew so fierce that an independent arbitrator was brought in. Late in 1924 he rendered a sweeping judgment, arguing that AT&T had no right to manufacture radio sets or even to engage in radio broadcasting. Shocked, top managers at AT&T plotted a countermove, hiring one of the nation's most prominent lawyers to render an advisory opinion on the arbitrator's ruling. This second opinion was even more dramatic, indicating that the entire patent pooling agreement—the technical and economic foundation of RCA—was itself illegal. All of the partners realized the potential impact of this finding: if it ever leaked into the public realm, it would seriously damage all of the erstwhile allies and ensure a hostile, and potentially devastating, conclusion to the FTC investigation.

It was at this point that Sarnoff struck. Why not, he argued, split the contentious broadcasting piece away from the larger and more powerful firms? Why not make broadcasting self-supporting, divorced from the equipment manufacturers who might otherwise be found in restraint of trade? With this shift, AT&T could be kept permanently out of the radio business and the reconstituted RCA would be free from the antitrust allegations that were otherwise likely to haunt it.

In January of 1926 a new company was formed, owned 50 percent by RCA, 30 percent by General Electric, and 20 percent by Westinghouse. Grandly called the National Broadcasting Company, it made its first move in November, purchasing AT&T's leading radio station (the New York–based WEAF) for a startling $1 million. It was a high price, but WEAF was more than just a station. It was instead the beginnings of a network, the largest in a chain of stations that AT&T had originally hooked together, each nominally independent but drawing from the same base of programming and advertisement. The idea behind this network—an idea that NBC quickly copied and expanded—was straightforward. First, a network of stations was inherently more orderly and rational than a cacophony of competing broadcasters. Each station that joined the network would reduce the chaos a little more, furthering Sarnoff's vision of a well-mannered, publicly committed industry.[65] Then, by pooling their programming—that is, by playing the same music, news, or featured entertainment—the stations could reduce their individual costs while still securing the most lavish and expensive programming. Or, as NBC stressed in its press release:

> *The purpose of [the] company will be to provide the best program available for broadcasting in the United States.*

The National Broadcasting Company will not only broadcast these programs through station WEAF, but it will make them available to other broadcasting stations

---

[65]Indeed, during this period, Sarnoff envisioned the radio network as a public institution of sorts, financed through a combination of public and private contributions rather than by advertising. For more on his original notion of network broadcasting, see Bilby, *The General*, pp. 68–71; and Archer, *History of Radio to 1926*, p. 343.

throughout the country so far as it may be practicable to do so, and they may desire to take them.

It is hoped that arrangements may be made so that every event of national importance may be broadcast widely throughout the United States.[66]

This, then, was the start of true network broadcasting in the United States. It was also the continuation of two common refrains in radio, echoes of the same connections that had built Marconi's messaging company and the early RCA. First, like Marconi's ships and RCA's patents, NBC's stations found power in numbers. By joining together around a central core, each piece became part of a network that was vastly more valuable than the sum of its parts. This was network economics long before the term was coined: using a common standard and common content to create critical mass for a new industry. Second, like its predecessors, NBC was cloaked at its birth in the language of national interest. Sarnoff wasn't building an empire just to increase his own personal fortunes, he was bringing order to a critical but unruly industry. NBC wasn't portrayed as an entertainment company, but as a voice of reason—a responsible means to channel education and information to an increasingly hungry mass audience. Clearly NBC stood to benefit from the role it would play in the burgeoning broadcast industry. But Sarnoff's ability to link this role to a broader public mission gave the new company prestige, support—and power. And the fact that this all occurred under RCA's auspices made it stronger still.

Regulation: The Radio Act of 1927

Meanwhile, the rest of the U.S. radio industry was rolling in the chaos that Herbert Hoover had unleashed. Once it became obvious that the government had precious little power to deny or even police radio licenses, stations simply jumped

[66]NBC Press Release, November 1926. Reproduced in Barnouw, *A Tower in Babel*, p. 187. Italics in original.

into the ether at whim. Many were stations in name only, often run by complete amateurs who shunned even the radio boys' informal norms of behavior. In one infamous example, a California evangelist named Aimee Semple McPherson decided to run a small station from her local temple. With no knowledge of radio mechanics, she roamed all over the dial, broadcasting from whatever frequency she happened to stumble upon. When Hoover's agents threatened to close her station, Sister Aimee wired the Secretary at once. "Please order your minions of Satan to leave my station alone," she demanded. "You cannot expect the almighty to abide by your wave length nonsense. When I offer my prayers to him I must fit into His wave reception. Open this station at once."[67] And so it was. Sister Aimee agreed to find a competent engineer, and Hoover reopened her station.

Other broadcasters were less emphatic, but no less troublesome. Throughout the summer and fall of 1926 they poured onto the airwaves in raucous gangs, jumping across frequencies and "stealing" spaces that had already been allocated under Hoover's old system. Quickly, all semblance of order broke down. In most metropolitan areas, where listeners could no longer even receive a consistent signal, sales of radio sets began a dismal slide. "Radio broadcasting," notes one account, "was in danger of destroying itself by the mad scramble of selfish interests."[68]

In the face of such chaos, the larger radio stations descended at last upon Congress. With no more reference to open markets or freedom of the airwaves, they begged legislators to play a more aggressive role in the radio industry: to stop the "wave jumpers" and "pirates" and to police a system of allocated frequencies. While there was a predictable amount of disagreement about the specifics of regulation, consensus—among the larger firms at least—was tight. As Hoover remarked, the radio industry by this point was "probably the only industry of the United States that [was] unanimously in favor of having itself regulated."[69] Between 1925 and 1927, eighteen bills for radio

---

[67]Quoted in George Douglas, *The Early Days of Radio Broadcasting* (Jefferson, N.C.: McFarland & Company, 1987), p. 94.
[68]Archer, *History of Radio to 1926*, p. 370.
[69]Quoted in Douglas, *The Early Days of Radio Broadcasting*, p. 93.

regulation were submitted to Congress, and even President Coolidge joined the chorus of supporters, arguing that "the whole service of this most important function has drifted into such chaos as seems likely, if not remedied, to destroy its great value."[70] It was ironic in some ways—this roughshod industry beseeching the government to impose order upon it. But there it was. Just as congestion on the wires had forced the early telegraph companies to mend their feuds and coordinate their actions, so did congestion on the airwaves compel the radio broadcasters to establish rules and divide their terrain. The difference, though, was that the telegraph wires had clear owners and property rights, while the airwaves were intangible and, in theory at least, commonly owned. Without an established system of property rights, the firms had little choice but to turn back toward the state.

The law that emerged, the Radio Act of 1927, was a concrete reflection of these demands. First, in a sharp correction of the 1912 Act, it clearly made radio the province of the federal government and its regulatory powers: henceforth, the U.S. government would "maintain control over all channels," granting licenses for the *use* of channels "but not the ownership thereof."[71] Thus there were property rights in the ether, but ownership lay strictly with the state. Second, the Act established a Federal Radio Commission (FRC), which, guided by standards of "public interest, convenience or necessity," was to oversee the administration of radio licenses.[72] There was to be no censorship of radio broadcasts, with the exception of "obscene, indecent or profane language," and no licenses were to be granted to any party found guilty of monopoly.

In effect, the 1927 Act made radio a kind of regulated utility. It served a public purpose of some sort; it was subject to the

---

[70]Quoted in Douglas, *The Early Days of Radio Broadcasting*, p. 95.

[71]Public Law No. 632, 69th Congress, 2d session, 1927.

[72]For arguments about how the FRC reached its decisions, see Hugh Richard Slotten, "Radio Engineers, the Federal Radio Commission, and the Social Shaping of Broadcast Technology: Creating 'Radio Paradise,'" *Technology and Culture* 36, no. 4 (October 1995), pp. 950–986; and Lawrence W. Lichty, "The Impact of FRC and FCC Commissioners' Backgrounds on the Regulation of Broadcasting," *Journal of Broadcasting* VI, no. 2 (Spring 1962), pp. 97–110. For a comprehensive review of the FRC's activities, see Thomas Porter Robinson, *Radio Networks and the Federal Government* (New York: Columbia University Press, 1943).

usual concerns over decency and fair play; and it was to be ruled by a dedicated authority. There was no more mention of the inherent freedom of airwaves; no hint that the ether was infinite or uncontrolled. Rather, the 1927 Act pulled radio harshly back to earth, forestalling the dreams of the radio boys and carving the ether into manageable chunks of space. It also paved the way for commercial radio and left the airwaves wide open to the emerging power of NBC. For although the Act prohibited monopoly, it didn't restrict either chain stations or commercial advertising. It also made no specific provision for small-time stations or public interest broadcasting. As a result, the FRC during its formative years edged closer and closer to the side of network broadcasting. Or, as one leading political scientist commented in 1935: "While talking in terms of the public interest . . . the commission actually chose to further the ends of the commercial broadcasters."[73] In the summer of 1927, the FRC reallocated all existing radio licenses.[74] It then held hearings on 164 stations, asking them to demonstrate how "their continued operation would serve public interest, convenience, or necessity."[75] Only eighty-one of these stations were eventually renewed, most with reduced broadcasting time and power. Many of the educational stations were eliminated in this process, or forced to share time with their commercial counterparts. Twenty-three educational stations thus disappeared in 1928, and another thirteen in 1929.[76] The larger stations, by contrast, fared eminently well. Of the twenty-five channels "cleared" by the FRC for more efficient broadcasting, twenty-one were network stations.[77] By 1929, as Table 3.1 indicates, NBC alone controlled sixty-nine stations.

[73]E. Pendleton Herring, "Politics and Radio Regulation," *Harvard Business Review*, January 1935, p. 173.
[74]See Federal Radio Commission, *Annual Report of the Federal Radio Commission for the Fiscal Year 1927*, p. 8.
[75]Federal Radio Commission, *Annual Report of the Federal Radio Commission*, p. 15.
[76]Barnouw, *A Tower in Babel*, p. 218. For more on this trend and its implications, see Tracy F. Tyler, ed., *Radio as a Cultural Agency: Proceedings of a National Conference on the Use of Radio as a Cultural Agency in a Democracy* (Washington, D.C.: The National Committee on Education by Radio, 1934).
[77]See the discussion in Barnouw, *A Tower in Babel*, pp. 211–19; and Hearings before the Committee on the Merchant Marine and Fisheries, U.S. House of Representatives, 70th Congress, 1st session, January 26—February 14, 1928 (Washington, D.C.: U.S. Government Printing Office, 1928).

TABLE 3.1 NBC Stations, 1926–1936

Year	Number of Stations
1926	19
1927	48
1928	56
1929	69
1930	72
1931	83
1932	85
1933	85
1934	86
1935	87
1936	103

Source: United States, Federal
Communications Commission, "Report
on Chain Broadcasting" (Washington,
D.C.: U.S. Government Printing Office,
1941), p. 15.

As this industrial shakeout progressed, the popularity of radio exploded. When radios first appeared on retail shelves in 1920, annual U.S. sales had totaled only $10 million. By 1929 they hit nearly $850 million, giving radio a growth rate that surpassed that of any other industrial product. As Table 3.2 demonstrates, between 1922 and 1929 radio sales increased by more than 1,300 percent—or roughly 46 percent a year.

Indeed, as many commentators argued at the time, radio in the 1920s grew faster than just about anything ever had: "The rapidity with which the thing has spread," claimed one contemporary article, "has possibly not been equalled in all the centuries of human progress."[78] At the decade's end, 75 thousand home receivers were being produced every month.[79] Meanwhile, in a trend reminiscent of the 1990s Internet stock craze,

---

[78] "'Listening In,' Our New National Pastime," *Review of Reviews*, 67 (January 1923), p. 52. Cited in Douglas, *Inventing American Broadcasting*, p. 303.
[79] All figures here are taken from Bilby, *The General*, p. 90.

TABLE 3.2 The Boom in U.S. Radio Sales

Year	Sales*	Average Price of Receiving Unit
1922	$60,000,000	$50
1923	$136,000,000	$27
1924	$358,000,000	$67
1925	$430,000,000	$83
1926	$506,000,000	$114
1927	$425,600,000	$124
1928	$650,550,000	$122
1929	$842,548,000	$136

*Includes receiving sets and tubes, as well as supplementary apparatus such as aerials and batteries.

Source: W. Rupert Maclaurin, *Invention & Innovation in the Radio Industry* (New York: Macmillan, 1949), p. 139.

RCA's stock followed radio sales to dizzying heights. From $85 in early 1928, it rose to over $200 by June of that year. In November it hit $400; it reached $500 by the summer of 1929. The stock was then split into five shares, each worth $101. At this stage, a $10,000 investment made in 1921 was worth over $1 million. And the company had yet to pay any dividends. F. Scott Fitzgerald invested part of the proceeds from *The Great Gatsby* in RCA stock; former president Calvin Coolidge and General John J. Pershing also became shareholders.[80] RCA was the Yahoo! of its day, the must-have stock for a newly eager investing public.

And then it crumpled. When the market crashed in 1929, RCA fell from a post-split high of $110 to lower than $20. Fitzgerald lost much of his fortune, as did Owen D. Young. (Sarnoff, remarkably, had sold all of his shares just months earlier.) Along with the rest of the industrial world, RCA was thrown in the 1930s into a period of commercial despair. Sales plummeted, workers were laid off, and even the company's bedrock messaging business suffered from a total contraction of world trade. But broadcasting, not surprisingly, flourished, making radio a permanent feature of American life and culture.

[80]Bilby, *The General,* pp. 90–91.

By 1931, the influence of NBC was well established. It had eighty-three stations clustered into two linked networks, a net income of over $2.6 million, and had been described by its own president, Merlin Aylesworth, as an "enormous power concentrated in the hands of a few men controlling a vast network of radio stations."[81] The impact of such power was immense, and it pushed radio along a very particular course, one that veered in many ways from the path scratched out during radio's boom days. With NBC and other "chain" stations on the air, for example, listeners no longer cared about the sounds of far-off stations, since so many of them carried the same programming. And so long-distance listening, a herald of the radio boys, dropped away. So, too, did the custom of free performances by artists who sang or played simply for the thrill of going "on air." Now that radio was a distinctly commercial enterprise, performers demanded their own share of the proceeds and joined talent agencies and unions, which eventually fixed their fees and helped negotiate contracts. This, in turn, further impeded the independent stations, which lacked skills and capital to deal with these more formal demands.

Less than two decades after its birth, then, the U.S. broadcast industry had settled into a distinctive pattern of conduct. No longer a free-for-all along the technological frontier, it was instead a stable and well-regulated market, complete with property rights (both in the atmosphere and in the music) and organizations to dispense these rights and patrol violations. There were notions of theft now, and codes of conduct, plus an active audience with its own interests in radio and its own means of influence. Despite a persistent raft of allegations, NBC never developed into a true radio monopoly. Indeed, as early as 1928 several other powerful radio networks, including CBS (the Columbia Broadcasting System), had emerged, along with a handful of strong independent stations. But the structure that NBC established—the rules, the rewards, the players—had become the dominant structure of commercial radio.

[81] Merlin Aylesworth, "Commission on Communications: Hearings before the Committee on Interstate Commerce, U.S. Senate, 71st Congress, 1929–1930, on S. 6. (Washington, D.C.: Government Printing Office, 1930), p. 1702.

Radio Europe

As radio was rolling out across the United States, a similar process was also occurring in Europe. Here as well, radio made a quantum leap in the 1920s from pure communication to media, creating a broadcast industry as strong as that which emerged in the United States and drawing an equally passionate audience. In Europe, as in the United States, the emergence of radio broadcasting was a major cultural and political event: a channel, potentially, for transforming nations, or uniting them, or projecting their images to far-flung lands. But in Europe, the early exuberance of radio was dashed much more quickly by the intervention of the state, and the broadcast market was tied from the outset much more closely to state demands. Rules came to European radio even before the market developed, creating an industry that was similar to its American cousin but considerably more orderly, measured, and well behaved.

In Britain, the path toward this relationship began in the opening days of World War I, when the Royal Navy abandoned its arms-length relationship with radio and took over production at the Marconi Works. For the next several years, Marconi employees worked directly for the British government, intercepting enemy transmissions and training soldiers in the mechanics of radio technology.

After the war, the military was keen to retain its control over radio, seeing it still as a communications device too powerful to let slip into public hands. But the Marconi Company was also licensed by the Post Office to set up two experimental broadcasting stations, one in Ireland and another in Chelmsford. Still linked to the state and its security interests, these stations were essentially just testing the technical possibilities of broadcasting. Engineers who played instruments would play; if they knew someone who sang, they would sing. Yet, as in the United States, even these crude transmissions rapidly developed a devoted following and an enthusiastic audience. Using homemade receivers, many early fans would spend hours listening to lists of train schedules or marine forecast reports, entranced by the airwaves and hungry for more.

As this enthusiasm mounted, the Marconi Company moved gingerly to meet it. In 1920 it broadcast a concert by the popular entertainer Dame Nellie Melba, a move that delighted its listeners but enraged prominent members of the British military, who regarded the broadcast as a serious abuse of technology. When subsequent concerts began occasionally to interrupt transmissions of weather or tidal information, the Post Office suspended all broadcasts from the Chelmsford station. This action transformed Britain's lonely bands of listeners into a united and angry coalition. In 1920 and 1921, the Wireless Society of London held a series of conferences that demonstrated radio's growing popularity and revealed a seething discontent with the government's restrictions. In response to these complaints, the Post Office relented a bit, agreeing in 1920 to distribute more licenses for radio purchases and grant broadcasting licenses to amateurs who could realistically promise not to interfere with official transmissions. It was a significant step toward recognizing radio's popular ambitions, but hardly a wholehearted one. Applicants for broadcast licenses still had to submit character testimonials to the Post Office, and their transmissions had to be dedicated to scientific research or public education. Such procedures whetted the appetites of radio enthusiasts but did not satisfy them. Within a year, the Wireless Society was pushing for a complete lifting of content restrictions and the establishment of regular popular programming.

Initially the government balked, maintaining its opposition to the popularization and commercialization of radio transmission. In an attitude that would be repeated throughout the coming decades, public officials pointed to the United States as evidence of market failure—of everything that could go wrong if the medium were let loose from government's reins. As one official at the time recalled, "The American experience provided a valuable lesson. It showed the dangers which might result in a diversely populated country of a small area like our own if the go-as-you-please methods of the United States were copied."[82] Chief among these dangers, it appears, was technical

[82]Arthur Burrows, BBC Director of Programmes, quoted in Asa Briggs, *History of Broadcasting in the United Kingdom, Vol. I, The Birth of Broadcasting* (London: Oxford University Press, 1961), p. 67.

chaos and the loss of quality content. British officials argued that the commercialization of radio in the United States had led to a jumble of overlapping frequencies and a proliferation of crass and vulgar programming, to content that was all about entertainment rather than national security or any attribute of public service. In America, the British stressed, the government had given up its control of the airwaves and gotten virtually nothing in return. This was not an experience that Britain was eager to emulate.

At the same time, though, it was becoming clear that neither could Britain confine radio to a purely military, wholly controlled sphere. The technology was simply too popular, too promising, and already too prevalent. The government needed some way to release this technology onto the market without letting market forces overwhelm its development. This policy compromise became the BBC.

Launched in 1923, the BBC (British Broadcasting Company, later the British Broadcasting Corporation) was established as a consortium of radio manufacturers. Formally independent, the BBC was nevertheless linked closely to the British government. Its license to broadcast came from the Post Office and its revenue from two novel taxes: a ten-shilling license fee paid to the Post Office by all new radio owners and a 10-percent royalty from the sale of all BBC-member radios. There was to be no advertising on BBC radio and thus, planners presumed, no pressure from commercial forces. There was also explicitly no competition for its services. Instead, all potential competitors (that is, all the radio manufacturers who had applied, or might apply, for a broadcast license) were brought from the outset into the BBC's consortium structure. In one move, therefore, the BBC was set up to solve the several problems that radio had simultaneously posed. It would provide the popular programming that enthusiasts demanded. It would prevent multiple broadcasters from crowding the spectrum and interfering with official transmissions. And it would create a broadcast system free from the chaos that had already engulfed the American airwaves.

From the start, these objectives carved a distinctive character upon the BBC. It was radio linked to the ideals of government, radio that was committed to providing public service in

addition to entertainment. For J. C. Reith, first general manager
of the BBC, radio was too important to be devoted to leisure,
too promising to be developed along laissez-faire lines. It was
instead a powerful tool for mass communication and for incul-
cating the British public with a socially desirable set of values. "I
think it will be admitted by all," he wrote, "that to have ex-
ploited so great a scientific invention for the purpose and pursuit of 'entertainment' alone would have been a prostitution of
its powers and an insult to the character and intelligence of the
people."[83] Rather, radio would be a tool of progress and democracy, informing and enlightening the public and preserving
a certain "high moral standard."[84] "Our responsibility," he
urged, "is to carry into the greatest possible number of homes
everything that is best in every department of human knowledge, endeavor and achievement, and to avoid the things which
are, or may be, hurtful."[85] As the BBC developed, both the company and the technology became nearly an object of worship for
Reith. "Broadcasting," he proclaimed in 1924, "is a servant of
culture and culture has been called the study of perfection. The
whole service which is conducted by wireless broadcasting may
be taken as the expression of a new and better relationship between man and man."[86] In marked contrast to his American
peers, Reith had absolutely no interest in giving audiences what
they might want; he gave them what he thought they needed.

Reith ran the BBC for sixteen years and left an indelible imprint upon it. By the time he resigned in 1938 the BBC was already seen as the voice of authority in Britain—fairer and more
informative than even the well-regarded print media. Under
Reith, it also perfected a distinctive balance between its public
face and political ties. Officially, the organization remained
closely linked to the British government. Funding for the service was channeled through the Post Office, and the Postmaster
General retained veto power over broadcast content.[87] In practice, though, the BBC operated not as a propaganda arm of the
government but rather as an independent, autonomous agency.

[83]J. C. W. Reith, *Broadcast Over Britain* (London: Hodder & Stoughton, 1924), p. 17.
[84]Reith, *Broadcast Over Britain*, p. 32.
[85]Reith, *Broadcast Over Britain*, p. 34.
[86]Reith, *Broadcast Over Britain*, p. 217.
[87]Briggs, *History of Broadcasting*, p. 358.

It was generally on the side of the government, but never in its pocket. This was Reith's argument, and perhaps his greatest legacy. As one historian has described it, the BBC under Reith shaped a relationship with the government based on "calculated imprecision." By maintaining quality standards and a deep-seated commitment to public service, the BBC won in return "all the liberty, independence, [and] autonomy that can be hoped for."[88] All of which, though, remained ultimately in the hands of the British government.

As an institution, the BBC had and retains a distinctive character. It is a symbol of Britain to the outside world and a dominant figure, as shall be seen in Chapter 4, of Britain's political and cultural landscape. The basic structure of the BBC, however, and the relationship between state and station that it embodies are hardly unique. Indeed, across the European continent governments greeted the broadcast era with similar strategies of regulation and control. After an initial period of "controlled freedom," for example, the French government moved rapidly and decisively into the radio industry. In 1926, it issued the Bokanowski decree, proclaiming:

> A country that is not equipped with a coherent, rational, and powerful system of radio is exposed to trouble from widespread, suggestive, and persistent foreign propaganda, and at the same time deprives itself of the most effective stimulus to artistic and intellectual development. If it receives foreign broadcasts without being able to broadcast the expression of its own genius beyond its borders, a country resigns itself to suffering serious damage.[89]

To ensure that radio served the national interest, the French established a public system of broadcasting, more open in some

---

[88]Tom Burns, *The BBC: Public Institution and Private World* (London: MacMillan Press Ltd., 1977), p. 21.

[89]Benjamin Huc and François Robin, *Histoire et Dessous de La Radio en France et Dans Le Monde* (Paris: Les Editions de France, 1938), p. 41. In French, the quote reads: *un pays qui n'est point parvenu à se doter d'un système de radio diffusion cohérent, rationnel, et puissant, est exposé à devenir tributaire de la propagande étrangère sous la forme la plus accessible, la plus suggestive, la plus persistante, en même temps qu'il se prive des moyens d'action, d'expansion intellectuelle et artistique les plus efficaces. Recevoir les émissions étrangères sans pouvoir diffuser au delà des frontières l'espression de son propre génie, c'est, pour un pays, se résigner à une grave diminution.*

ways than the British—private stations were permitted—but
tightly controlled in terms of access and content. Only French
citizens were allowed to broadcast; broadcasters were held to
strict rules of censorship; and the largest player in the industry
was the government's own extensive chain of stations. In
France, claimed one of the country's leading politicians, "it
seems impossible . . . that radio not be controlled by the state. It
is an organization that involves the national security and the
National Defense."[90]

A similar attitude prevailed in Germany, where the government moved from the outset to control what it saw as a potent
and potentially disruptive technology. In 1922, authority over
the radio industry was split between two ministries, the Ministry of Post (RPM) and the Ministry of Interior (RMI). Although independent production companies were permitted to
provide the state-run stations with content, the RPM retained a
majority of voting shares in each of these companies and local
censorship bodies oversaw all programming. This structure became even tighter over the course of the 1920s and 1930s, until
the government eventually owned all of the equipment in German radio stations, sanctioned all the news that was broadcast;
and even determined who in the country was allowed—or not
allowed—to listen in.[91] Nearby, meanwhile, governments in Italy,
Austria, and Czechoslovakia controlled their budding radio networks through a series of ostensible private monopolies that
were effectively linked to the state; while newly communist Russia, having no need to pretend, simply placed all radio broadcasting under state control.

By the time World War II crashed through Europe, therefore,
the continent was crossed by an extensive web of state-run, or at
least state-linked, radio networks. As in the United States, broadcast radio had become a powerful social factor—a technological

[90]Edouard Herriot, quoted in Huc and Robin, *Histoire et Dessous de la Radio,* pp. 41–42.
For more on this period, see also Rene Duval, *Histoire de La Radio en France* (Paris: Edi-
tions Alain Moreau, 1979); and U.S. Department of Commerce, *Radio Markets of the
World, 1932* (Washington, D.C.: U.S. Government Printing Office, 1932).
[91]For more on the early development of radio in Germany, see Rudolf Arnheim, *Radio,*
trans. Margaret Ludwig and Herbert Read (London: Faber & Faber, Ltd., 1944); and
Kate Lacey, *Feminine Frequencies: Gender, German Radio, and the Public Sphere, 1923–1945*
(Ann Arbor: University of Michigan Press, 1996).

breakthrough that seized consumers' imagination and created a massive and untapped demand for news, information, and entertainment. In contrast to the United States, however, radio in Europe did not experience a period of chaos or any ambiguity over property rights. Instead, there were rules from the start in European radio that created technical and content standards and established broadcasting as a tool of the national interest. In the United States, the consolidation of the radio industry meant the growth of NBC and then eventually CBS: powerful broadcasters that carved out their own rules and engaged—slowly, painfully, but ultimately successfully—with the regulatory regime that emerged around them. In Europe, the consolidation of the same industry saw the emergence of the BBC and its counterparts: equally powerful entities, equally engaged with their own governments, but tied here much closer to the state, and regulated from birth by the rules that created them.

Ruling the Waves

The story of radio by no means ends in the 1930s. On the contrary, radio spans the remainder of the century, gaining in popularity as the decades wear on and affecting both business and politics in profound ways. In the United States and Europe, radio gave birth to distinctive modes of entertainment: the variety show, the news program, and the serial. Shows such as *Amos 'n Andy* and *The Rise of the Goldbergs* became parables for their time, capturing slices of cultural life and then shaping in turn the attitudes and experience of their listeners.[92] American radio also pioneered a distinctive mode of business, one based not on messaging or public service, but rather—fully

[92]*Amos 'n Andy,* a parody of black life in America, was wildly popular in the early 1920s despite its deeply racist overtones. *The Rise of the Goldbergs,* which aired for thirty years, traced the story of an immigrant Jewish family in New York. For more on the cultural impact of radio see for instance Carolyn Marvin, *When Old Technologies Were New* (New York: Oxford University Press, 1988); Michael Brian Schiffer, *The Portable Radio in American Life* (Tucson: University of Arizona Press, 1991); Andrew F. Inglis, *Behind the Tube: A History of Broadcasting Technology and Business* (Stoneham, Mass.: Focal Press, 1990); Hilmes, *Radio Voices,* and Douglas Czitrom, *Media and the American Mind: From Morse to McLuhan* (Chapel Hill: University of North Carolina Press, 1982).

and profitably—on advertising. During the Great Depression, radio stations, like many businesses, were extraordinarily hard-pressed for cash. Hundreds of independent stations were forced to shutter their operations, while others cut back drastically on their services or programs. Still others began to trawl for new sources of funding, trying to find some means—any means—of keeping their finances afloat. Eventually, they came to what seems in retrospect an obvious solution. They began to expand relations with the firms that were already sponsoring some of their programs, selling them an allocated block of time—thirty seconds, usually, or a minute—in which to tout their product. This was the start of broadcast advertising, and of commercials as we know them today.[93]

Politically and culturally, meanwhile, radio's reach grew even wider than many prophets had predicted. Radio became a tool for political communication, epitomized by Franklin D. Roosevelt's "fireside chats," and a fiery weapon of propaganda. Hitler used radio extensively during World War II, as did Winston Churchill. After the war, radio became a means of enforcing control within the communist bloc, and a way for Western states to try to break the communists' monopoly on information. The U.S. government, for example, used Radio Free Europe and Radio Marti to broadcast political information to Eastern Europe and Cuba; while the British relied on the BBC World Service to reach interested listeners around the globe. As shall be seen in Chapter 4, therefore, radio was the political precursor to satellite television, a way of propelling information across international boundaries and around the tentacles of national governments. At the same time, as will be discussed in Chapter 7, it was also the catalyst for a whole new industry—the recorded music industry—and for transforming music's position within popular culture. Jazz grew up in the radio stations of the 1920s, and rock 'n' roll erupted there in the 1950s.

For our purposes, though, the story of radio really does end in the 1930s, for it is here that the frontier gives way to order and stability. Between 1895 and 1930, radio was a technology in

---

[93]For more on the development of radio advertising, see Barnouw, *A Tower in Babel*; Barnouw, *The Golden Web*; and Douglas, *Inventing American Broadcasting*.

flux, a technology driven by impassioned inventors and flights of scientific fancy. Mystical to many, radio seemed during this time as ethereal as the waves that carried it. It couldn't be controlled, people thought. It couldn't be owned; it could never be regulated. By 1930, though, these beliefs had been swept away, along with many of the people who had staunchly avowed them. Radio had become by that time a tool of both big government and big business—ruled, regulated and armed with a distinctive set of property rights.

The forces that led to regulation, however, were different in radio than they had been in either telegraphy or ocean trade. They had mostly to do with coordination and congestion, and they emerged in two consecutive waves. During the first wave, the technical beauty of wireless was its ability to communicate at sea and to drop the wires, poles, and cables that cluttered the telegraph industry. The problem, however, was that wireless systems, like telegraph systems, had to be compatible in order to be useful. For ships to communicate either with the shore or with one another, they needed—quite literally—to be on the same wavelength and using similar equipment. Because telegraphy was already well established by the time wireless rolled in, the early users of wireless simply adopted Morse code as their common language, solving one problem of compatibility without even perceiving it as an issue. Marconi then solved the next problem by fashioning a seamless web of compatible ships and a network of wireless tied to a common standard. In a common refrain of the technological frontier, though, this single firm solution created its own predictable problems. Governments feared Marconi's power while other firms resented his commercial reach; and the two forces moved before long to break his stranglehold. In Europe, the replacement for Marconi was a series of state-linked companies overlaid, by 1906, with a formal set of common standards. In the United States, it was RCA.

During the next phase of radio, wireless morphed into broadcasting and code was replaced by voice. This transformation brought radio into the popular consciousness and created a whole new realm of cultural and commercial pursuits. In the United States, however, it also led directly to chaos, as hundreds and then thousands of would-be broadcasters streamed across the airwaves. The problem during this stage, therefore, was congestion, and the solution lay with government intervention. By 1927, an initially reluctant U.S. government had created a formal regulatory structure for radio. Government officials established a clear set of property rights in the once-ethereal ether and a set of formal laws to accompany them. They enforced the rights and laws through both legislation and the court system, and they topped it all off with a newly crowned regulator. All of these moves were initiated and supported by the private sector, or at least by the dominant private firms. Could the U.S. radio industry have solved its congestion problem without government regulation? Perhaps. In a 1959 article, the economist R. H. Coase (who would go on to win the Nobel Prize in 1991) suggested that it might have been possible to leave spectrum allocation to the market, letting demand for the airwaves determine who received them, and at what price.[94] Yet while such an outcome seems theoretically reasonable (firms could have bid for the spectrum, or divided it amongst themselves), the pattern of events denies its feasibility. For in the 1920s, it was private firms that expressly denounced a market solution and instead petitioned the government for formal allocation of the spectrum. It was private firms—and particularly the large and successful ones—that asked the Commerce Department to redefine the whole notion of the ether, transforming it from a common and inchoate space into something that could be owned and managed and controlled. The reasons for this behavior are obvious. If broadcast networks such as NBC were to flourish in the new world of radio, they needed to secure a stable system of property rights. They needed consistent access to their segment of the spectrum and a reasonable guarantee that neighboring stations or unsophisticated amateurs would not come barging into

---

[94]R. H. Coase, "The Federal Communications Commission," *Journal of Law and Economics* 2 (October 1959), pp. 1–40. For subsequent arguments that build upon Coase's work in this area, see Ithiel de Sola Pool, *Technologies without Boundaries: On Telecommunications in a Global Age* (Cambridge, Mass.: Harvard University Press, 1990); and William H. Meckling, foreword to *A Property System Approach to the Electromagnetic Spectrum*, by Arthur S. De Vany, Ross D. Eckert, Charles J. Myers, Donald J. O'Hara, and Richard C. Scott (San Francisco: Cato Institute, 1980). For a critical review of this literature, see Hugh G. J. Aitken, "Allocating the Spectrum," pp. 686–716.

their broadcasts. If the number of radio stations had remained small and homogeneous, it is possible that the players could have negotiated the terms of engagement by themselves. But once the number grew and commercial stations found themselves jostling with the likes of Aimee Semple McPherson, the likelihood of private regulation diminished. By this stage, the only solution that served the commercial stations was a system of property rights and formal allocation. And the only entity that could preside over such a solution was the state.

Against this backdrop, it is interesting to see how fears of monopoly played a different role in radio than they had in telegraphy and would in software and satellite television. As in all these industries, there was a certain tendency in radio toward monopoly, a certain advantage that could be gained through common standards and a network of interlocked players. Marconi clearly saw these advantages and acted upon them. Explicitly, he strove to create a monopoly in radio communications and to fashion the industry's rules around his own commercial position. He almost succeeded. But in the end governments were unwilling to cede this power to Marconi and each, in its own way, attacked him: Britain through nationalization; France and Germany through the establishment of competing technologies; and the United States with the creation of RCA and its absorption of the American Marconi Company. In Europe, the reaction to Marconi then flowed seamlessly into the broadcast era, and states carved out national broadcast monopolies before a commercial market could even develop.

In the United States, by contrast, the government sat further from the broadcast industry and the monopoly question was more complicated. In 1930, the Justice Department brought suit against RCA and its allies, charging them with anti-competitive behavior in the radio industry. The charges were almost certainly accurate and they could have been devastating to RCA and NBC since, under the 1927 Radio Act, any station found guilty of monopoly was liable to have its license revoked. In 1932, however, the charges were dropped, and NBC continued to exert its powerful influence over the American broadcasting industry. How are we to explain this apparent inconsistency? Why did the U.S. government attack Marconi (and Western

Union, and AT&T) and leave NBC unfettered for so long? Part of the answer may be sheer luck: the radio suit came in the midst of the Depression, when the government would have been hard pressed to penalize one of the few American companies that was actually making money. Part of it, too, was related to the inherent qualities of radio and a lingering unwillingness to subject broadcasting to the kind of chaos that had flourished in the mid-1920s. But much of the explanation falls back to David Sarnoff, a brilliant negotiator and uncanny reader of the political situation around him. When Sarnoff learned of the Justice Department's antitrust suit, he deftly rearranged RCA once again, divorcing it formally from General Electric and Westinghouse.[95] After two years of fragile negotiations, RCA emerged with the alliance's radio manufacturing facilities, GE and Westinghouse received debentures and some real estate, and NBC became a wholly owned subsidiary of RCA. The Justice Department withdrew its suit, leaving the broadcasting industry to follow once again in NBC's footsteps. Later Sarnoff would boastfully recall, "They [the government] handed me a lemon, and I made lemonade out of it."[96]

In the end, allegations of monopoly did catch up to RCA. In 1942, the Federal Communications Commission (heir to the Federal Radio Commission) forced the giant radio concern to sell off one of its twinned networks, creating what would eventually become the American Broadcasting Company (ABC).[97] Yet even this measure did not really affect what RCA did or how it operated. RCA remained a dominant producer of radio equipment and NBC, together with CBS and the new ABC, continued to provide commercial broadcasts in a stable and profitable market. Throughout its commercial life, therefore, RCA was something of a golden child, linked to government and regulated by it, but never truly punished.

To understand this privileged position, it useful to consider what kind of rules RCA wanted and how they overlapped with the U.S. government's political aims. For essentially, what RCA

---

[95]Details of this negotiation are available in Barnouw, *A Tower in Babel,* pp. 252–68.
[96]Quoted in Bilby, *The General,* p. 109.
[97]The process of separation began in 1942, although NBC's Blue Network wasn't formally sold until 1943, when the American Broadcasting Company (ABC) was formed.

had wanted since 1920 was a particular set of rules: rules that protected ownership in the ether, prevented interference, and ensured its own technological edge. Clearly, the objective of these rules was to protect and advance RCA's commercial position. Their practical impact, however, was also to further the government's own interest in an orderly and technically proficient radio sector. In the U.S. radio industry, therefore, there was an almost perfect alignment between what RCA wanted and what the government thought necessary. This alignment didn't completely protect RCA from political attack—witness the constant stream of antitrust investigation—but it certainly helped.

In the end, few of radio's technical pioneers realized the commercial promise of their work. Marconi built his empire only to see it, toward the end of his life, absorbed into Britain's behemoth Cable and Wireless. Popov's work disappeared into the chaos that became the Soviet Union; Slaby's, into a Telefunken tied increasingly to the German state. In the United States, Fessenden was actually kicked out of the radio market by a string of patent liability decisions, and De Forest suffered several bankruptcies and a stock fraud indictment before abandoning radio in pursuit of other technologies. Edwin Armstrong, who invented the superheterodyne set (a vastly superior receiver) and developed FM radio, plunged from a thirteenth-story window in 1954, pushed to despair over patent litigation with RCA. Meanwhile, of course, the radio boys' hopes vanished into the ether from which they had sprung. Radio became instead a stable and heavily regulated field, parsed and divvied as property and shaped by the likes of David Sarnoff and J. C. Reith, men who wrote the rules of radio and made them stick. This pattern persisted throughout radio's own technological advances and then slipped indistinguishably into television, the next major development along the technological frontier.

CHAPTER 4

BSkyB and the
New Wave of Television

> Broadcast information is too important to be left to
> market pressures alone.
>
> RUPERT MURDOCH

In 1983, Rupert Murdoch's News Corporation bought a floundering British company called Satellite Television and renamed it Sky. The transaction garnered very little attention at the time, and those who did notice it were inclined to scoff. Satellite television, they said, was an absurd idea: too expensive, too far-fetched, too unnecessary. And Murdoch, they continued, was even worse. He was a newspaperman, and an Australian. Sky, most observers clucked, would be gone overnight—a painful example of pushing technology where markets did not go.

For nearly a decade, it looked as if the observers would be proven right. Sky lost hundreds of millions of pounds; it was reviled by the British broadcast establishment; and in 1988 it engaged in a knockdown battle with British Satellite Broadcasting (BSB), a hastily assembled consortium of Britain's media and commercial elite. This time around, many more people were watching, most of whom still gleefully predicted the imminent downfall of Sky. Yet, by 1989, Sky had trounced BSB in the marketplace. In 1990 it effectively acquired its well-heeled competitor, and by 1997 it was the most profitable broadcaster in the United Kingdom. In less than fifteen years, Sky had created a brand new market in Great Britain and undeniably grown to dominate it. In the process, it changed the rules of British television and brought Britain's regulators to a power-defying standstill.

The tale of Murdoch's Sky Broadcasting is but a fragment of the broader story of satellite television. In the 1980s, satellite broadcasting surged around the world. Employing technologies that were first developed in the late 1950s, a host of companies followed on the path laid down by Sky.[1] They launched independent satellite networks, circumventing government networks in many cases and testing governments' ability to regulate information flows. In many places, including China, India, and the Soviet Union, the advent of satellite television raised wrenching political questions. It allowed broadcasters to perch for the first time above a sovereign state, relaying news or shows or music that might not fit the state's notion of what was appropriate or even correct. It created the phenomenon of instant, round-the-clock news—of broadcasters such as CNN (the Cable News Network), which used a network of satellites to depict graphic, real-time news and often became a critical information point for the very people featured in the stories. During the Gulf War, for example, Libyan leader Muammar Qaddafi called CNN when he wanted to release his plans for peace, and both George Bush and Saddam Hussein acknowledged that they had watched the upstart station's coverage.[2] Satellite providers also helped to

---

[1]The idea for satellite communications is much older, going back to the science fiction writer Hugo Gernsback, who proposed something along these lines in 1911. The first experimental communication satellite, however, was not launched until 1958. Development proceeded throughout the 1960s and commercial applications—mostly in long-distance telephony—began to emerge in the 1970s.

[2]See John Lippman and William Tuohy, "Rivals Challenge CNN's Global Grasp," *Toronto Star*, December 24, 1992, p. A17; Thomas Rosenstiel, "CNN: The Channel to the World," *Los Angeles Times*, January 23, 1991, p. A1; and Bill Thomas, "The Bad News Bearers at CNN Have Whipped the Networks in the Gulf War, But Will They Get Zapped When the Scuds Stop Flying?" *Los Angeles Times*, March 3, 1991, Magazine Section, p. 16.

prod the balance of power within authoritarian states, showing citizens images of their own country—such as the 1989 massacre in China's Tiananmen Square or the 1991 attempted coup in Moscow—that the state refused to depict.[3]

Sky's story is more subtle than these. Certainly the company did not change the logic of politics in the United Kingdom or the structure of political power. It didn't bring a new kind of news into the country or a substantially different menu of entertainment. And yet Sky still typifies in many ways the evolution of satellite broadcasting and the satellite television industry. In less than ten years, Sky opened the British television market to competition and creative anarchy. Led by Murdoch and a handful of brilliant lieutenants, the company seized a technology that others had disdained and constructed a whole new industry upon it. Using this technology, it skirted the boundaries of existing regulation and defied the establishment that sought to constrain it. Sky did not invent the technology of satellite broadcasting; it didn't even pioneer the commercial model of satellite television. But it took this technology and this business model and thrust them into a nearly perfect environment, breaking rules and reconstructing them all along the way.

For nearly two decades, Sky presided unchallenged over the market it had created. Like Sir Francis Drake (who was a pirate long before he became a Sir), Murdoch-the-pirate morphed seamlessly into Murdoch-the-pioneer, and Sky joined the elite it had once sought to trample. Eventually, though, both technology and politics caught up with Sky. Gradually, regulators in both the United Kingdom and the European Union began to grapple more systematically with satellite technology and the market it had wrought. Social concerns and social issues slowly pushed their way through the political maze, as did a chorus of would-be competitors. And in the meantime, the technology

---

[3]For further treatment of these developments, see Ralph Negrine, *Satellite Broadcasting: The Politics and Implications of the New Media* (London: Routledge, 1988); Jonathan F. Galloway, *The Politics and Technology of Satellite Communications* (Lexington, Mass.: Lexington Books, 1972); Johanna Neumann, *Lights, Camera, War: Is Media Technology Driving International Politics?* (New York: St. Martin's Press, 1996); and Thomas E. Skidmore, ed., *Television, Politics, and the Transition to Democracy in Latin America* (Baltimore: The Johns Hopkins University Press, 1993).

that drove Sky's business evolved as well, pushing Sky from the cutting edge of the technological frontier and raising the possibility of new kinds of rules and shifting tactics of regulation.

At the turn of the twenty-first century, British Sky Broadcasting was still a very powerful company. It had more than eight million subscribers, £850 million in revenues, and an enviable position in Britain's new media market.[4] But it also had a whole new range of competitors and a much more solid, less ambiguous set of rules. Just as with telegraph and radio, satellite television had become by this point a stable, well-regulated industry. Sky, to be sure, had helped to shape this regulation and—just like Western Union, the Marconi Company, and RCA—was powerfully placed to benefit from it. Yet the frontier in the new millennium was already starting to advance, raising new commercial and political issues and jostling the balance of power once more.

Ground Zero:
British Broadcasting and the BBC

In Britain, the market for television did not come naturally. Rather, television was for many years solely the province of the state, linked to government and its politics through a complex web of finances, friendships, and ideology.[5]

The roots of this relationship began in 1926, when an entrepreneur named John Baird approached the BBC to discuss his plans for broadcasting visual images. At this time, of course, the BBC was still a fledgling public service corporation, prodded by J. C. Reith's vision of progress and devoted to the preservation of a "high moral standard." It wasn't clear in this context just what visual images could offer, but the BBC allowed Baird to set up a small shop in Broadcasting House and experiment with

---

[4]The figure of eight million includes both direct subscribers to Sky and cable customers who subscribe to some menu of Sky programming.

[5]The following account of the BBC's early history draws on Asa Briggs, *The History of Broadcasting in the United Kingdom, Volume I: The Birth of Broadcasting* (London: Oxford University Press, 1961), pp. 1–134. Briggs's five-volume history is the definitive work on the history of the BBC.

transmissions from time to time. In 1933, the BBC struck a similar deal with EMI, an electronics firm that dreamed of developing a mass market for television sets.[6]

By this point, it was already clear that television in Britain would be tied in some ways to the BBC and its public service mandate. But just as the details of this connection were being sketched out, World War II intervened, thrusting the BBC into the arms of Britain's new Information Ministry and shoving television back to the technological sidelines. This neglect lingered until 1945, when the British government decided that television would henceforth be developed solely by the BBC, using technology developed by EMI. There would be no commercials on BBC Television and no competition. Instead, revenues would come from a mandatory license fee to be paid annually by all British television viewers. It was Reith all over again, only this time with pictures.

Accordingly, by the early 1950s, BBC Television had joined BBC Radio as the voice of Britain and the symbol of its psyche. Television sets were selling at a rate of forty thousand a month in 1951, and the average British viewer was slowly becoming attached to shows such as *What's My Line?* and *Animal, Vegetable, Mineral?* For these viewers, television was simply synonymous with the BBC, so much so that the BBC was widely referred to as "Auntie"—a gentle, somewhat bossy presence that resided quite comfortably in the living room. True to her radio roots, Auntie was somewhat liberal in her political preferences but devoutly balanced in her political portrayals. She was also rather proper, as might be expected, and refrained from broadcasting any material that wasn't educational or uplifting or informative.

As Auntie settled down to her postwar prosperity, though, complaints about her power began to emerge. The BBC, critics argued, was too dominant in British life, too complacent and narrow-minded. Within the Conservative Party, a number of MPs began to grumble publicly about the need to "set the

---

[6]For more on the early days of British television, see Asa Briggs, *The History of Broadcasting in the United Kingdom, Volume II: The Golden Age of Wireless* (London: Oxford University Press, 1965), pp. 525, 561, 577.

people free," eliminating the BBC's monopoly and its stranglehold on the public's imagination. These grumbles became louder and more acceptable in 1951, when Winston Churchill and the Conservatives regained power. In May of the following year, the Conservatives launched their first salvo—a white paper that agreed to renew the BBC's charter for ten years but simultaneously urged that competition be introduced into television as soon as it was fiscally possible.[7]

For the next several years, Britain's political elite aligned themselves on the edges of an increasingly sharp divide. On one side, Conservatives formed the Popular Television Association to win support for a new, more commercial television system. On the other, Labour leaders and people connected to the BBC formed the National Television Council, focusing on the evils of commercialization and publicizing the dire social consequences of allowing "non-British" content onto the airwaves. In one pamphlet, the Labour Party sketched out a particularly ominous scenario:

> Parents all over Britain switch on [Children's Hour] confident that the programme will be properly presented and planned in the interests of children and of no one else. This is because the BBC is a Public Corporation and has a proper sense of public and social responsibility. Supposing that responsibility were removed? . . . In the fierce competitive struggle for the largest audience, all today's good standards would be abandoned in favour of the "sensational" and "stunt" type of programme.

The pamphlet went on to cite a recent report describing horrified American mothers who had tallied a total of 104 gun shootings during a half-hour TV show.[8] The same fate was sure to await British children, argued opponents. The Conservatives, however, were nonplused. "We," they stated, "are a much more mature and sophisticated people."[9] British television could go commercial, they believed, without ever sinking that low.

[7]Asa Briggs, *The History of Broadcasting* (London: Oxford University Press, 1979), pp. 423–425.
[8]Asa Briggs, *The History of Broadcasting* (London: Oxford University Press, 1979), p. 897.
[9]Asa Briggs, *The History of Broadcasting* (London: Oxford University Press, 1979), p. 431.

In the end, the government struck a familiar sort of compromise. In 1954 it created the Independent Television Authority (ITA), a regulatory agency that would henceforth license a string of new broadcast firms. This meant competition in the television market, and commercialization—but only of a very limited sort. For regulatory purposes, the ITA split England into three regions (service to Wales, Scotland, and Northern Ireland were established later) and awarded two broadcast licenses in each—one for weekday viewing and another for the weekend. Unlike the BBC, these new broadcasters were permitted to advertise and to earn profits, but they were still closely tethered to the ITA and its notions of what constituted "responsibility," "decency," and the proper degree of "British" programming. Composed of politicians, Lords, academics, artists, and the like (Britain's "great and good," as critics sneered), the ITA retained considerable control over the operations of the independent stations, or ITVs, as they came to be called. The ITA could tell the ITVs what programs to cancel, what kind of advertising to show, and even when to schedule certain shows. Any station that fell foul of the ITA would simply not have its license renewed.

It was an odd relationship in some respects, but also an eminently successful one. The ITVs grew closer and closer to the ITA over the next few decades, locked in a bond of both formal controls and informal links. They developed a unique brand of programming—high quality and "British" like that of the BBC, but lighter-hearted and more popular. They also drew a significant share of Britain's television viewing audience and made a lot of money. After initially grabbing a shocking 80 percent of the total audience in 1957,[10] the ITVs receded to a more comfortable 50 percent in the early 1960s, as the BBC lightened its own fare somewhat and added more "realistic" shows to its educational mission. This 50/50 split soon became the norm, with the ITVs jostling the BBC from time to time with particularly popular shows and the BBC roaring back whenever news stories became more prominent. The BBC continued to shun any

---

[10]Asa Briggs, *The History of Broadcasting in the United Kingdom, Volume V: Competition* (London: Oxford University Press, 1995), p. 21.

commercial content and lived happily (if frugally) off its license fee, while the ITVs consolidated and grew rich. By the mid-1980s, a handful of ITVs dominated the market.

This was the world of British television—quiet, staid, high quality, and comfortable. There was competition, but only between the BBC and a single licensed regional provider. And there was commercialism, but only under the watchful eye of the ITA, and in a market that essentially still lacked competition.

But that was before Rupert Murdoch came along.

The Coming of Sky

Like many successful firms, Sky began its commercial life with a bright idea and a financial disaster. Born as Satellite Television (SATV), Sky was the brainchild of Brian Haynes, a television journalist who had once done a documentary on Ted Turner and his Cable News Network. After completing his research, Haynes had become entranced with Turner and with Gerald Levin, who had introduced Home Box Office (HBO) into the U.S. market at about the same time that Turner launched CNN. Haynes figured that if these programming innovations could take root in the relatively competitive U.S. market, they could achieve even more dramatic results in Britain's environment, allowing the new service to broadcast literally above the regulators' heads.[11] Like Turner and Levin, therefore, Haynes decided to become a satellite broadcaster.

It was a radical idea in many ways, but also a logical one. Technologically, satellite broadcasts had been possible since the late 1960s, when scientists first succeeded in launching data transmission satellites into orbit. Essentially, these are transmitters that hover exactly 22,300 miles above the earth, the point at which their revolution precisely matches the earth's rotation. From this point, the satellite's transmission can consistently reach a specified region of the earth—the satellite's "footprint"—twenty-four hours a day.

[11]See Michael Schrage, "Murdoch Reaches for Sky in European TV Battle," *The Washington Post*, March 3, 1985, p. F1.

In an effort to extend and control the commercial use of satellites, a multilateral organization known as the World Administrative Radio Council had parsed out the world's known satellite space in 1977. Each of the participating countries received five channels on existing broadcast satellites at that time, plus a certain amount of space for future satellites. While most European governments wrestled with the allocation of these channels, Britain moved particularly slowly. And thus the field was wide open in 1981, when Haynes set his sights on the skies and began to construct Britain's first direct-to-home (DTH) satellite system. Rather than petitioning for a piece of Britain's allotted space, he rented space on a low-powered satellite owned by European postal agencies and received a license that essentially gave him a footprint that stretched from Finland to Malta. Eventually, Haynes hoped to beam his satellite programming directly into viewers' homes. But initially he would concentrate on Europe's emerging cable industry, providing them with the content they needed and replicating in the process the success of CNN and HBO.

In the European market, and particularly in Britain, this was a pathbreaking idea. It was a way of combining a proven business model with cutting-edge technology, of circumventing regulation by leaping to a mechanism that had, so far, no rules. The problem, though, was that Haynes's audacious venture demanded equally audacious amounts of capital investment, which Haynes himself did not have. Instead, like many entrepreneurs, he was trying to sell his idea simultaneously to both the consumer and capital markets, raising investment capital as he slowly proved the commercial viability of SATV. Unfortunately, such a strategy was inherently unsustainable, at least in this industry and at this time. With only a small budget behind him, Haynes could not afford to make or buy the programming that might lure subscribers or cable operators to his new channel. And without a solid subscriber base, he could not earn sufficient revenues to fund his operations or his programming. It was a classic catch-22, and Haynes was caught. By 1983, with no revenues in sight, early investors in SATV declared their intention to sell.

It was at this point that Rupert Murdoch's News Corporation became involved. Already, Murdoch was one of the most colorful figures in British media. Australian by birth, Murdoch was a self-made magnate, the son of a small-time publisher who, through guile and wit and financial wizardry, had built one of the world's largest and most controversial media empires. He was particularly controversial in Britain, where his News Corporation had purchased the family-run *News of the World* in 1969 and the London *Sun* in 1970. Both papers were quickly brought into Murdoch's editorial orbit, and the *Sun* was transformed into a full-fledged tabloid, complete with bare-breasted women on page 3 and a decidedly conservative political tilt. Subsequently, News Corp had also purchased the *Times* and *Sunday Times,* making Murdoch one of the most powerful men in British news. He was also one of the most reviled. According to his many critics, Murdoch was no more than a pirate, a scheming lowbrow publisher who catered to his readers' basest desires and had no concern for serious journalism. He was "ruthless" and "cynical"—the "Mephistopheles of popular journalism."[12] And with several newspapers and millions of pounds behind him, he was also extremely powerful.

In June of 1983, News International (News Corp's British arm) purchased 65 percent of SATV for £5 million. Less than a year later, with SATV's chairman cautioning that "any anticipation of significant profits must still be regarded as highly speculative,"[13] News Corp underwrote an additional £5.3 rights issue, which effectively transferred a further 17.6 percent of equity into News Corp's control. By the end of that year, News Corp was in full control of SATV and had already renamed the company Sky. Haynes and the original investors were gone.

For the next several years, Sky essentially bled money. News continued to invest £3 to £4 million a year in the venture, but proved unable to boost either the audience for Sky or its

[12]"Rupert the Ruthless," *Economist,* October 7, 1978, p. 52; David Alpern, "What Makes Rupert Run?" *Newsweek,* March 12, 1984, p. 70; and "Mr. Murdoch's Poodle?" *Economist,* March 20, 1982, p. 15.
[13]Quoted in Raymond Snoddy, "News International Underwrites Satellite TV Rights Call," *Financial Times,* April 17, 1984, p. 35.

advertising revenues. Instead, the channel filled a desultory niche in the European market, providing rehashed American content to a handful of hotels and expatriate viewers. Jeered at by Britain's television elite, Sky had accumulated losses of over £28 million by 1987.[14]

In the meantime, though, many of these elites had also slowly come around to the idea of satellite television. After years of debating what to do with its slice of the satellite spectrum, the British government finally arrived at a solution in 1986—a particularly British solution, in fact, and one that screamed of the regulatory preferences that would dominate television developments in the decade to come. First, two of the slots were given directly to the BBC.[15] Then, under the regulatory auspices of the IBA (the regulatory body that had succeeded the ITA), the government declared that the three remaining satellite frequencies would be given as a bundle to the most competitive bid. The winner would be able to establish a new commercial television service in Britain, subject only to the IBA's usual rules concerning quality, decency, amount of advertising, and the like. The winner also had to employ a cutting-edge satellite transmission standard known as D-MAC, whose virtue lay in the ability to deliver a very high-quality picture to a very small dish.

The structure and implications of this policy are highly revealing. Essentially, what the government did was to take a hugely valuable public commodity and split it into two tiers of users. One was reserved solely for the BBC, the government's guarantor of culture, decency, and access; the other was given to a private firm or consortium, free to "compete" with the BBC yet linked back to the government's preferences through the regulatory reach of the IBA. In effect, the system incorporated a monopolistic exchange: in return for compliance with its regulatory agenda, the government ceded a private monopoly in a highly lucrative market. It was the old structure of broadcast television, recast now in the new space of satellite.

---

[14]Richard Belfield, Christopher Hird and Sharon Kelly, *Murdoch: The Great Escape* (London: Warner Books, 1994), p. 187.
[15]The initial decision to give these slots to the BBC came somewhat earlier, but the BBC had not yet made any plans to use its allocation.

Predictably, News International submitted one of the five bids for Britain's satellite spectrum.[16] Just as predictably, it lost. The bid went instead to British Satellite Broadcasting (BSB), a consortium that had been the clear front-runner since the beginning of the bidding process. Composed of Pearson, Richard Branson's Virgin, Granada, Anglia Television, and Amstrad, BSB was the heavy-hitting choice, an "establishment thang" in the words of one observer, and a group with both substantial television experience and deep financial pockets. Which, as it turned out, was a very good thing.

British Satellite Broadcasting would eventually become the second most expensive commercial launch in British history, coming only after the Channel tunnel. Between December 1986, when its bid was formally accepted, and 1990, the group would invest roughly £1.25 billion. And it never even really got off the ground.

Befitting its new media, BSB started with ambitious plans. Soon after winning the license in 1986, the company announced its intention to launch a four-channel satellite service by August 1989. It expected to attract four hundred thousand viewers in the first year and break even in 1993. Its hope was to reach half of all British households in fifteen years, when its satellite franchise was set to expire.[17] To build this lofty vision, BSB hired some of the best names and most well-connected talent in British television. They also began to work furiously on the technical side of their project, adapting the D-MAC standard into a commercially viable system. Finally, realizing the critical importance of new content, they began to court the Hollywood studios assiduously, hoping that the purchase of big-name movie rights would ensure their breakthrough in the British market.

Meanwhile, as BSB scurried to shape its business and build a satellite system, Sky was undergoing its own transformation. On June 8, 1988, Murdoch announced that Sky, too, would

---

[16]It was part of a consortium including Celltech, a biotechnology company, British and Commonwealth Shipping, Ferranti, Sears, and Cambridge Electronic Industries. See Matthew Horsman, *Sky High: The Inside Story of BSkyB* (London: Orion Business Books, 1997), p. 35.

[17]Belfield, p. 184.

soon offer a direct-to-home (DTH) satellite service, complete with its own small receiving dishes and popular new programming. To transmit the DTH channels, Sky would lease four transmitters on a communications satellite launched by Société européenne de satellites (SES), a small private company based in Luxembourg. It was only a medium-powered satellite by emerging standards, but it was private, foreign, and officially a communications, rather than a broadcast, satellite. Britain's regulators had no jurisdiction over anything transmitted from Luxembourg's satellite space.

In retrospect, the genius behind Sky's move came from the simultaneous grasping of a commercial opportunity and a regulatory gap. In 1989, the evolving European Commission had determined that the appropriate jurisdiction for any satellite broadcast would be the "uplink" rather than the "downlink" country.[18] In other words, satellite broadcasters would be regulated by the country they broadcast *from* and not the countries they broadcast *to*. It was this directive that enabled Sky to service the British market out of tiny Luxembourg, a country that, not coincidentally, had virtually no regulation over satellite broadcasting.

BSB's management saw instantly the threat that Sky posed and railed against the unfairness of the situation. As Anthony Simmonds-Gooding, chief executive at BSB, complained:

> The government insisted on BSB marketing a pioneering technology, complex, high risk, but if successful, of great long-term benefit, not only to the consumer but also to the European manufacturing and retail community as a whole. The government then allowed BSB to be bypassed by a powerful competitor unregulated, with no technology demands, and promoted by the most powerful media group in the UK.[19]

Sky, of course, took a wholly different tack, painting itself as the underdog and attacking BSB as part of Britain's cozy and

---

[18]The key piece of legislation here was a European Commission Directive entitled "Television Without Frontiers." See Council Directive 89/52/EEC of October 3, 1989.
[19]Quoted in Belfield, pp. 186–87.

anti-competitive environment. Sky's service, according to the Murdoch, would be the start of television's new age. "We are seeing the dawn of an age of freedom for viewing and freedom for advertising," he insisted. "Broadcasting in this country has for too long been the preserve of the old Establishment that has been elitist in its thinking and in its approach to programming."[20]

As the shape of their impending battle became clear, both sides rushed directly into conflict. BSB pulled rank and the levers of power within its reach; Murdoch used his newspapers to tout Sky and dismiss any prospects for BSB's success. Both sides also raced to Hollywood, where they quickly ignited a bidding war for film rights. Finally, both BSB and Sky launched massive marketing efforts, trying to convince Britain's untutored viewers first, that satellite television was worth paying for, and second, that their satellite service was superior. It was classic and massive push advertising, attempting to create demand for a service that most potential customers had only recently heard of and had never missed. To jump-start its efforts, Sky even resorted for a while to old-fashioned door-to-door sales. It sent a thousand sales representatives into people's homes, pitching them on Sky and arranging for rapid installation of the necessary equipment. To sweeten the deal even further, Sky heavily subsidized the purchase price of the dish and offered new subscribers free maintenance of their system.[21]

In the end, Sky won. It lured a respectable base of subscribers; wrestled the most attractive movie deals; and launched its four-channel service in February of 1989. BSB, meanwhile, fell farther and farther behind, eating into its capital as it tried desperately to iron out technical problems with the untested D-MAC format. When BSB finally did launch, in April of 1990, Sky already had the market essentially sewn up. Those who were eager to try this "new" television had jumped in with Sky, lured by its offerings and cheap receivers, and were reluctant to switch providers. Those who hadn't were mostly content, it appears, to watch the BBC and Britain's other free programs.

---

[20]Quoted in William Shawcross, *Murdoch* (New York: Simon & Schuster, 1992), p. 344.
[21]Richard Lander, "The Satellite War: Low Tech against High Tech," *The Independent,* March 25, 1990, p. 10.

Furious at Sky's position, BSB waged a public and political battle. In a highly visible campaign, BSB management charged that Sky had circumvented British regulation only through "technological wizardry," using its perch above Luxembourg to exempt its programming from British regulation. They also lobbied relentlessly for new legislation that would extend Britain's existing prohibitions on newspaper and television cross-ownership to the field of satellite. At the core of their campaign was a deeply held belief that Murdoch and Sky had somehow, repeatedly, broken the rules; that they had taken the market that British regulators had so carefully carved out and run roughshod over it. In the process, Britain's concerns for preserving its own brand of television were under attack from Murdoch's lowbrow "American" fare—and BSB had lost a reported £1.25 billion.

As events unfolded, Sky continued to triumph, both in the market and along the corridors of power. With subscriptions running far below their expected level, BSB was quietly forced to concede defeat by the fall of 1990, when it lost approval for a last, critical, tranche of financing. Meanwhile, backed by Margaret Thatcher's Conservative government, the terms of the pending 1990 Broadcast Act turned sharply in Sky's favor. While cross-ownership was indeed prohibited among the holders of television, newspapers, and high-powered satellite services, satellites that employed non-U.K. frequencies were distinctly excluded from the ban. Incensed by Murdoch's Houdini-like ability to circumvent British law, many observers at the time saw the 1990 Act as proof that Prime Minister Thatcher was clearly beholden to him for the support his papers consistently lent her government. Yet participants in the legislation continue to insist that any relationship between Thatcher and Murdoch had more to do with ideological compatibilities than personal favors. Thatcher, they argued, wanted desperately to break the monopolistic hold that prevailed in sectors such as television. Murdoch was simply doing her job for her.

Where political antagonism ran deeper was in the regulatory agencies themselves, the organizations that had structured the same stability that Thatcher disliked and established policies such as the 1986 satellite bid. The IBA, in particular, was generally aghast at Sky's doings and eager to find some way to quash

Murdoch's growing influence. Yet they were essentially caught by the regulatory twist of satellite. Under prevailing law—both EC and British—they simply had no jurisdiction over Sky.[22] As Andrew Neil, then executive chairman of Sky, recalls:

> The IBA never really understood the implications of satellite television... The idea that you could confine satellite transmission to national boundaries, to only a few extra channels, and that no one else would try to muscle in on your act was fatally flawed from the start.[23]

Or as one participant recalls, more poignantly perhaps: "How could we regulate what wasn't even there?"[24]

In any case, by the end of 1990, BSB had failed to stop Sky's advance. Both companies were also scarred from two years of fierce competition. After a brief and intensely secret round of negotiations, they agreed in November to merge into a single company named BSkyB. Though all parties presented the negotiations as amicable, the structure of the new firm made clear where the power lay. With 50 percent of BSkyB, News Corp was the controlling force in the company, and in British satellite television.

Once again, controversy roiled through the industry and across the British government. The chairman of the IBA, who first heard of the merger only an hour before its signing, declared the deal "illegal and brutal." Canceling the BSB's formal contract to transmit, he proclaimed, "It is clear to the IBA that the completion of the merger, for which the IBA's consent was neither obtained nor sought, gave rise to a serious breach of BSB's program contract."[25] Echoing these sentiments, the Labour Party's broadcasting spokesperson asserted, "This merger is a skyjack. We are totally opposed to a satellite monopoly, particularly when controlled by a non-EC national."[26]

---

[22]There are a few minor exceptions to this claim. In an unofficial arrangement, Sky did agree to be governed by the IBA's regulations regarding content and acceptable advertising.
[23]Quoted in Belfield, p. 187.
[24]Interview with author, London, October 1998.
[25]Quoted in "UK: The Marketing Story," *Marketing*, November 8, 1990.
[26]Quoted in Ibid.

The Office of Fair Trade, Britain's anti-monopoly watchdog, declared its intention to investigate whether having a single satellite provider was anti-competitive—despite the fact that the country's official policy had in fact been structured to produce precisely that result.[27] Even the usually pro-business *Economist* joined the anti-Sky fray, calling the deal a "Wapping in the Air" and referring to Murdoch's previous success in breaking the once-mighty British printing union.[28]

Yet the newly formed BSkyB remained largely unaffected by the uproar. Since one of the provisions of the 1990 Broadcast Act was to transform the IBA into a new regulatory agency, threats from the current IBA carried virtually no weight. And Thatcher's government remained strongly supportive of the Act and of BSkyB, despite personal attacks launched on the prime minister's conduct. Thatcher acknowledged having a conversation with Murdoch two days before the announcement of the BSkyB merger, but insisted that no specifics of the deal had been discussed, much less approved. In the end, no formal proceedings were ever raised against the merger, and BSkyB set out alone to create Britain's satellite television industry.

Consolidation and Control

Over the next ten years, BSkyB grew to dominate and define the British subscription television market. From a company that had been first ignored and then scorned by the media establishment, Sky became the very center of its industry, pushing the margins of change and forcing all of the other players—including even the venerable BBC—to watch and respond to its moves. At the same time, Sky (as the company was generally still known) became as well the subject of intense commercial and political scrutiny. The furor that had surrounded the merger of Sky and BSB was, as it turned out, only the beginning of BSkyB's tussle with Britain's regulatory au-

---

[27]Andy Fry and Mat Toor, "Sky's the Limit for Satellite TV," *Marketing*, November 8, 1990.
[28]"Broadcasting: Wapping in the Air," *Economist*, November 10, 1990.

thorities. Again and again, the same pattern unfolded: Sky would recognize how some emerging technology could enable the firm either to leap through a regulatory barrier or consolidate a competitive foothold; then its managers would move to grasp this technology, develop it, create whatever standard might be necessary, and thus control its usage by any other parties. This was how Murdoch had first used satellite technologies to circumvent the otherwise solid net of British regulation and it was how Sky operated throughout the next decade.

From the perspective of Sky and its investors, the pattern was flat-out brilliant. It enabled Sky, in an amazingly short period of time, to rustle the entire British television market and rewrite the rules of what had been a plodding and predictable game. It also was the conduit through which the company introduced and mastered a wide range of cutting-edge digital technologies. For the regulators and other television firms, however, Sky was simply maddening. Every time they turned around, there was Sky again: slipping through regulatory crevasses, controlling new technologies, and carving up a market nearly before it had even been born. Moreso perhaps than any other firm of the information age, BSkyB knew how to write the rules of its own game.

Sky's takeoff phase began almost immediately after the merger. At that point, the new company was in horrendous shape. Animosities were still running strong between the two sides of the new venture, and weekly losses amounted to roughly £14 million.[29] Even more ominous, despite months of intense commercial squabbling, neither Sky nor BSB had conceived of a truly viable commercial model. Rather, both companies were still hobbled to a large extent by Haynes's old problem: they simply did not have enough subscribers to entice the advertising revenues that they needed to cover their content costs. Blessed with almost obscenely generous backers, BSB and Sky had, unlike Haynes, been able to invest considerable sums in

---

[29]Estimates of this amount vary considerably. The figure of £14 million comes from Horsman, *Sky High,* p. 76. Belfield et al. claim that, as of June 1991, BSkyB had accumulated operating losses over the past nine months of £650 million, or £17 million a week. Interest charges of another £115 million accounted for another £3 million a week. See Belfield, p. 215.

programming. Yet their spending, which included a combined $1.2 billion to lock up Hollywood's most promising film rights, had merely raised the stakes in the same old equation. To cover these costs, the satellite stations would need to generate huge advertising revenues. And even with their blitzkrieg efforts—even with nearly giving the receiving equipment away—neither firm had enough subscribers to lure advertisers away from more conventional media. Instead, they had merely upped the ante, raising the number of subscribers and level of advertising they needed to cover their costs.

The man who inherited this unhealthy balance was Sam Chisholm, a prickly New Zealander who had risen to prominence by transforming Australia's Channel 9 to the country's most popular station. He had been whisked to London at Murdoch's personal request and would soon become intimately intertwined with Sky and its rising fortunes.

Known for his late hours and brusque ways, Chisholm lost no time in transforming the disorganized scramble that Sky had become. He slashed costs, fired staff wildly, and instilled a sense of discipline that many described as dictatorial. Desperate to renegotiate the $1.2 billion that Sky and BSB had together contracted to pay to the major U.S. film studios, he also spent large chunks of 1991 in Hollywood, trying to winnow these costs down to a somewhat more manageable size. It was in the course of these negotiations that Chisholm hit upon the dual strategies that would come to define BSkyB.

CONTROLLING CONTENT

The first piece was already inherent in BSkyB's approach to satellite television. But Chisholm refined it, magnified it, and gave it the twist that made it fly. Content, of course, is the central element of any commercial television venture. Without compelling content, there simply is no market—especially in a place like Britain, with a rich palate of free-to-air offerings. It was this search for content that drove Sky and BSB to Hollywood in the first place, and that compelled them to pay such hobbling prices. Chisholm pushed this strategy several notches higher. Obtaining content was never sufficient for Chisholm.

He needed instead to control it, with a nearly religious drive. He realized the importance of lock-up—of preventing any competitor or potential entrant from getting any programming that might pull viewers away from Sky. Or as a sometime rival described it: "Sam came out of a truly competitive television environment. The whole of his strategy was that if any piece of programming looked like it might have any value at any time in the future, you should buy it and grow it or buy it and kill it."[30]

In Hollywood, this strategy was already largely in place, since the competition between Sky and BSB meant that their combined acquisitions covered nearly all of the available motion picture industry: Sky had contracted with Orion, Touchstone, and Murdoch's own Fox Studio; BSB had pledged more than $800 million to secure films from Paramount Pictures, Universal Studios, Columbia, and MGM. Chisholm's success here lay in forcing the studios, after months of negotiation, to drop their movie package prices by more than 30 percent.[31] This was extremely significant, not just because it meant weekly savings for Sky of a reported £500,000, but also because it signaled the company's coming of age and Chisholm's ability to negotiate with the industry's heaviest hitters.[32] Indeed, in forcing the studios to capitulate, Chisholm honed many of the tactics that would come to characterize his reign at Sky. First, he clearly played for broke, refusing to back down or concede negotiating points. Second, he used a divide-and-conquer strategy, letting the first concession ease the way toward others' capitulation. Here, it was Columbia that backed down first, largely because Sky was already deeply in debt to them.[33] And finally, Chisholm, like Murdoch, also made ample use of whatever legal channels he could find. When UIP (a consortium of Universal Studios, Paramount, and MGM) refused to renegotiate, Sky's lawyers went to court, claiming that UIP's structure was essentially

---

[30]Adam Singer, quoted in Mathew Horsman, "Rupert's Sam Missile," *The Guardian*, November 10, 1997, p. 2.
[31]Lisa O'Carroll, "USA: Satellite Broadcaster BSkyB Persuades Hollywood Studios to Cut Cost of Movie Packages," *Broadcast*, November 29, 1991.
[32]The figure of £500,000 was never officially reported, but is based on industry estimates. See O'Carroll, "USA."
[33]Horsman, *Sky High*, pp. 86–87.

anti-competitive and thus in violation of European Union law. In the end, Paramount, Universal, and MGM decided to renegotiate rather than fight.

Chisholm's stay in Hollywood thus left Sky in a distinctly stronger position, both financially and in terms of industry perception. More important, though, it also underscored a subtle but critical shift in Sky's business model. At this point, Sky was still fumbling for a business based on some combination of advertising revenues and subscription sales. The studio renegotiations forced Sky closer to the subscription side of the business. Which, as it turned out, was the right place to be. To ensure that their movies did not wend their way out of Britain or into nonsubscribing homes, the studios demanded that Sky provide foolproof encryption of all movies transmitted, and that they manage their subscribers as closely as possible. These demands strengthened the bonds between subscriber management and revenue.

Whereas the old Sky model, like older-style television, was based on providing the content free of charge and using advertising to generate the channel's revenue, this new model was based on a fundamentally different relationship between the customer and the channel provider. In order to guarantee that the Hollywood studios retained control of their intellectual property, Sky had to stop giving its content away, and instead persuade customers to pay for it. Then it had to ensure that everyone paid for exactly what they watched and that no one was able to get the content for free. Essentially, Sky had to take what people had thought was free—television—and turn it into private property. It also had to create a system for matching viewers to payment and preventing theft, not an easy task for a product that was "sold" from twenty-three thousand miles up in the air.

By the time Chisholm came to Sky, much of this infrastructure was already in place. The movie deals cemented the focus, giving Sky highly desired content and forcing the channel to strengthen the link between specific viewership and revenues. Indeed, part of Chisholm's renegotiation apparently linked Sky's payments to its subscriber base: the more people signed on to watch Disney movies, the more Sky paid back to Disney. What Chisholm added, though, was key. He realized that movies, despite their appeal, were not quite powerful enough to create the mass subscriptions that Sky would need. There had to be some other draw, something to persuade Britons to buy a dish, contract with Sky, and pay for various viewing options. That something turned out to be football.[34] And it saved Sky.

Up until this point, sports television had been the sole province of the BBC and ITVs. In 1992, though, just as Sky was trawling for new content, some of Britain's sports franchises were beginning to realize their economic potential. With the football teams in particular grumbling about breaking away from their traditional arrangements, Sky moved in for the kill. In May of 1992, against the advice of its financial advisers, Sky offered £304 million for a five-year contract with the Premier League, a newly formed assembly of Britain's twenty-two most popular football clubs.[35] For the first time, a popular sporting event had been transferred from the public realm to the private.

This transfer let loose a howl of outrage. Critics screamed that Sky had once again usurped the public space, stealing privileges that constituted some core piece of British culture and society. Calling for a government investigation into News Corporation's media holdings, the head of ITV Sport argued that the Premier deal "ensured that many poor and old people will never see a live Premier League game in the country."[36] The chief executive at Channel 4 was equally outraged. "Is this a market operating in the best interests of viewers as a whole?" he asked rhetorically. "I think not."[37] Even politicians got into the act, with Labour MP Bryan Davies denouncing the deal as "absolutely appalling."[38]

Sky, however, just held its ground, quietly scrambling to turn its contractual victory into a revenue stream. This wasn't hard. Once viewers knew that Sky had rights to the most popular

---

[34]Soccer, in American parlance.

[35]Under the terms of the deal, the BBC retained some peripheral rights, enabling them to continue to broadcast *Match of the Day,* a popular show of football's highlights.

[36]Georgina Henry, "ITV Chief says BBC Little More than Poodle of Murdoch," *The Guardian,* June 24, 1992.

[37]Quoted in Michael Leapsman, "Grade Fears Satellite Monopoly on Sport," *The Independent,* May 30, 1992, p. 6.

[38]"UK: ITV Fury at BSkyB Premier Football League Win," *Marketing Week,* May 22, 1992.

football, they flocked to the channel in droves. Within two days of the deal, Dixons (a major electronics retailer) reported a 20 percent rise in dish sales.[39] By August, nearly one million viewers had signed on, agreeing to pay £5.99 a month to subscribe to Sky's sports channel. Revenues rose accordingly, hitting £380 million by the end of 1993 and putting Sky, for the first time, on stable financial footing.

CONTROLLING ACCESS

During these early years of Sky, it was movies and sports, along with Chisholm's blustery style, that drew popular attention. Papers continued to rail at the scrappy station, at Chisholm's destruction of the old BSB management team, and at Sky's supposed assault on British cultural values. Yet the core of this growing enterprise was quiet, nearly invisible to public view. It lay in an anonymous building in Kent and in a sprawling service center in Scotland. It lay, more precisely, with the systems Sky developed to control access to its growing stock of content.

As Sky's managers rapidly understood, the world of satellite television is fundamentally different from that of its more traditional predecessor. It demands not only a new model of revenue generation, but also a wholly different use and view of technology. In traditional broadcast television, technology works essentially as a conduit for programming and a source of quality enhancement. The model is a simple one. Stations like the BBC (or NBC, or CBS) create content and broadcast it as an analog signal, a radio wave, from a central tower. The signal, which can only travel across a standard line of sight, is received by any antenna in this area that is tuned to the particular frequency, known generally as a channel. It then travels from the antenna to a receiver inside the television box, where the waves of information are turned back into electronic signals and translated into the individuals dots of color (pixels) that compose the television picture. Note that in this model, the key relationship is between the television station and the viewer: technology is just the means that brings them together. From the viewer's per-

---

[39]Cited in "Satellite Sales Boom Kicks Off," *Electronics Times,* May 28, 1992, p. 3.

spective, moreover, the key piece of technology is the box itself, a product purchased outside the relationship between the station and the viewer.

In the world of pay television, technology plays a very different role. For once programming is transmitted only to a select group of customers, the technology of transmission becomes decidedly more complex. Rather than just sending the signal "free to air," providers must direct it to particular receivers; as their mix of programming gets more complex, they must also segment their signals between customers, ensuring that each one gets the precise channels or programs for which he or she has paid. On the flip side, moreover, the television provider must also ensure that all customers do, indeed, pay for their viewing, and that no one "steals" the signal as it travels invisibly along the airwaves.

Note that in this model, property rights are paramount and technology becomes a strategic asset rather than just a means of transmission. To make money in pay television, providers such as Sky need not only to control the content that will lure viewers to their channels, but also to erect the technical systems that will prevent piracy and guarantee payment. They need to direct the signal, encrypt it, and then track its usage. For Sky, these demands were particularly intense, since both its Hollywood and Premier contracts relied to some extent on the number of viewers that came to the channel. If nonpaying viewers managed to capture the signals on their own, both Sky and the content providers would lose revenue. This was a dominant concern for the Hollywood studios, since their deals with Sky had only covered rights for Britain and not the rest of Europe.

The introduction of pay television thus forced providers such as Sky to wrestle with some fundamental questions about the conduct of trade in the age of information. Ultimately, all that Sky sells is content: information transmitted through the ethereal reach of air. The product is intangible and the transmission invisible. The sale, moreover, is far removed from the transaction itself, since viewers do not actually pay for each episode of "The Simpsons" as they watch it. Making this business work, then, entails creating new means for controlling property and property rights. It means crafting some system that lets the owner of this property reap the full rewards of its sale, even if the property is but a wave of information and the sale occurs invisibly.

Sky's management realized this relationship early on. In 1990, executives from the company traveled to Israel to meet with Adi Shamir, a well-known mathematician who had been one of the creators of the RSA encryption algorithm, the world's most respected means for scrambling complex streams of data. After protracted negotiations, News Corp eventually partnered with Shamir and his Israeli-based research institute and established a new subsidiary, News Datacom (NDC), that would receive exclusive license to some of the Israelis' key encryption algorithms. Officially, NDC was wholly separate from Sky. With its headquarters in Maidenhead, it was legally an independent firm, with its own management team and board of directors. In practice, though, it functioned largely as the technical arm of Sky. Or as Stephen Barden, the general manager of Sky who become chief executive of NDC in 1992, recalls: "It was a company that had one product and one customer... News Datacom was completely, totally, utterly dependent on Sky."[40] Another industry executive is even blunter: "It was Rupert's empire," he suggests, "and nothing more."[41]

NDC's sole commercial objective was to develop the technical means for controlling Sky's transmission. It achieved this, rapidly and unobtrusively, by developing several overlapping layers of technology. First, all programming transmitted by Sky's satellite transponders was encrypted using specially designed versions of the RSA algorithm. Simply put, the signal was mathematically scrambled, in a process directly akin to the encryption described in Chapter 5. In order to receive the signal, viewers had to have not only a basic receiving dish, but also a set-top box that de-scrambled the signal and let them watch a particular show. Meanwhile, all interaction between the customer and Sky was conducted through a subscriber management service, a sort of high-technology credit card center. Viewers would phone the center (located in Livingstone, Scot-

---

[40]Interview with author, London, December 1998.
[41]Interview with author, London, October 1998.

land) to subscribe to Sky or change their programming requests. The center would then process this information, encrypt it, and send it to a broadcast receiver. This receiver, in turn, would then send another scrambled signal directly to a "smart card" located in the viewer's set-top box. At this final link, the signal would tell the card which portions of the Sky broadcast to receive and decipher.[42]

Technologically, the News Datacom system was quite sophisticated. It used cutting-edge algorithms to protect the entire length of the broadcast and communication chain, making it difficult for any would-be pirate to steal Sky's content.[43] It also gave Sky the ability to track precisely its viewers' preferences and payments. Most important, though, the conditional access system gave Sky an extremely powerful competitive edge. The power lay, oddly enough, not in the encryption per se, or in the transmission mechanism, but in the set-top box. Together, NDC's systems for conditional access and subscriber management gave Sky, as former chief executive Barden notes, "its segmentation, its gateway, its collection system and an enormously powerful marketing tool."[44] Once viewers had succumbed to Sky's charms, in other words—to its soccer, its movies, its marketing blitz and discount equipment—they were essentially hooked.

Sky-wards

As soon as Hollywood and football and News Datacom were in place, Sky soared through the British market. Between June and December of 1992, the company gained 290,000 new subscribers. By the end of that year, Sky Sports alone claimed 1.1 million paying subscribers. While these figures were somewhat lower than analysts' predictions,

---

[42]See "Murdoch's News Datacom at Cutting Edge of Digital Broadcasting," *Jerusalem Post*, October 21, 1996, p. 3.
[43]In practice, theft did occasionally occur and some hackers claimed even to have broken NDC's underlying algorithm. True to form, though, NDC retaliated against any theft, bringing suit against several pirate rings.
[44]Interview with author, London, December 1998.

they nevertheless served to push Sky's finances, for the first time, into the black. In 1993, Sky reported operating profits of £62 million on revenues of £380 million. In 1994, revenues rose to £550 million and profits hit a rather dazzling £170 million.

As usual, each round of Sky's success set off another round of criticism and demands from various quarters that this "brutal monopolist" be put under some restraint. Uncannily, though, Sky continued to escape, pushing the technological frontiers beyond the regulators' grasp and writing the rules on its own.

In 1992, the ITV sued the Football Association for granting the Premier League contract to BSkyB and called upon the Office of Fair Trading (OFT) to examine whether media concentration had "distorted competition in this important sector."[45] In 1993, the Labour party vowed to have BSkyB referred to the Monopolies and Mergers Commission, arguing that News Corp's dual control of newspapers and television made the corporation—and Murdoch—too powerful a force in Britain. Both complaints languished for a while, and eventually disappeared.

A more serious threat came in 1994, when Britain's leading cable companies launched their own legal and regulatory battle against Sky. The claim here was of monopoly practices, and the evidence was quite strong. Ever since Sky had mastered its conditional access system, other industry players had chosen to piggyback on Sky's system rather than recreate their own. It was a logical, though often begrudging, decision. With Sky's growing base of subscribers and set-top boxes, other content providers realized that it would be nearly impossible—or at least exceedingly costly—to compel subscribers to switch to any kind of separate system. Instead, it just made more sense for these providers to add their content to Sky's existing system, letting Sky handle the transmission, the encryption, and the billing procedures. By 1992, most of the major content providers had made some kind of arrangement along these lines. UK Gold, a service of Thames Television, was an early ally of Sky, adding its package of upmarket British reruns to Sky's basic package. Flextech (linked to the U.S.-based TCI) came on board with a package of channels, and Viacom's MTV and Nickelodeon fol-

---

[45]"BSkyB Deal Attacked," *The Times*, May 30, 1992.

lowed shortly. For Sky, of course, such arrangements were extremely attractive. They gave the firm more content to offer viewers, and more flexibility in terms of both packages and price. For the providers, however, this piggybacked relationship effectively left them at the mercy of Sky. So long as Sky controlled the gateway, it had the power, potentially at least, to act as a monopolist.

The situation was particularly galling to Britain's cable companies, the recipients of these piggybacked schemes. To maximize its own revenues and keep other providers from simply abandoning the satellite market, Sky had decided early on to sell its content as bundled channels. Thus, instead of choosing menus of specific programming, viewers were offered a handful of packages: a basic channel of news and weather, for instance; a sports channel; a movie channel, and so forth. For the viewers, who were already accustomed to channels of mixed content, this arrangement was just fine. For the cable companies, however, which purchased Sky programming and redistributed it to their own cable customers, it rang of monopoly. Bundling, they charged, was fundamentally anti-competitive. It forced the cable companies to buy product they did not want at prices they could not afford, usually 59 percent of the retail price paid by Sky's own subscribers.[46] Commenting on this pricing structure, one cable executive complained, "My typical cost of sales is 65%. If I am paying 59% to Sky, where do I make my money? I still have to pay to transmit. They are a brutal monopolist."[47] Because Sky was the only serious provider, though, the cable companies had to accept these terms—or risk losing customers who were not willing to go without their football, or movies, or daily dose of "The Simpsons."

Frustrated, the cable companies took their complaints to Britain's various regulatory agencies. It was a complicated lobbying process from the start, since no one seemed quite sure which the appropriate regulatory agency was and what rules should be applied in reviewing Sky's conduct. After a fair

---

[46]The price was based on a so-called "pay-to-basic" ratio, where cable operators paid between 50 and 60 percent of the retail satellite price, depending on the percentage of their customers that purchased the premium channels.

[47]Interview with author, London, June 1997.

amount of nosing around, however, the industry at last won
some support from the Office of Fair Trading (OFT), Britain's
anti-monopoly watchdog, which agreed in 1995 to launch a full-
scale review of Sky's commercial conduct. For several months,
it appeared as if OFT might turn the tables of competition
against Sky. But when the agency completed its review, it concluded that although Sky did indeed have a leadership position
in all elements of its pay television business, its leadership did
not constitute an abuse of power.[48] As the decision was announced, the Director of OFT even praised Sky for "creating a
market which has greatly enhanced consumer choice." "I believe it is important," he continued, "that companies which
show enterprise and flair are not sent the wrong signals by regulators as soon as they move into profit."[49]

From Sky's perspective, this string of regulatory victories
was simply the evidence of its success. Indeed, both publicly
and in private, Sky's management maintained that the company
had never bent any laws nor been handed any political favors.
On the contrary, Sky remained the scrappy outsider, the firm
that the BBC, the ITVs, and the media establishment were still
trying to condemn. It had succeeded on wits and savvy—with a
fair dose of Murdoch's money thrown in for good measure.
Sky's opponents, of course, saw a different picture. In every regulatory victory, they saw hints of backdoor dealings, scraps of
Murdoch's political ties to Margaret Thatcher and the Tory
party.

Implicit in the criticism of Sky was the familiar argument of
nonlevel playing fields. Essentially, Sky's critics felt that the
company was competing on unfair terms—that because of
News Corporation's position in the British newspaper market,
Sky was permitted to play by different rules in the television
market. To some extent, this criticism rings true. Sky did play
by different rules, and during the early 1990s British regulators
demonstrated a distinct unwillingness to rein in the renegade

[48]Office of Fair Trading, "The Director General's Review of BSkyB's Position in the
Wholesale Pay TV Market," London, December 1996.
[49]Cited in Christopher Marsden, "Multimedia Multinationals and the Regulation of
UK Digital Television," Ph.D. thesis, Faculty of Law, London School of Economics,
1998, p. 311.

broadcaster. Yet this special treatment was not necessarily the result of special favors. Indeed, a more plausible explanation is that Sky was treated differently because it *was* different. Sky was neither a traditional broadcaster nor a telecommunications firm. It was operating in a market that it had itself created and that was not even technically located within Britain. Thus the rules that prevailed in Britain simply did not apply to Sky. And the regulators were not necessarily either in Sky's pocket or blind to its emerging market power; they were just stymied.

A case in point is competition law. Under prevailing British doctrine, anti-competitive behavior is defined as conduct that has the effect of "restricting, distorting or preventing competition" in a given market.[50] What is critical here for Sky's case is the delineation of the market. According to its critics, Sky clearly had a controlling position in the pay television market: it controlled 100 percent of the premium film channels in the United Kingdom, all but one of the premium sports channels, and roughly 30 percent of the basic channels available from either cable or direct-to-home subscription services.[51] Yet, as Sky argued, the pay television market is but a tiny piece of the total television market, in which Sky had only a 3.9 percent share.[52] Which is the relevant market? It depends on one's view. But in any case, it is not entirely clear. Sky essentially controlled a new market that it had created on the back of a much older and larger one. Does that constitute anti-competitive or pro-competitive behavior? In the mid-1990s, Britain's competition agencies were not well equipped to answer that question. So they dodged it, and simply concluded that Sky's "powerful position" in the pay television market had not been detrimental to the public interest.[53] Like other pioneers, then, Sky was able to compete on a tilted playing field because it had created the playing field itself.

[50]A short description of British law is available at http://www.mmc.gov.uk/ancomp.html. For more on the prevailing doctrine, see the Competition Act of 1998, available at http://www.oft.gov.uk/html/comp-act/index.html.

[51]Office of Fair Trading, "The Director General's Review," pp. 36–37.

[52]From Office of Fair Trading, "Director General's Review," p. 27.

[53]The Office of Fair Trading did find pay television to be a "clearly differentiated" market. The confusion lay in determining whether or not it was "differentiated enough" to constitute a separate market. See Office of Fair Trading, "Director General's Review," p. 30.

The result of these advantages was a business that seemed both unstoppable and untouchable. In 1996, BSkyB reported profits of £315 million on revenues of just over £1 billion. Between 1993 and 1996, the company's profits had grown at an astonishing compound annual growth rate of 72 percent and it was widely regarded as the most profitable pay-TV broadcaster in the world. In 1997, BSkyB was making £10 of profit every *second*.[54]

Going Digital

By this point, both Sky and Murdoch looked invincible. But then, just as satellite television was settling into its commercial space, the technological frontier lurched forward again, opening the possibility of an even more radical form of broadcasting and a more frantic rush for real estate. The innovation in this round was digital television, and its implications—both commercially and politically—were profound.

Technically, digital television (like all digital applications) involves the transmission of data in a simple binary form. In traditional television, analog signals are broadcast to a receiver that translates the waves into electrical impulses and displays them as individual dots of color. In digital television, by contrast, images and sound are recorded not as analog waves but as digital streams, the same flows of ones and zeroes that characterize computer processing and flood the Internet.

The differences between the two formats are considerable. In analog transmission, the wave configuration of the signal takes up a fair amount of space in the radio spectrum. Thus, in any given geographical area, only a certain number of channels can coexist. Moreover, because analog signals are sent as waves, certain bits of information must be transmitted more than once in order to ensure that the full picture is received accurately. If two characters in a soap opera, for instance, are seated before a stone fireplace, the whole fireplace image must be retransmitted each time the characters move, even if the fireplace stays pre-

---

[54]Mathew Horsman, "The Rise of UK Pay," *Broadcast*, February 5, 1999, Sky Tenth Anniversary Issue, p. 3.

cisely the same. This is known as "information redundancy." In digital transmission, by contrast, the fireplace is digitally coded and need not be retransmitted for each frame. Digital coding, moreover, means that each transmitted image is a precise replica of the original, with none of the fuzziness that often creeps into analog transmission. Even the audio component of digital television is markedly better, with the same accuracy and clarity as a compact disk recording.

The most dramatic aspects of digital television, however, have nothing to do with the quality of digital broadcast. Rather, they come from the vast multiplication of content that digital transmission allows. In the analog world, the number of channels is physically constrained by the width of the available radio spectrum. Each country has a certain amount of bandwidth, which it parcels out to various functions and users. Some is reserved for military communication; some for emergency services, such as fire and ambulance; some for radio and television. Given that certain frequencies can be used only to carry certain kinds of transmission, the result in most countries has been room in the spectrum for roughly three to five television stations. Thus we have the BBC and a handful of ITVs in Britain; ABC, CBS, and NBC in the United States, and so on.

Digital transmission, however, can shatter these physical limits and the commercial structure that has grown upon them. With digital signals transmitted in a compressed form, television frequencies can carry five to ten times as many channels as they could in the analog world. So can radio frequencies, shortwave radio frequencies, and even fiber-optic cables. This means, of course, a vast new space for commercial transactions and the possibility of the much-lauded "500-channel universe." But it also translates—somewhat counterintuitively—as an expansion of government power. For in the short run at least, the multiplication of channels means that governments, which already control the spectrum, now control a considerably bigger, considerably more lucrative piece of real estate. They are the ones, in this case, who physically control the frontier and who get to carve it up and hand it out. And thus while satellite television was somehow made for renegades, digital television was a strangely conservative creation, tied innately to the property rights that governments had established long ago over the radio spectrum.

As one might expect, different governments dealt with this development in rather different ways. In the United States, for example, the Federal Communications Commission (FCC) blessed a new technical standard for digital television, allocated the expanded spectrum space to the existing broadcasters, and then simply urged these broadcasters to convert by 2006 to a fully digital format.[55] In Germany, the government was even less intrusive, intervening only—and somewhat paradoxically— to prevent the nation's largest media firms from combining their forces in the digital realm.[56] And in the United Kingdom the government hastened to create a whole new system of distribution, a system that would thrust British television into the digital age while still maintaining the shadow of its analog structure. What role Sky would play in this new system was uncertain. But one thing was clear: this time, the cycle of technology had actually shifted the balance of power back toward government.

[55]Formally, policy in the United States had a chicken-and-egg quality to it. Broadcasters were obliged to return the additional chunks of the spectrum made available by their transition to digital, but were not required to do so until 85 percent of all television viewers had switched to digital TV. These provisions were codified in the U.S. Telecommunications Act of 1996, P.L. 104–104, 110 Stat. 56. See "Charting the Digital Broadcasting Future," Final Report of the Advisory Committee on Public Interest Obligations of Digital Television Broadcasters, Washington, D.C., December 18, 1998; Stephen Labaton, "The Battle of the Bandwidths," *New York Times*, August 11, 2000, p. C1; Michael James, "HDTV or Bust," *Baltimore Sun*, October 16, 2000; and Joel Brinkley, "Digital Era Still Remains out of Reach," *New York Times*, August 7, 2000, p. C1. For earlier descriptions of the standard setting and license allocation processes, see Edmund L. Andrews and Joel Brinkley, "The Fight for Digital TV's Future," *New York Times*, January 22, 1995, section 3, p. 1; Liza McDonald and Ben Hammer, "FCC Approves Digital-TV Standard," *Seattle Times*, December 26, 1996, p. D1; and "Future of TV: FCC Approves Digital Licenses," *Minneapolis Star Tribune*, April 4, 1997, p. 1A.
[56]In two key instances, competition authorities in Brussels prohibited German mergers that were intended to create new ventures in the field of digital television. In 1994, for example, Karel van Miert, the EU's competition commissioner, blocked a proposed alliance between Bertelsmann, Kirch (Germany's leading television firms), and Deutsche Telekom, the state telephone company. In 1998, he forbade a digital television alliance between Bertelsmann and Kirch. Concerns in Brussels were also shared by Germany's own cartel office. See "Satellite Television: The Revolution that Could Bring Viewers 1,800 New Channels," *Financial Times*, April 25, 1996, p. 13; Samer Iskander, "Pay-TV Groups Fight to Avert EU Veto," *Financial Times*, May 27, 1998, p. 2; and "German Television: Karel's Service," *Economist*, May 30, 1998, p. 62.

The Broadcast Act of 1996

Initially, no one in Great Britain had seemed to know how digital television could be made commercially feasible, or what its development would mean. Nontraditional broadcasters such as Sky were trumpeting their movement to the digital realm; traditional broadcasters such as the BBC were mumbling their interest but essentially sticking to their analog ways; and the government was struggling to understand the implications of digitization. It was a classic situation of regulatory gap: for several years in the early 1990s, British regulators and the politicians who directed them simply did not know how digital television would develop and what they should do about it. Over time, however, a slow consensus of objectives began to form. First, there was a general consensus among British policymakers that the country should use television as a portal to the digital age. Already, several British firms had taken a global lead in crafting digital television technologies. There was News Datacom, of course, which ran one of the world's most sophisticated systems for conditional access, and NTL (National Telecommunications Ltd.), a cutting-edge provider of digital television services. There were also strong hardware producers such as Amstrad and Pace, and a host of innovative programmers. Britain's software and Internet sectors, by contrast, were relatively unsophisticated, pale cousins of their Silicon Valley counterparts. If Britain were to compete with the United States in the field of digital technology, television was an intriguing place to start and the government was eager to support it.

A second motive revolved around the tremendous financial windfall that digital television created. In the past, the British government (like nearly all governments) had split the television portion of its spectrum into a handful of frequencies, reserving some for government-linked broadcasters such as the BBC and then licensing the remaining space to privately run networks. Now, though, the advent of digital compression meant that five or ten channels could be squeezed into the space formerly reserved for one, allowing the government, potentially at least, to collect five to ten times its old revenue. Or it could push a single station into a fraction of its former space and then auction the remaining spectrum to other digital users, such as mobile phone companies. In Britain, where cellular phone usage was already extremely high, the government estimated that a re-allocation of the spectrum could yield as much as £5 billion in additional revenue per year.[57]

A final motive was much more subtle and discrete. By the mid-1990s, virtually all of Britain's television elite were convinced of the possibilities of the digital age. All of the ITV regulators, the BBC management, and the politicians backing and staffing the Department of National Heritage acknowledged that digital technologies had the power to revolutionize broadcasting. Yet they also understood what revolutions meant: they destroyed the existing order and replaced it with something new. And in this case, most of Britain's policymakers did not want to replace the kind of television that the country had known and nurtured for so long. They did not want their television to be a series of talk shows and low-budget comedies, nor did they want to replicate what many derided as "American-style" entertainment. They wanted instead to maintain the Britishness of British television, with its own distinctive styles and formats. And thus there was a certain conflict inherent in any digital policy. Policymakers wanted to embrace the technology, but they did not want to change the basic face of British television or the role it sustained in Britain's public discourse.

The task of assembling these jumbled objectives fell initially to Stephen Dorrell, a young, fast-rising member of the Conservative Party who was appointed Secretary of State for National Heritage in July of 1994. Dorrell, who had previously served as Treasury Financial Secretary, had virtually no experience with the television industry. But he was a businessman at heart and, according to most observers, a fair and efficient policymaker. After months of consultation and debate, Dorrell released a white

---

[57]This figure of £5 billion was reported to the author during interviews conducted in London in late 1998 and early 1999. In fact, when the British government auctioned just five new licenses in April 2000, it reaped a staggering £22.5 billion. See Alan Beattie and Alan Cane, "Treasury Rings up £22 Billion Windfall," *Financial Times,* April 28, 2000, p. 1.

paper on broadcasting, which—after more consultation and debate—eventually became the Broadcast Act of 1996. In its final form, the Act essentially liberalized Britain's media sector, negotiating the labyrinth of rules that had grown out of the 1990 Act and allowing more room for television, radio, and newspaper companies to move into related industries. It also committed the British government to a far-reaching agenda for digital terrestrial television—a more ambitious agenda, indeed, than that of nearly any other industrialized country. Rather than waiting for market forces to move of their own accord to digital transmission, and rather than pushing existing broadcasters to move toward digital technologies, the British government effectively compelled this move by the granting of wholly new, fully digital, licenses. The details of this new license procedure were technically and legally quite complex—so complex, indeed, that they barely drew attention from the British public or even the usual media watchdogs. Yet buried in the details were the means for transforming British television while, miraculously perhaps, recreating it.

Under the terms of the 1996 Broadcast Act, the British government agreed to establish six new digital "multiplexes." These were to be frequency channels, licensed to "multiplex operators" who would use them to transmit digital television programming and its related services—things such as electronic program guides (EPGs), subscription data, or interactive games. As was the case with traditional analog licenses, these new multiplexes would fall under the regulatory authority of the ITC, which would award the licenses and track the operators' compliance with existing standards for diversity, decency, acceptable advertising, and so forth. Of the six licenses, one was to be allocated, free, to the BBC. Another was reserved for the digital services of channels 3 and 4, that is, to the digital services of the existing ITV license holders. And one half of the third multiplex was set aside for Channel 5. (In Wales, the remainder of this multiplex was reserved for S4C, the Welsh fourth channel.) The three-and-a-half remaining multiplexes were to be put up for public auction. Bidders (who had to meet certain requirements and pay a nonrefundable application fee of £100,000) could apply for either individual multiplexes or a package of all four.[58]

As a policy document, the 1996 Act is dense nearly to the point of incomprehensibility. It deals not only with digital television, but also with digital radio, setting out a similar array of new licenses and a similarly intricate weave of requirements, conditions, and restrictions. As a political document, however, the Act is rather brilliant. In one legislative swoop, it simultaneously propels British television into the technical future and cements it into the regulatory past. Under the Act, the BBC will move, at no cost, into the world of digital television. So will the ITVs, in order that diversity and quality will be upheld in the digital age and British culture preserved. The ITC, meanwhile, will resume its position as the watchdog of quality, while the market will function as markets should, with the best bidder offering a fair price for the increasingly valuable slices of government-allocated frequency.

Moreover, by structuring the digital bids as it did, the 1996 Act essentially used market forces to jumpstart the public portions of digital television. Recall that in Britain, as elsewhere, the policy rub lies in the leap to digital. Without programming, no one will invest in a digital television set. But without an installed base of sets, there is little incentive for broadcasters to switch to digital transmission. This is a game that theoretically could drag on forever, especially if any of the broadcasters (like the BBC) are obligated to ensure that everyone in the country can receive their programming. To navigate through this problem, the Broadcast Act offered a seductive deal: it gave one private firm, or perhaps just a tiny handful of firms, an extraordinarily valuable piece of real estate and a monopolistic stake along the emerging frontier. All they had to do in exchange was to make digital television work: to create the content, subsidize the television sets, and trumpet the benefits of digital reception to a comfortably analog world. It was a rather remarkable piece of legislation, a clear regulatory push clothed in the language

---

[58]For a full description of the 1996 Broadcast Act, see Helen Burton, "Digital Broadcasting in the United Kingdom," *Computer and Telecommunications Law Review* 3, issue 1 (February 1997), pp. 33–42.

and practice of the market. Even more remarkably, it worked. In 1998, digital television burst onto the British market. And all of the expected players were there.

The Quest of Marco Polo

On the south bank of the Thames, a glass-clad building rises awkwardly among warehouses and loading docks. This is Marco Polo House, a monument to a certain ill-fated breed of modernism, one that floundered briefly in the late 1980s before succumbing to other, more amenable forms. It is also, some might say, a monument to other forms of folly, and other short-lived and ill-fated pursuits. For it was here at Marco Polo House that BSB was launched and killed. And it was here again, from the very same glass box, that Britain's second round of television wars was waged.

This one began more quietly, without the fanfare that had attended the creation of BSB or the derision that greeted the initial launch of Sky. It began in the final weeks of 1996, when Sam Chisholm sat down with Michael Green, chairman of Carlton Communications. By all accounts, it was an unlikely meeting. Sky was still the scrappy outsider of British television, the company whose startling ascendancy the 1996 Broadcast Act was supposed to slow. Carlton, by contrast, was the one of the country's leading ITV firms and thus one of the presumed beneficiaries of the 1996 Act. Both Carlton and Green were voices of the establishment: respectable, influential, unimpeachable. They weren't supposed to work with Sky. But there was a reason for the two companies to cooperate, and they both knew it.

By itself, Sky could never win endorsement from the newly empowered ITC, the agency charged with dispensing the multiplex licenses. Sky was the firm that the ITC had been trying to restrain for all these years. It was the dominant pay television provider and the only British broadcaster beyond the government's regulatory reach. It was—well, Sky. Carlton, arguably, was in a much better position to grab the lure of digital television. It had the support of the ITC and the respectability that came with a decade of top-quality broadcast television. Unfortunately, though, Carlton had no access to key sporting events or movie rights, since Sky had long since locked these up. It also had no experience with pay television, and it certainly had never had to *sell* programming. Green reportedly realized these deficiencies and chose to act upon them. Thus, when Sam came calling, he listened. And when Sam proposed an alliance, he agreed.

Before long, this circle of not-quite-friends was expanded to include Granada, another leading ITV and a longtime associate of Sky. The three firms formed a venture that they soon dubbed BDB—British Digital Broadcasting. With a total planned investment of £300 million, BDB asked the ITC to be considered either for all three available digital multiplexes, or just for one.

Given its surprising lineage, BDB from the start was a rather strange child. It was obviously created through a marriage of convenience, and it involved partners who were not known for their easygoing ways. It also was in many ways a subtle attack on the stated purpose of the digital scheme. Digital television, according to the 1996 Act, was supposed to create diversity and competition in the broadcast sector, widening the field of providers and expanding viewer choice. BDB was essentially the same old thing all over again. It involved the same broadcasters, showing the same programs, presumably to the same viewing audience. The only difference was that this time, Sky was part of the establishment. And this in fact was the brilliance of the BDB scheme. Sky could get what it wanted: a place in the digital arena and an end, perhaps, to the regulatory sniping that had surrounded the firm since its inception. Granada and Carlton could piggyback on Sky's marketing expertise and programming rights to carve out their own digital position, and the government could use the combined weight of the BDB partners to push British viewers into the digital spectrum. An odd child indeed, but one that made a great deal of commercial and political sense. Once BDB declared its intentions, it instantly became the front-runner for the digital bid. Its only competitor was Digital Television Network (DTN), a company owned by NTL, the cable and broadcasting services firm spun off from the Independent Broadcast Authority. Both firms submitted bids to the ITC in January of 1997. Nearly all bets were on BDB.

Then something unusual happened. On June 3, Karel van

Miert, the European commissioner in charge of competition issues, formally expressed his concern about the BDB bid. Noting that "there could be an enhancement of an already dominant position" in the pay television market, van Miert made it clear to the ITC that he wasn't pleased with BDB, and particularly with BSkyB's participation in it.[59] Over the next three weeks, a furious round of conversations ensued. The members of the BDB consortium met with each other, with the regulators, and with their lawyers. Sometimes they went together. Other times it appears Carlton and Granada made separate visits, particularly to the ITC. Then the partners responded formally to the regulators' request.

On June 24, the ITC announced that BDB would receive all three of the digital multiplexes. The central condition of the award was that BSkyB withdraw from the consortium but continue to provide it with a full slate of programming.[60] It did, and the following week the BDB partners signed a new agreement. Under its terms, Carlton and Granada together paid £75 million to repurchase BSkyB's shares. They also contracted to purchase from Sky all the programming that had been included in the original BDB bid. This was a critical piece of the agreement, since without Sky Sports and Sky Movies the ITC might well have rejected BDB's application. It was also predicted to contribute roughly £300 million a year to Sky's operating profits by 2005.[61] The partners parted amicably.

But the deal left many players frustrated and unhappy. DTN, of course, was angry at having lost to BDB and announced its intent to appeal the ITC's decision to the Office of Fair Trading in London and EU competition authorities in Brussels. OFTEL, the telecommunications regulator, also expressed its unease, stating that BSkyB's relationship to the consortium still raised the possibility of "substantial competition

[59]Neil Buckley and Raymond Snoddy, "EU Raises Doubts on Digital TV Licence Bid," *Financial Times,* June 4, 1997, p. 11; and Andrew Garfield, "Did BSkyB's Top Executives Jump Last Week or Were They Pushed?" *Scotland on Sunday,* June 22, 1997, p. 5.

[60]"ITC announces its decision to award multiplex service licenses for digital terrestrial television," ITC News Release, 24 June 1997.

[61]Based on Morgan Stanley estimates. See "Defensive exchanges," *Financial Times,* June 22, 1997; and "Digital TV," The Lex Column, *Financial Times,* June 21, 1997, p. 20.

concerns."[62] Not even the winners were happy. At Carlton, executives recall that they "couldn't quite believe" what was happening. The European Commission, they felt, had made a horrible mess, engaging in what Nigel Walmsley, chairman of Carlton Television, quickly dubbed "pre-regulation." Rather than waiting to see if BDB was indeed anti-competitive, they had sprung far too quickly, imposing a "nexus of regulation" around a market that did not yet exist.[63]

Perhaps. Clearly there were tensions in the ITC process, and clearly the competition authorities in Brussels had acted in advance of any anti-competitive actions. But there was also a certain disingenuousness in complaints issuing from BDB, for essentially the outcome of the ITC's decision was a terrific windfall for the consortium. Under the terms of the multiplex bid, Carlton and Granada received a digital license good for twenty-four years. The BBC would be in the digital terrestrial spectrum, of course, and so would channels 3 and 4, but a significant chunk, roughly half the total digital terrestrial spectrum, had just been handed to BDB. And Sky, whose programs were critical to the BDB bid, was no longer there. It was, as even one of Carlton's top executives acknowledges, an "absolute steal."[64] Meanwhile, the ITC had also accomplished precisely what the policymakers wanted: private firms competing in the market for digital television, driving a consumer demand that would eventually allow the government to "switch off" analog transmission.

Round Two

For Sky, the changes came at a particularly unfortunate time. Just weeks after the company was dismissed from the BDB bid, Sam Chisholm announced his intention to resign from Sky, citing complications from the asthma that had plagued him for years. David Chance, Chisholm's deputy and confidant, followed shortly thereafter; word of his resignation hit the press

[62]"Advice to ITC on Bids for DTT License," Office of Telecommunications Press Release, June 24, 1997.
[63]Interview with author, London, December 1998.
[64]Ibid.

just hours after Chisholm's. Sky's stock began plummeting. By the end of the month, shares were selling for just 453 pence, down from 666 pence earlier in the year. More than £3.5 billion in total market value had been lost.[65]

For the first time in years, media pundits began to speak about the possible demise, or at least the probable decline, of Sky. While "this is not the end of BSkyB's dominance," the *Financial Times* intoned, "it is probably the beginning of the end."[66] "Suddenly the Sky seems to be falling," read a headline in *The Guardian*, noting an investor's concern that "the aura of invincibility has begun to disappear."[67] What complicated the picture and piqued critics' interest even further was the simultaneous ascendancy of a new manager. In March of 1996, Elisabeth Murdoch, Rupert's twenty-seven-year-old daughter, had been appointed general manager at Sky. No one wanted to say it very loudly, but speculation within the industry was rife. Elisabeth, most insiders thought, was being groomed to take her rightful position within the Murdoch empire; Sam left because he didn't want to play second fiddle to Murdoch's own offspring, and things would never be the same at Sky without Chisholm at the helm.

Sam or not, though, it was soon clear that Sky would not go easily. Quashing fears that Elisabeth would rise too quickly at Sky, Murdoch imported Mark Booth from News Corp's Japanese operations and named him the new chief executive. By the end of 1997 the furor that had surrounded Sky had subsided, replaced once again by growing fears of its dominance. The second battle of Marco Polo had begun.

THE AIR WAR

In September 1997, officials from BSkyB watched as Société européene de satellites launched a new digital satellite over the skies of French Guyana. Sky had leased half the satellite's capacity,

---

[65]Raymond Snoddy, "BSkyB Chief to Explain Strategy," *Financial Times*, June 30, 1997, p. 24.

[66]"Beady-eyed BDB," Lex column, *Financial Times*, June 25, 1997, p. 30.

[67]Maggie Brown and Lisa Buckingham, "Suddenly the Sky Seems to be Falling," *The Guardian*, July 7, 1997, p. 10.

meaning that it now had the ability to broadcast somewhere between two hundred and five hundred *digital satellite* channels.
Recall that BDB had only won the rights for digital *terrestrial*
programming—for a slice of the spectrum that physically covered the United Kingdom. Sky, which had been bounced yet
again from the formal bid, had retaliated by returning from
whence it came: 23,000 miles up. Only this time, it was going
digital.

Even before the actual launch, Sky began to promote its new
service, called SkyDigital. It was to have up to sixty channels of
movies alone, with staggered starting times to increase viewers'
flexibility. There would be separate channels for each Premier
football match, with viewers paying separately for each game
they chose to watch. There would be more sports, dedicated
documentary channels, four music channels, and forty audio
channels offering digital-quality radio broadcast. Sometime in
the not-too-distant future there would also be interactive services, allowing viewers to do their banking or purchase their
groceries directly from the television screen. And all of this, Sky
promised, would be devastatingly simple. All customers needed
were a new set-top box and an updated satellite receiver.

Sky planned to launch SkyDigital in the autumn of 1997.

THE GROUND WAR

As Sky mustered its forces, the remaining partners of BDB
began to plot their own digital launch. They peppered the company with executives borrowed from Carlton and Granada;
named Stephen Grabiner, formerly an executive director with
United News and Media, as chief executive; and ensconced
themselves—amazingly—in Marco Polo House. It looked just
like a replay of the BSB/Sky rivalry. Which wasn't particularly
good news for the Marco Polo crowd.

True to form, BDB decided to compete directly and explicitly with Sky. It, too, would offer more choice, better quality,
and ease of access in a newly digital world. It would ask customers to buy a set-top box (no satellite dish necessary) and
then choose from a menu of subscription offerings. It would include basic and premium channels, with a heavy dose of movies

and sports. It looked—well, an awful lot like SkyDigital. Except for the numbers: where Sky was talking grandiosely about starting with two hundred channels and working its way up to five hundred, BDB was considering launching with thirty, only a handful of which were exclusive to the platform. Half of BDB's channels would be free, composed mostly of material that was already available on basic broadcast television. The other half would consist primarily of Sky channels: the movies and sports that Sky was obligated to provide to BDB under the terms of the digital license. The only unique content would be a series of new offerings such as Public Eye, a Granada-produced crime program; Carlton Cinema; and the Carlton Food Network.

It didn't look like much of a fight. Sky had more channels, more programming, more experience, and much deeper pockets. It also knew how to make and manage set-top boxes. But executives at BDB were nonplused about their prospects. Sky, they thought, simply couldn't attract that many more viewers to its "brash and American" offerings; it couldn't convince too many more people to install rooftop dishes that remained "ugly and common." Where Sky was aggressive and "naughty," BDB, according to Carlton's Nigel Walmsley, would present itself as friendly and comfortable; it would be high-technology television from "those awfully nice people you've let into your homes all these years."[68] BDB would be all about "quality" and "British content," the very characteristics that Sky openly disdained. BDB also repeatedly belittled Sky's size. "There will be people who want 200 channels," Grabiner acknowledged, "But these are sad unhappy people who live in lofts."[69] For the rest of Britain, BDB was prepared to offer "manageable choice"— fewer channels, to be sure, but better ones. Sky was not impressed. "Manageable choice," scoffed Ray Gallagher, Sky's director of public affairs, "is like getting only one kind of peanut butter, even if it is the best kind."[70] Or as chief executive Mark Booth expressed it, with customary post-Chisholm reserve:

[68]Interview with author, London, October 1998.
[69]Quoted in Peter Thal Larsen, "Can Digital TV Convert the Dish-hating Viewer?" *The Independent,* July 29, 1998, p. 17.
[70]Interview with author, London, October 1998.

"Less for more is not a powerful business proposition . . . We're not too worried about the Carlton Food Network."[71]

The battle began in earnest on October 1, 1998, when Sky-Digital began transmission. With 140 channels, it was the first digital broadcaster in the United Kingdom, and one of the most significant in the world. Six weeks later, BDB followed suit. Newly renamed ONdigital, the service boasted roughly thirty channels, with a mix of free-to-air, primary, and "premium" offerings.

Though executives from both companies continued to insist that there was "room in the market" for several players, they sprang almost instantly into headlong competition. Seizing once again the advantages of deep pockets and rapid speed, Sky promised to offer "much cheaper entry prices" and to spend up to £480 million to subsidize consumers' purchases of its new set-top box.[72] It announced new channels and a tiered pricing scheme that would enable viewers to sign on to SkyDigital for as little as £8 a month. "I don't want to sound too aggressive," Mark Booth proclaimed, "but we have the opportunity to establish the benchmark in digital television. We are going to set that bar very high indeed."[73] ONdigital volleyed back. "Mark Booth," thundered Stephen Grabiner, "is trying to create an illusion that he can finish off ONdigital at birth. He can't do that."[74]

As of February 1999, Sky's digital service had signed up 350,000 customers. ONdigital had 70,000.[75]

REGULATORY SKIRMISHES

Commercially, Sky seemed in 1999 to be poised on the brink of another success. Even if ONdigital managed to encroach upon its traditional dominance in the pay television market, Sky was likely to remain on top. It had the programs, the penetration

---

[71]Interview with author, London, December 1998.
[72]"BSkyB Throws Down Gauntlet to ONdigital," *Cable Europe*, August 5, 1998.
[73]Quoted in Neil Bennett, "The Dash for Digital," *Sunday Telegraph*, August 2, 1998.
[74]Quoted in Bennett, "The Dash for Digital."
[75]Jemimah Bailey, "Family Finance," *Sunday Telegraph*, February 14, 1999, p. 9. Note, however, that many of Sky's "new" subscribers were existing analog customers who had simply switched to the digital offering.

levels, and the ability to serve a growing base of customers across a range of delivery platforms. When Mark Booth confidently predicted that 50 percent of British homes would have digital television by 2003 and that Sky would own half this market, no one in the industry saw reason to disagree.

Quietly, though, the political situation had started to evolve, and the charges that had long been lobbied against Sky began, at last, to stick. The first blow came in June of 1998, when the ITC ruled that content providers such as BSkyB could no longer bundle together a mandatory package of premium and basic channels. From here on, viewers had to be able to pick and choose as they wanted, rather than being forced to buy a whole bundle of channels in order to receive one particular favorite.[76] Although the decision affected a range of companies, and although Sky accepted it with a certain aplomb, "unbundling" was nevertheless a direct attack on Sky. It was the first time in a decade that British regulatory authorities had decided against a Sky practice and in favor of competitors' claims.

A second blow came in 1999 and was in many ways more damaging. It involved sports, the centerpiece of Sky's content arsenal and a major point of loyalty among its viewers. For several years, Murdoch's News Corp had been acquiring not only sports rights around the globe but also, increasingly, the sports franchises themselves: the Los Angeles Dodgers, Australia's Super League Rugby, and interests in New York's Knicks and Rangers.[77] The strategy was bold and completely logical. Realizing that content was the ultimate driver of success in the media industries, Murdoch saw the control advantages that full vertical integration could bestow upon his growing empire. By owning the most popular teams, Murdoch could benefit from both sides of the equation. He could enhance the teams' visibility by

[76]"ITC Confirms End of Minimum Carriage Requirements," ITC News Release, 26 June 1998.

[77]For more on these acquisitions, see Alan Deutschman, "The Mogul," *New York Times,* October 18, 1998, section 6, p. 68; Conal Urquhart and Martin Flanagan, "First Baseball, Now Murdoch Pitches for European Football," *The Scotsman,* September 7, 1998, p. 3; Michael White, "Entertainment Giants Are Playing Ball," *Minneapolis Star Tribune,* May 30, 1998, p. 1D; and Lisa Olson, "Murdoch: The Thunder Down Under," *Seattle Times,* October 12, 1997, p. C11.

assuring their coverage from News Corp media, and he could protect the media by ensuring that his stations secured viewing rights to the games and other events. Even if the relationship between the teams and the stations remained at arms' length, the arms, presumably, were still part of the same body.

In the United Kingdom, this logic had eventually brought Sky to Manchester United, one of the country's most popular and profitable football clubs. In September, Sky bid £623 million to purchase Manchester United. Almost immediately, criticism hit the streets and the press. Many fans were outraged at this obvious commercialization of their beloved sport, and painted Murdoch as a commercial bogeyman. They claimed that Sky had no understanding of Manchester's deep-seated traditions and jeered over the fact that Mark Booth could apparently not name the club's left back.[78] In the City and Parliament, meanwhile, other voices worried less about tradition and more about competition. Once again, the familiar threat of control emerged: critics claimed that the acquisition would put Sky on "both sides of the table," allowing the holder of Premier League television rights also to be one of the leading players within the League.

Although Sky's proposed takeover was admittedly unique, Sky executives were fairly confident that the uproar would die down and that the bid would be approved. Thus they cooperated fully with the competition authorities and simply waited, as usual, for the complaints to go away. As one Sky insider recalled, "We lifted our skirts and let the competition police take a good look."[79] Only this time, for the first time, the "police" decided against them. On April 9, Stephen Byers, the U.K. Trade and Industry Secretary, accepted a harsh recommendation from the Monopolies and Mergers Commission to forbid Sky's takeover of Manchester United.

Sky was shocked. This was a major decision against the company, and a largely unexpected one. Because the final decision had come directly from the highest levels of government, it also suggested that Sky's comfortable symbiosis with Downing

[78]See Tim Allan, "View from the Dugout," *The Guardian*, April 12, 1999, p. 3.
[79]Quoted in Allan, "View from the Dugout," p. 3.

Street might be in danger. Murdoch had actively courted Tony Blair for years, and his newspapers' glowing coverage of the Labour politician was widely seen as an important plank in Blair's election and his popularity. The Manchester United decision called this budding relationship into question. It indicated that the political winds might be shifting against Sky and that its sporting rights—the central jewel of the realm—might eventually be called into question.

The third blow was the subtlest, but in some ways the most telling, of the lot. In March of 1999, the Department of Trade and Industry (DTI) released a relatively obscure document entitled *Advanced Television Standards Regulations (1996): Revised Guidance*. The paper, which appeared without any particular prompting from the Parliament or regulatory agencies, updated the provisions that the DTI had issued three years earlier with regard to advanced television standards. The revisions were slight and technical, barely noticed by the press or viewing public. For Sky, though, they were critical.

Under the revised guidelines, any television set that contained an integrated digital decoder had to include a "standardized socket," a technical interface that allowed conditional access systems or other software to be plugged directly into the set. The socket itself had been required by the first, 1996, guidelines. The revisions, though, were more precise. They required a particular kind of socket, one that Sky had explicitly not employed.

Known generally as idTVs, integrated digital decoder sets are widely seen as the next wave of television. While all of the sets existing as of 1999 needed some kind of additional box in order to receive and de-scramble digital signals, idTVs are fully and internally digital: the box, in other words, is already in the box. Thus, if digital television really does replace analog, idTVs will no doubt become the appliance of choice. Accordingly, during the mid-1990s, electronics firms had scrambled along with Sky and the ITVs to manufacture the best idTVs they could, at the lowest possible price. Like many things digital, the race had a certain chicken-and-egg quality to it: in order to make idTVs attractive to the mass market, the price had to be reasonable—not much above the price for a traditional analog set. But in order to produce these more complicated sets at this kind of price, manufacturers needed the scale economies of a mass market. Standardization, therefore, was once again key.

As usual, Sky had decided early on to set the standard in this market, rather than waiting for it to evolve or joining with the ITVs in a common technological front. In 1998 it had contracted with the Korean manufacturer LG Electronics to manufacture digital television sets capable of receiving Sky's new digital service.[80] By 1999, LG had reportedly produced about fifty thousand of these sets, for a total value of roughly £25 million.[81] Under the new rules, all these sets were actually illegal.

For Sky, therefore, the ruling was a rout. Either the company had to give the sets away or swallow the £25 million loss. In either case, it would have to revamp its production strategy, abandoning its proprietary software in favor of a more open standard. Its sets would have to be compatible with ONdigital's broadcasts—and with the BBC, cable companies, and whomever else might venture along the digital path.

All in all, then, the spring of 1999 was a rocky time for Sky. The company, to be sure, was still undeniably strong, both as a content provider and as a distribution platform. It had a devoted following of viewers, some of the most valuable sport and movie rights in the United Kingdom, and access to the highest levels of government. But somehow, subtly, the balance of power seemed to have shifted. Sky was a formidable contender in the digital realm, but it was no longer alone. Cable was there, and the ITVs were there, and so, increasingly, was the BBC. In April, chief executive Mark Booth left BSkyB to run another part of the Murdoch empire.

The Empire Strikes Back

There are several ways of interpreting the events that surrounded the launch of digital television in the United Kingdom. One is that they simply represented the market

---

[80]Other manufacturers signed on as well. See Ben Potter, "BSkyB Signs Digital Deal with TV Makers," *Daily Telegraph,* July 30, 1998, p. 33.
[81]Chris Barrie, "Row over Sky: Digital Sets Face Ban," *The Guardian,* April 23, 1999, p. 23.

catching up with Murdoch. In this view, the rise of Sky was a technical anomaly, a blip of satellite technology that threw riches at the company but disappeared once a new and more attractive technology came along. Sky, in this view, was the lucky beneficiary of one technological gap and the victim of another. Once digital television came on stream, Sky lost the cocoon that satellite had momentarily provided it and became more of a normal company, with competitors and regulators and market shares to ponder. In this politics-as-usual view, there really are no politics, just the normal course of business and the normal ebb and flow of technological change.

A second and more cynical view favors politics above all else. Here, the dominant shift is not from satellite to digital but rather from Margaret Thatcher to Tony Blair, from a government that was ideologically aligned with Murdoch and beholden to News Corp to a more freewheeling, liberal, and skeptical regime. Certainly this view lines up neatly with the chronology of events and with the opinions of those who had bemoaned for a decade the purported Murdoch-Thatcher link.

A third view is decidedly more complicated, pairing technology with a messy agglomeration of politics and business and a host of deep-seated relationships. When Sky Broadcasting first burst onto the British scene, it simultaneously circumvented one set of rules and established another. Officially, Sky was not transmitting under British jurisdiction: it was beaming from Luxembourg and subject only to that country's minimal set of rules. And even on the ground, where Sky was subject to British law, the novelty of the company's practices made it very difficult to lodge legal complaints against it. Was Sky a monopolist? It is hard to say. Did it behave anti-competitively? Hard again. It was a classic case of creative anarchy, with technology opening spaces and creating markets that defied existing regulation.

During this early period, Sky did not need much in the way of external, formal governance. Unlike radio's pioneers, for example, it did not need to resolve issues of crowding or congestion. Since the satellite spectrum had already been assigned by an earlier international accord (the 1977 World Administrative Radio Council), and since Sky was the only service transmitting across its bit of the heavens, there was no congestion in the skies of Britain, and thus no need for any formal resolution. Unlike telegraph's pioneers, meanwhile, Sky also had no problem with connection, or with compatible standards. Rather, from the start, the broadcaster insisted on setting its own standards, both technical and commercial, and then using the force of its customer base to drive competitors and suppliers to play by its own rules. Thus, it was Sky that set the technical standard for its transmission, choosing indeed an older technology rather than the government-approved D-MAC. And it was Sky that established standards for encryption and conditional access and subscriber management and then imposed these systems across the entire pay television industry. In some instances, Sky even manipulated the formal standard-setting process to its own advantage, "helping," for example, to craft a European-wide standard for conditional access that clearly favored its own system.[82] Finally, whereas many pioneers ultimately need governments to protect their property rights, Sky was remarkably able to protect itself. It had secure access to its satellite space; it used advanced encryption systems to protect the intellectual property—the movies and sports—it conveyed; and it bolstered the whole enterprise by using its own proprietary systems and software. In its earliest days, then, the relationship between Sky and the British government was characterized by at least two levels of separation: the government couldn't get Sky, and Sky didn't need the government.

As satellite technology gave way to digital, however, the contours of this relationship began to change. Suddenly, regulators had new, highly lucrative territory that they alone were able to distribute and rule. Suddenly, there was room in the market for new kinds of television and a sense that government had to reshape this market to make it grow. None of these developments meant, necessarily, the demise of Sky or an end to its preeminence. But they engendered a different sort of market and a different set of rules. Meanwhile, of course, the sheer passage of

[82]See the descriptions of this process in David A. Levy, *Europe's Digital Revolution: Broadcasting Regulation, the EU and the Nation State* (London: Routledge, 1999), pp. 70–79; and David A. Levy, "The Regulation of Digital Conditional Access Systems: A Case Study in European Policy Making," *Telecommunications Policy* 21, no. 7 (1997), pp. 661–76.

time also meant that some of the old rules of the game—and particularly rules of competition—were finally able to work themselves through the political system and address the issue of Sky. It is still unclear, of course, whether Sky ever was a monopolist or whether it behaved anti-competitively. But because Sky was so big, and because it had imposed its own rules so successfully, its competitors were eventually able to force the political system to respond. This is a common phenomenon along the technological frontier, one described already in the radio and telegraph industries and again in Chapter 6's story of Microsoft. When the pioneer gets too big and too powerful, society frequently forces government to respond. What is almost lyrical about Sky's case is that the first blow from the competition authorities came out of football, the very game that established Sky's dominance in the first place.

In May of 1999, just as the BBC was gearing up for its full-scale digital launch, BSkyB and ONdigital entered a new and particularly expensive phase of combat. One month after the Manchester United decision, Sky made headlines again, announcing that it would give away free set-top boxes to all new digital subscribers. It was a bold move, even for Sky. At £199 the boxes were already heavily subsidized; for free, they were going to cost the company upwards of £400 million. At this point, ONdigital really had no choice. Almost immediately after Sky's announcement, the rival digital provider made a nearly identical offer: free set-top boxes for all new subscribers to ONdigital's pay services. Both companies made light of the situation, describing the giveaway in promotional terms and stressing its impact on new sales and customers. "These are not just short term promotions," stressed Ian West, Sky's entertainment managing director. "This is a new strategy that will help Sky rise to the challenge offered by its competitors."[83] Ashley Faull, executive director of broadcasting at ONdigital, was perhaps more direct, noting wryly that "free's been fairly popular."[84]

Clearly, though, "free" was exceedingly costly for both companies, especially when they were already spending millions of

[83]"Home News," *Press Association Newswire,* May 5, 1999.
[84]Interview with author, London, June 1999.

pounds on digital transmissions systems and expanded programming. In the summer of 1999, Sky reported annual pre-tax profits of only £73 million, down from £271 million in 1998; it also expected to take an exceptional charge of £315 at the end of the year.[85] ONdigital was in an equally treacherous state, with expected losses for the year of £100 million.[86] The upshot of these losses, though, was that digital television began seriously to take root. By the summer of 2000, an estimated 20 percent of Britain's households were receiving digital television and the government was already discussing whether it might switch off analog service earlier than originally had been planned.[87] In the United States and Germany, by comparison, digital take-up was less impressive: in the United States, only about forty thousand digital television receivers had been sold by the end of 2000; and in Germany, the country's leading digital network seemed stuck with a subscriber base of just 1.8 million viewers.[88] Sky, by comparison, had five million digital subscribers in a considerably smaller market, and ONdigital boasted one million.[89]

At the century's end, it seemed clear that digital television would indeed be the wave of the future and that Britain's digital market would be both robust and competitive. Three companies (Sky, ONdigital, and the BBC) had already launched digital television services as of 1999, and several others (Cable and Wireless, Telewest, and NTL) were rushing to join them via the country's cable infrastructure.[90] Where British television

---

[85]The full amount of the exceptional charge was £461.5 million. The associated tax relief, however, brought the figure to a net charge of £3315. See "BSkyB Profits Eclipsed by Cost of Digital Moves," *New Media Markets,* August 12, 1999.

[86]Roger Baird, "ONdigital Fall Guy?" *Marketing Week,* July 22, 1999, p. 20.

[87]Jonathon Carr-Brown and Nicholas Hellen, "Brown Eyes £50 Billion Digital TV Auction," *Sunday Times,* August 27, 2000, p. 10.

[88]Joel Brinkley, "Digital TV Era Still Remains Out of Reach," *New York Times,* August 7, 2000, p. C1; and "DTV in Germany—Does Anyone Care About Burn Rates?" *New TV Strategies,* September 30, 2000, p. 6.

[89]Put somewhat differently: Germany had 11 percent digital penetration among household viewers in 2000. Italy had 6 percent and the Netherlands 9 percent. The next best market in Europe was Denmark, with a digital penetration level of 16 percent. See "European Digital TV: Markets and Platforms," Jupiter Communications, Online Intelligence, 2000, vol. 1.

[90]Note that although the BBC was set to be a major player in the digital market, it remained a channel provider rather than a platform operator.

had been a monopoly and then a duopoly, it was now a vibrant, rivalrous market, full of jostling competitors and cutting-edge technologies.

The irony of all this was Sky. For it was Sky, after all, that had pushed British broadcasting out of its complacency; it was Sky that had pioneered the technology that caused the market to explode and the rules to change. It was Sky that brought digital television to Britain and convinced Britons to pay for it. And it was Sky that was likely to get lost, or at least minimized, in the final rounds of change. Like other companies that operate along the technological frontier, Sky was a true pioneer. It deployed the new technology and broke the old rules. For a while, it even had pieces of the government on its side, joined by a mutual desire to break out of the existing system and reshuffle both political and commercial power within it. So long as this mutual desire remained, Sky was able to reap all the benefits of a pioneer, and even of a monopolist. It owned the market it had created and seized all the profits that emerged. But monopolists often don't last. Over time, governments reenter the picture and redefine the new market so that it fits once again within their regulatory reach. This, at least, appears to be the story of Sky: a brilliant ascent, a phenomenal period of high profits and few rules, and then, eventually, a demise into normalcy. It is indeed the path of the technological pioneer. But as one official at the Department for Culture, Media and Sport acknowledged somewhat ruefully, "Over the long run, pioneers do get killed."[91]

[91]Interview with author, October 1998.

CHAPTER 5

Last Stand of the Cypherpunks

> A man is crazy who writes a secret in any other way
> than one which will conceal it from the vulgar.
>
> ROGER BACON,
> *Secret Works of Art*
> *and the*
> *Nullity of Magic*

Just outside of Washington, D.C., stands a complex of buildings that few people have ever seen. Heavily fenced and consciously obscure, it is the National Security Agency (NSA), a central figure in the U.S. defense establishment. For the NSA is the home of all U.S. signals intelligence (or "sigint," as the pros call it), the electronic brains behind both the U.S. military and the Central Intelligence Agency. Inside the building, officials manage a web of high-technology eavesdropping devices: the satellites and listening stations that the United States uses to track its enemies around the world. They also write the codes that power these various devices and protect official information flows. Whenever a general in the field transmits a message to headquarters, it is disguised (or "scrambled"); whenever a CIA operative signals or an embassy relays information back home, this too is encoded. The process that allows for these messages to be scrambled and received is known generally as encryption, and the mathematicians who practice it refer to themselves as cryptographers.

Until the advent of widespread computing, encryption was an obscure art. The masters all worked at the NSA or its equivalents around the world, where they spent their lives perfecting mathematical algorithms and developing new methods for breaking their rivals' codes. There were also some amateurs who fiddled with the math from time to time, and a small academic community. But the core of cryptography remained—heavily guarded and highly confidential—within just a handful of government agencies: the NSA, Britain's Government Communications Headquarters (GCHQ), and Russia's Federal Government Communications and Information Agency (FAPSI).[1]

As computing expanded and the Internet developed, however, cryptography rapidly went mainstream. Suddenly, there was data flowing across thousands, and then millions, of unseen computers: financial data, commercial data, and personal data. This was the great beauty of the Internet, after all, and the great promise—that data could flow across an impersonal network, shredded into innocuous bits and employing whatever pathways were quickest and most efficient. Yet such openness also created hazards. How, for example, could firms protect internal communications from interference, or even theft? How could consumers trust their personal information or credit card numbers to such a diffuse and free-flowing network? The answer, it quickly appeared, was through encryption—through the same kind of math and codes that had long protected MI5 operatives and far-flung ambassadors. By encrypting their transactions, firms and consumers could gain the security that cyberspace demanded. And with security, the Net was poised to explode.

In the mid-1990s, therefore, a new market for encryption emerged along with the commercial Net. Dozens of firms rushed to offer "security solutions" to a newly eager public, and

[1]For more on the little-known FAPSI, see "Heirs of the KGB—Russia's Intelligence and Security Services," *Jane's Intelligence Review,* September 1, 1998, p. 15.

hundreds of mathematicians left the obscurity of the NSA to ply their trade for substantially higher profits. It was the usual scramble along the technological frontier, the usual race to stake a claim in a fevered and highly promising market. By 1995 the market for encryption products was already nearly $1 billion, and industry analysts were predicting an "explosive" path of growth.[2]

There was only one problem. In the United States, at least, encryption was still considered a military technology. There were formal limitations on what kinds of encryption could be developed, even stricter limitations on what could be exported, and a general presumption that cryptography did not belong in private hands. Or, as one government spokesperson explained, "Our feeling is that cryptography is like nitroglycerin: Use it sparingly then put it back under trusted care."[3] As soon as the private market for encryption developed, therefore, it quickly ran afoul of these restrictions and ignited a long-standing battle with the U.S. security establishment. On one side were the NSA, the CIA, the FBI, and several defense agencies, all of which saw commercial encryption as an illegal venture and a threat to national security. On the other were a growing legion of high-technology firms, grouped somewhat awkwardly with the "cypherpunks," private cryptographers who saw encryption as a tool of social empowerment and a way for average citizens to protect themselves.

It was a classic standoff, and a critical one. For according to U.S. officials, high-end encryption software was simply too powerful to be left to market forces. If anyone could buy it, they argued, then everyone would—including Mafia hit men, foreign terrorists, and money launderers, all of whom would now be able to erect an impenetrable wall of secrecy. In the past,

---

[2]See U.S. Department of Commerce and the National Security Agency Working Group on Encryption and Telecommunications Policy, "A Study of the International Market for Computer Software with Encryption" (Washington, D.C.: 1995), III–1. For a related estimate, see A. Michael Froomkin, "It Came from Planet Clipper: The Battle over Cryptographic Key 'Escrow,'" 1996 University of Chicago Legal Forum, p. 15.

[3]NSA spokesman, quoted in Joe Abernathy, "Promising Technology Alarms Government: Use of Super-Secret Codes Would Block Legal Phone Taps in FBI's Crime Work," *Houston Chronicle*, June 21, 1992, p. A1.

U.S. officials (like most government officials) had employed a whole range of techniques for tracking hostile groups and communications: they intercepted telegrams, monitored radio broadcasts, and wiretapped phones. If encryption became the online norm, it would mean that a vast and promising new realm of communication could be bolstered against interception. Even worse, because the Net was inherently an international medium, this cloak of secrecy would extend across the world, giving all sorts of potentially hostile groups the ability to operate without fear of U.S. detection. Terrorists, for example, could trade bomb-making secrets in a secure chat room; drug dealers could launder money on their laptops. None of this activity, of course, was new. But what *was* new, and what alarmed the U.S. law enforcement establishment, was that the Internet could simultaneously multiply and hide criminals. By opening the floodgates of electronic communication, it made it easier for criminals to communicate with one another and harder, ironically, for anyone to catch them. Or, as the director of the FBI warned in 1997: "Unbreakable encryption will allow drug lords, spies, terrorists, and even violent gangs to communicate about their crimes and conspiracies. . . . Our national security . . . will be jeopardized."[4]

To the cypherpunks and interested firms, though, concern over law enforcement paled before the promise of encryption and the seamless world of privacy that it offered. Yes, they acknowledged, encryption would enable terrorists to shield their messages more completely. Yes, it might give drug dealers more opportunity to move their money around. But the rewards of anonymity were worth the costs. And in any case, they stressed, it was already too late. Aided by the Net itself, encryption software was already in the public domain, spread by pirates and well-wishers who had distributed key algorithms in the early 1990s. Higher-end encryption (longer and more complex algorithms) was already under development in places such as Moscow and Tel Aviv, where even the long arm of U.S. law enforcement was unlikely to penetrate. It was a new world, the

---

[4]Louis B. Freeh, quoted in Justin Matlick, "Encryption Debate Ignores the Obvious," Editorial, *The Seattle Times*, October 2, 1997, p. B5.

cypherpunks argued, a world where information flowed of its own will and where the old rules, bound as they were to wires and borders, no longer applied.

And thus in the late 1990s the battle raged—between the cypherpunks and the NSA; the government and private firms; the bearers of invention and of order. It wasn't a highly visible battle. Indeed, most Americans at the time probably didn't even know what encryption was, much less what the fight was all about. They hadn't heard of pioneers such as Phil Zimmermann and Whitfield Diffie; they didn't understand the complicated jargon that surrounded encryption fights. Yet, despite its relatively low public profile, the battle over encryption was one of the first and most significant contests of the Information Age. For by embracing encryption as a private good, the cypherpunks and their supporters had taken something—security—that used to belong to governments and thrust it into private hands. They had used technology—specifically, the combination of encryption and the Internet—to attack politics, arguing that technological advance had fundamentally changed the rules of the game. And when the politicians struck back, the cypherpunks used this same technology to circumvent the state's now-rusty laws and proclaim a new order.

In less than a decade, then, the advent of the Internet seemed to shift power away from the state. Pushed by technology that was evolving much faster than policy, national governments abandoned their restrictions on high-powered encryption or, as in the U.S. case, weakened them substantially. Quietly, these governments let one aspect of control slip away from their grasp: they let the cypherpunks win. This was a major concession, and evidence of the Net's radical political promise. And yet the last stand of the cypherpunks is not necessarily the last word on encryption. For the issues that drove the debate—and in particular the issues of law enforcement and national security—remain very much alive, lurking in the background until a public outcry brings them to the fore. And the very force that weakened governments' ability to regulate this area—the international spread of encryption technologies—may, ironically, be the same force that paves the way to a tighter and more all-encompassing regulatory regime.

Keeping Secrets:
A Brief History of Encryption

Compared with digital music or satellite TV, encryption is an ancient art. It has been used presumably since the dawn of communications, ever since one person or group tried to keep secrets from another.

In its most primitive form, encryption is the simple act of taking a message, changing it so that the message is no longer comprehensible, and then using some kind of decoder to replace the now-garbled message with the original text. For instance, let's suppose that two friends agree to play cards after dinner. They don't know quite what time that will be, so Frankie agrees to call Louie when he's ready. Neither boy's parents, though, want their son to be playing cards, so when Frankie calls, he loudly proclaims, "Okay, Louie, I'm finished eating now. Let's *do our homework* at 8:00." That's encryption. Louie and Frankie have agreed on a simple code—that "doing our homework" means "playing cards"—and they have sent their message in that encoded form. Frankie coded, or encrypted, it as he spoke; Louie decoded, or decrypted, it as he listened.

If the boys had wanted to send a more detailed message, they might have used a slightly more sophisticated form of encryption.[5] Suppose now that Frankie wants to tell Louie precisely what game they'll play, and where. Suppose, too, that his parents have caught on to the "homework" message, forcing Frankie to rely on notes ferried by his little sister, Olive. Frankie's message this time is: "Let's play gin at the baseball field." Since he doesn't trust his little sister not to read the message, he writes it in a basic code, substituting each letter of the alphabet for the one next to it:

A B C D E F G H I J K L M N O P Q R S T U V W X Y Z
B C D E F G H I J K L M N O P Q R S T U V W X Y Z A

Using this pattern of substitution, Frankie's message would read "MFU'T QMBZ HJO BU UIF CBTFCBMM GJFME."

---

[5]Formally, the use of whole words to substitute for other words is known as a code, while the use of letter substitutes is known as a cipher. For purposes of simplicity, I refer to both these methods as codes.

When Louie receives the message, he will use his version of the substitution code—the "key"—to put the message back into a readable form and then meet Frankie at the field, confident that no one will be able to figure out where they are.

But suppose that Olive is on to the boys' scheme and keen to find some form of leverage against her older brother. She wants to crack the code and catch Frankie. How can she do it? Two ways. First, while Frankie is out of the house, she can rummage through his room and look for the piece of paper that now holds the key, the letter-by-letter substitution laid out above. If she finds that, she can read the message as easily as Louie did and then track the boys down. Second, even if she can't find the key, she can try to decode the message herself, which, in this case at least, isn't all that hard. After just a few minutes of peering at the message, for instance, Olive would probably note the double Ms at the end of the second-to-last word. Realizing that there aren't that many letters that typically appear in this formation, Olive could guess that M in this code is probably either L or E or S. She could also guess that the T after the apostrophe is most likely an S, and that the BU combination probably stands for IS, AT, ON, or some other common word. With enough time and practice, Olive can crack this code fairly easily. Which means that the code is compromised, and the boys are caught.

This relationship embodies the dynamic of cryptography. Once people move beyond a simple word-for-word substitution (*homework* for *card game*) or the use of secret languages, they typically adopt some kind of letter-for-letter or number-for-letter solution. These codes are far more nimble and expansive than the word-for-word kind, but they also create a distinct set of vulnerabilities. Most critically, they demand some kind of code book, or key, which can then be either cracked or stolen, destroying the very secrecy that the code was designed to protect. People can hide their keys more carefully, of course, and they can make keys that are more and more complex, but the basic problem remains: with time, money, and patience, nearly any key can be either stolen or cracked.

The history of encryption is marked, therefore, by an ongoing, constantly accelerating race. In the Middle Ages, for example, dueling monks used to communicate with a form of encryption known as a nomenclator. A combination of word- and letter-based codes, nomenclators concealed all kinds of religious and political communication for several centuries, and were widely believed to offer nearly perfect concealment. In the middle of the fifteenth century, however, an Italian architect named Leon Alberti was prompted by a friend, a Vatican secretary, to come up with some "new ideas" regarding encryption. Alberti, who had already designed the Pitti Palace and written the first printed book on architecture, was apparently intrigued. And so for the next several years he worked on codes and codebreaking, stumbling eventually onto what is now known as frequency analysis. Essentially, Alberti's frequency analysis was just a more sophisticated form of Olive's guesswork above: a way of breaking codes by charting the frequency with which certain letters or pairs of letters appear.[6] Or, as Alberti explained in an essay written around 1467:

> First I shall consider the number of letters and the phenomena which depend on the rules of number . . . Here the vowels claim first place . . . Without a vowel there is no syllable. It follows that if you take a page of some poet or dramatist and make separate counts of the vowels and consonants in the lines, you will be sure to find the vowels very numerous . . . If all the vowels of a page were put together, to the number of, say, 300, the number of all the consonants together will be about 400. Among the vowels, I have noticed that the letter *o*, while not less frequent than the consonants, occurs less often than the other vowels.[7]

And so on. Through analysis, Alberti showed how any nomenclator, or indeed any system of single-letter substitution, was readily breakable. So long as there was a pattern in the substitution, a clever cryptanalyst (or codebreaker) could always

---

[6]Frequency analysis was not completely new, having already been developed by ninth-century Arab scholars. Alberti was the first Westerner to rediscover this breakthrough.
[7]Quoted in David Kahn, *The Codebreakers: The Story of Secret Writing* (London: Weidenfeld & Nicolson, 1967), p. 127. Kahn's book is the classic work on the history of encryption and provides background for most of this section.

crack it. The only way to defeat the codebreaker and truly hide a message was to destroy the pattern by using multiple alphabets—replacing each letter with a substitute derived from a different alphabetic code. This way, there would be no pattern and no way to chart frequencies. Using two circles cut from copper, Alberti thus proposed a new form of encryption based on a constantly shifting match of letters. Each circle was inscribed with the letters of the alphabet and the two circles, one larger than the other, would then be attached through the center with a small needle. Each time that the dial of the inner circle was spun, the letters would line up differently, suggesting a wholly new substitution code. This, claimed Alberti, would produce encryption "worthy of kings"; it would be unbreakable.

In one form or another, Alberti's "cipher disk" was to dominate European encryption for the next four centuries. Working from Alberti's central insight, generations of cryptographers took the cipher disk and laid layers of intricacy upon it, culminating in a sixteenth-century code known as the Vigenère cipher.[8] In this system, twenty-six separate alphabets are laid out in a giant table, each line a new alphabet that begins one letter later than the prior one. To navigate through the table, users would employ a key word or phrase that would point to the particular line in the table used to decipher each letter of the message. It was a painstaking process, but also a remarkably complex and secure one. Because the solution depended entirely on the key word, and because these keys could be changed with each message sent, the Vigenère cipher was considered unbreakable. Writing in the era of its creation one contemporary cryptanalyst proclaimed, "The key cipher is the noblest and the greatest in the world, the most secure and faithful that never was there man who could find it out."[9] Nineteenth-century firms and governments embraced the Vigenère cipher (or *le*

---

[8]The cipher is named after Blaise de Vigenère, a sixteenth-century French diplomat and scholar who was originally credited with its founding. More recent scholarship has shown that the real breakthroughs in the cipher were conducted by three earlier scholars: Leon Alberti, a German abbot known as Trithemius, and an Italian nobleman named Giovan Batista Belaso. Vigenère, however, continues to get the credit of his name. For a full account of the cipher's development, see Kahn, *The Codebreakers*, pp. 130–56.

[9]Quoted in Kahn, *The Codebreakers*, p. 150.

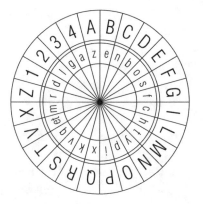

*Leon Alberti's
Cipher Disk*

chiffre indéchiffrable) as the best means for encoding their telegraphic communication and as late as 1917, *Scientific American* still portrayed it as "impossible of translation."[10]

Unbeknownst to *Scientific American*, though, the Vigenère cipher was not only breakable but indeed already broken. In 1854, an eccentric British inventor named Charles Babbage decided to crack the apparently uncrackable code.[11] Babbage was no stranger to painstaking work, having already spent years toiling over what, in retrospect, was the world's first programmable computer. Determined to find some way into the Vigenère cipher, he began to probe slowly and meticulously for a weakness—for some pattern that would let him unlock the underlying code. That pattern, not surprisingly, turned out to be the key word. Recall that the Vigenère cipher is based on an agglomeration of various alphabetic codes, each of which can encrypt the separate letters of messages. What gives the cipher its power is the key—the word or phrase that directs users to a particular line in the table. By applying a sophisticated form of frequency analysis, Babbage was able to read the key and thus unlock the underlying message. The math he used was quite complex, and far too tedious for the average eavesdropper to employ. But the

---

[10]Quoted in Kahn, *The Codebreakers*, p. 148. Kahn points out that *Scientific American* was actually describing a simpler version of the cipher, one that had been incorrectly attributed to its founder over the preceding centuries. For present purposes, though, the distinction is not critical.

[11]For more on the story of Babbage and the Vigenère cipher, see Simon Singh, *The Code Book* (New York: Doubleday, 1999), pp. 63–78.

principle he demonstrated was critical.[12] In order to encode
messages with any degree of secrecy, users need to employ multiple alphabets, or multiple kinds of signals. Then, in order to
decode these messages, they need to agree upon some kind of
key—a map of sorts that navigates through the various possible
signals. The key, therefore, is critical to secure encryption but
also its weakest link. If an outsider can decipher or obtain the
key, then she or he, like Olive or Babbage, can rapidly decipher
any communication encoded with this key. Finding keys is incredibly difficult, but so long as there is some underlying structure to the key itself—a repeated word, a logical phrase—it can
still be cracked. This turns out to be a fundamental principle of
encryption and a driving issue in current debates.

It is ironic that the greatest breakthrough in cryptanalysis (or
code breaking) occurred at the turn of the twentieth century.
For it was just at this time that the demand for encryption
boomed. Suddenly, the advent of telegraphy and then radio
meant that there were thousands of new clients for secrecy; millions of messages whirring unprotected along the wires and
across the airwaves. People wanted to disguise these messages,
and encryption offered an obvious means to do so. And thus the
growth of telegraphy ignited a race for new kinds of codes and
new forms of concealment. Though both governments and the
telegraph companies tried to prevent the use of code, they were
powerless before an onslaught of circumvention: foreign languages, made-up languages, simple substitution, and prefabri-
cated phrases. Most of these methods were fairly simple, but
they at least served to protect messages from curious clerks and
casual onlookers. Similar methods were applied in the early
days of radio, when operators would customarily cloak their
Morse transmissions in increasingly complicated codes.

The real boon to cryptography, however, occurred during
World Wars I and II, when the race to protect and decode messages assumed a supreme importance. Throughout these wars,
messages flowed endlessly: from commanders to their troops,

---

[12]Babbage himself never published his work on encryption, so his discovery was not recognized until the twentieth century, after scholars had pored through his notes. In 1863,
however, Friedrich Wilhelm Kasiski, a retired Prussian army officer, developed a similar cryptanalytic technique, which he published as *Die Geheimschriften und die Dechiffir-
kunst*. See Singh, *The Code Book*, p. 78.

from nations to their allies, from distant battlefields to regional commands. Such communication, of course, is a constant in war and may explain why Julius Caesar employed a basic form of encryption (known now as the Caesar shift) as early as 50 B.C. What changed in World Wars I and II, though, was the range of the fighting and the immediacy and magnitude of the information flows. With the advent of submarines and airplanes, battles now raged over vast and widely separated fields; with the advent of telegraph and radio, commanders could be in almost constant communication with both their troops and their superiors. It was a heady combination that changed the face of war.[13] It also meant that both encryption and decryption became vital elements of warfare, as nations strove to read each other's communications and protect their own. As mentioned in Chapter 3, for example, the German military relied heavily during World War I on a powerful cipher known as ADFGVX, which the Allies eventually managed to crack. Between 1914 and 1919, British cryptanalysts intercepted and deciphered an estimated fifteen thousand German messages, using the information to foretell German troop movements and meet German U-boats on a predicted course of attack.

During World War II, the race became even more fevered. In the waning days of World War I, a pair of German engineers had begun to tinker with an electrical encryption machine, a mechanical version of Alberti's original disk. The logic here was obvious: to use electrical power to generate more complex combinations of letters and more frequently changing codes. The result was ENIGMA, a German code machine that would baffle the Allies throughout the course of World War II. Built around a keyboard and a series of rotating disks, ENIGMA could encrypt any message automatically, without the sender having any knowledge of the code. By switching between settings, it could also reproduce 17,576 different scrambling arrangements, and move back and forth across them continually.[14] It represented a massive increase in encrypting power and left

---

[13]For an excellent account of how telegraph and radio affected warfare and international relations, see Daniel R. Headrick, *The Invisible Weapon: Telecommunications and International Politics 1851–1945* (New York: Oxford University Press, 1991).

[14]With modifications and additional features, ENIGMA was actually capable of generating approximately ten quadrillion possible keys. See Singh, *The Code Book*, p. 136.

the Allies desperate for an entry point. But even ENIGMA was eventually cracked, as was Japan's so-called PURPLE code.[15] In both cases, it proved once again a matter of time, patience, and keys. For ultimately, even the most ornate and rapidly changing code relied on some underlying pattern; and with sufficient repetition, the pattern could be discerned.[16]

As World War II faded into the Cold War, the field of encryption slouched back toward obscurity. Although World War II had produced true cryptanalyst heroes such as the small band of Britons who succeeded in breaking ENIGMA, most of these men and women went back to unrelated peacetime jobs and were forbidden or reluctant to speak of their work. Others clustered in the national agencies that were now devoted to the full-time pursuit of encryption and cryptology: the United States' National Security Agency and Britain's Government Communications Headquarters. There, under some of the world's most secure conditions, they began to push again at the boundaries of scientific secrecy, relying this time on the growing power of computers and a small stream of mathematical breakthroughs.

Over time, this secluded band of experts evolved into an elite fraternity of sorts. Most people who went to work for either the NSA or the GCHQ worked there for life. They were cutting-edge mathematicians, devoted to the complex task of unraveling bits of code and piecing them back together. Many played a formative role in the birth of computing, since the very basis of encryption—substituting one set of symbols for another—is also the backbone of computer programming, or "code." Indeed, recent historiography suggests that the world's first electronic computer was not in fact the ENIAC, unveiled in 1945 by scientists at the University of Pennsylvania, but rather the

[15]The story of how these codes were cracked has generated a substantial literature, nearly all of which lies beyond the scope of this chapter. Interested readers should see, for example, David Kahn, *Seizing the Enigma* (London: Arrow, 1996); F. H. Hinsley, *British Intelligence in the Second World War: Its Influence on Strategy and Operations* (London: HMSO, 1975); Andrew Hodges, *Alan Turing: The Enigma* (London: Vintage, 1992); and Józef Garlinski, *The Enigma War* (New York: Charles Scribner's Sons, 1979).
[16]According to Simon Singh, the one wartime code that was not broken was the Navajo language. Because this was a real language and did not therefore rely on artificial patterns, it remained unintelligible to all but a small band of native speakers. See Singh, *The Code Book*, pp. 191–201.

Colossus, a highly secret machine that Britain's wartime code-breakers devised in order to crack ENIGMA.[17] Because the cryptographers labored under such secrecy, however, their work and triumphs were rarely recorded. Instead they were a nearly invisible lot, shielded from public view and cloistered in an intense, almost entirely self-referential world.

As the Cold War heated up and the power of computing grew, the work of the NSA became more and more technical and the cloud surrounding it grew thicker still. In 1966, the NSA spent an estimated $1 billion and housed an estimated fourteen thousand people (precise figures are not available).[18] The NSA employed more mathematicians than any other organization in the world in the early 1990s, and was a major buyer of computer hardware.[19] Throughout this period, the agency presided over a secretive empire of spy planes and satellites, massive antennas and sophisticated listening devices. Yet hardly anyone outside the defense community even knew what it was: the centerpiece of U.S. intelligence efforts and the biggest site for masking and surveillance—code-making and code-breaking—in the world. There were, of course, other uses for encryption during this period. As banks, for example, embraced electronic money transfers in the 1960s and 1970s, they quickly developed (or bought) encryption software to protect these transactions. So did corporations that employed computer databases or maintained highly confidential computerized files. There were private firms that offered encryption programs during this time, and there was even the adoption, in 1976, of a formal standard of encryption: the Data Encryption Standard, or DES. But the vast bulk of encryption, and certainly the most high-powered encryption, still occurred under the watchful eye of the NSA. It was a realm of secrecy itself shrouded in secrecy—silent, staid, and invisible. And then the Internet came along.

---

[17]See Singh, *The Code Book*, p. 244.
[18]These estimates are from Kahn, *The Codebreakers*, pp. 677–84. See also James Bamford, *The Puzzle Palace: A Report on America's Most Secret Agency* (Boston: Houghton Mifflin Company, 1982).
[19]Reported in Singh, *The Code Book*, p. 248. See also Richard Lardner, "The Secret's Out," *Government Executive*, August 1998; and Dan Beyers, "After Decades of Secrecy, the NSA Lifts a Corner of Its Veil," *Washington Post*, August 25, 1991, p. B1.

Secrets in Cyberspace

In the United States, the Internet was born not far from the NSA. It was born, that is, as a child of the Cold War era and a tool of the U.S. defense establishment. Indeed, the early Net was in many ways the mirrored twin of the NSA, designed to ensure the sanctity of U.S. communications links while the NSA trawled to find weaknesses in others'.

The idea for this "network of networks" came initially from the U.S. Department of Defense and its Advanced Research Project Agency (ARPA). In the early 1960s the Cold War was at its peak, and the entire defense establishment was mobilized on its behalf. Reeling still from the Soviets' 1957 launch of Sputnik, U.S. officials were trying desperately to beat the Soviets on the scientific front and protect U.S. territory against what was still seen as a highly probable attack. These efforts led in many directions: to the space program, of course; to advanced research on missiles and launch technology; and to next-generation satellites. They also led, quietly and without any great fanfare, to experiments with network computing and a hope that the growing legion of military and research computers might be linked into some kind of overarching system. The logic behind these experiments was twofold. First, if the computers could be linked together, then scientists across the country would be able to share their knowledge efficiently and inexpensively. No university or research site would be forced to replicate the work of another, and the free flow of information would drive innovation and discovery. Second, if information on this network could be spread among a number of computers and research sites, then the entire network—and indeed the entire backbone of high-level communications—could be made impervious to nuclear attacks or natural disasters. By spreading information around, the network could protect it.

Before long, ARPA's experiments with such "internetworking" led to the creation of ARPANET, a series of physical links between various computer networks. Using ARPANET, a physicist at, say, the University of Chicago could transmit a document directly to a colleague in New Hampshire, and a missile base in New Mexico could receive electronic instruction from the Pentagon. The great beauty of the system was that no single route was either fixed or dominant. Instead, messages passed freely and invisibly across multiple pathways—sometimes traveling via a computer in Arizona to reach from Chicago to New Hampshire; sometimes going through Atlanta or New York; and often split en route into separate and discrete packets of information, each of which traveled along its own unique course. There was, as a result, no fear that information on the ARPANET would ever be destroyed, since the loss of any particular link would simply shift traffic onto another. There was also no great need for security, since the users of this system were all members of the same elite corps of researchers. Indeed, in the late 1960s and early 1970s, before the advent of personal computers, it was reasonable to conclude that anyone with the technical means to access ARPANET would be from the very scientific or engineering community that the system was designed to serve.

For roughly twenty years, this community quietly flourished online. Expanding rapidly from just four host computers in 1969 to nearly two thousand by 1985, ARPANET changed its name to Internet and became a common mode of communication among university researchers, government scientists, and a handful of outside computer engineers. Funding for the creation and maintenance of the system's infrastructure came from the National Science Foundation (NSF), which assumed this responsibility from the Department of Defense in the 1980s and explicitly discouraged the use of the Net for any "nonscholarly" purpose.

As computer use expanded and then exploded during the 1980s, though, the old rules of this community yielded to a surge of outside interest. Suddenly, people from beyond the world of research wanted in; and because the Net was so open and inclusive by design, because it was built around shared knowledge rather than hoarded information, there was no effective way of stopping them. Accordingly, by the late 1980s, the NSF had abandoned both its policy of scientific focus and its financial support of the Net. In 1989, commercial service providers were allowed to offer Internet access to paying customers, and in 1990 the Net was officially opened to commerce.

These reversals transformed the Net. In less than two years,

the community of researchers was recast as a commercial frontier and cyberspace was awash with "newbies"—new users, new
firms, and whole new ideas about what this space was and how
it might be used. In the early 1980s, the Internet community
consisted only of about twenty-five linked scientific and academic networks. By 1995, it had grown to encompass over forty-
four thousand networks, 160 countries and twenty-six thousand
registered commercial entities.[20] Users by this point totaled
somewhere between forty and fifty million and were growing by
an estimated 10 to 20 percent a month. Luxuriating in this new
space, the prophets were exultant: "Cyberspace," proclaimed in-
coming House Speaker Newt Gingrich, "is the land of knowledge. And exploration of that land can be civilization's truest,
highest calling."[21] Vice President Al Gore was equally effusive.
"Empowered by the movable type of the next millennium," he
prophesied, "we can send caravans loaded with the wealth of
human knowledge and creativity along trails of light that lead to
every home and village."[22]

Once the initial rush of excitement had passed, however, it
became clear that not all of the Net's underlying traits were particularly well suited to this new cast of users. And chief among
these mismatches was secrecy. On the original Net, secrecy was
not highly prized; on the contrary, the very purpose of the system was to share information and distribute it widely across the
network. There was no fear of interception, and thus no need to
cloak messages or restrict them. Once the Internet was open to
commercial ventures, though, and once it spawned a bevy of
consumer-focused businesses, the concern for secrecy grew. It
grew so much, in fact, that the lack of secrecy, the lack of privacy, became a major impediment to several emerging lines of
Internet trade. Banks, for example, were eager to replace costly
tellers and deposit windows with electronic transfers and online
banking. But how could they convince their customers to trust
their money, and information about their money, to an open

[20]Philip Elmer-Dewitt, "Welcome to Cyberspace: What Is It? Where Is It? And How Do
We Get There?" *Time*, March 22, 1995, p. 4.
[21]Quoted in Elmer-Dewitt, "Welcome to Cyberspace," p. 4.
[22]Quoted in Bill Tammeus, "Surfing for Meaning," *Kansas City Star*, November 10,
1996, p. L1.

and unrestricted system? Retailers, likewise, were anxious to ply their wares in cyberspace but wary of losing frightened customers. How could they convince a computer-hating fifty-year-old to release her credit card or shopping desires to some anonymous machine? The closer these firms got to the manna of online retailing—to matching customers with their particular preferences, for example, or showing them advertisements geared to their specific needs—the larger the issue of secrecy loomed.[23]

There were several ways, of course, for both firms and their customers to address this problem. They could, for example, handle only part of the transaction online; they could create powerful barriers around their internal communications; they could agree to release only certain bits of information. But the real solution to their problem was the same as Frankie and Louie's: it was to find some way of making their communication utterly secret, of disguising the information so that no one except the intended recipient could decode it. The solution, in other words, was encryption—high-powered, unbreakable encryption. And the best way to get this kind of encryption was to employ one of the true breakthroughs in cryptography, a recent innovation known as public key.

Cypherpunks and Public Keys

In 1974, Whitfield Diffie, a thirty-year-old cryptographer, got into his car and drove across the country to see Martin Hellman, a twenty-eight-year-old professor of electrical engineering whom he had never met. Through the small cryptographer's grapevine, Diffie had just found out that he and Hellman shared a common obsession. Both men were trying to

[23]The pursuit of privacy has long been a concern among American citizens. For more on the history of privacy in the United States, see David J. Seip, *The Right to Privacy in American History,* Harvard University Program on Information Resources Policy Publication P-78-3, July 1978; and Daniel J. Knauf, *The Family Jewels: Corporate Policy on the Protection of Information Resources,* Harvard University Program on Information Resources Policy Publication P-91-5, June 1991. For a specific examination of the role of privacy in cyberspace, see Whitfield Diffie and Susan Landau, *Privacy on the Line* (Cambridge, Mass: MIT Press, 1998).

do what most cryptographers deemed impossible: they were trying to hide the keys.

Recall that all encryption methods, from Frankie's note to Alberti's disk and the Germans' ENIGMA, share a common vulnerability. They all rely on some kind of a key—on a password or code phrase that tells the sender how to navigate through the code and where to begin decoding. Theoretically, keys allow encryption to become more and more complex; they allow users, for example, to employ the kind of polyalphabetic tables that Alberti pioneered, and to change the coding alphabet with every letter of a message. In practice, though, keys also create their own weakness, since any key can itself be stolen or discerned. And once the key is cracked, the entire coding system is laid bare. A critical component of any encryption system, therefore, is simply how to get the keys from one place to another—how, in the jargon of cryptography, to ensure secure key exchange.

After generations of attempts, most cryptographers had conceded that the problem was unsolvable. Short of physical contact, they reasoned that there simply was no way to guarantee secure exchange, and that the next best solution was the tried-and-true method of constant updating, constant shifting, and constantly staying just one turn ahead of the would-be interceptor. Diffie, though, was determined to break the cycle. In the early 1970s, just as ARPANET was gaining steam, he became convinced that the electronic network would shortly grow beyond its official boundaries and into the public realm. When this happened, he further reasoned, people would need encryption—not casual encryption or government-sponsored encryption, but high-powered, easily accessible encryption. They would need some way to protect their privacy in an electronic world, some way to encode their electronic messages and safely distribute the keys that would unlock them. "I was always concerned," Diffie later recalled, "about individuals, an individual's privacy as opposed to Government secrecy."[24] And so this is what Diffie, a passionate believer in privacy, set out with Martin Hellman to do.

---

[24]Quoted in Steven Levy, "Battle of the Clipper Chip," *New York Times Magazine*, June 12, 1994, p. 46.

For two years, the mathematicians tinkered with the problem from every conceivable angle. Diffie moved to Stanford, registered as a graduate student so that he could work with Hellman, and then enlisted the help of Ralph Merkle, another Stanford graduate student. After countless dead ends, the group stumbled upon a solution, creating the foundation for what is now known as public key encryption.

To grasp the essence of public key, it is useful to return one last time to our ill-fated Frankie and Louie. We left Frankie holding his message, desperate to get it to Louie but with no safe means of transmission. Now imagine that Frankie has a plan. He puts his message in a suitcase, secures it with his bicycle lock, and then asks Olive to deliver the package to Louie. When Louie receives the suitcase, he secures it again with his own lock, and then sends the twice-locked package back with Olive to Frankie. When Frankie receives it, he unlocks his own lock, but leaves Louie's secured. The exhausted Olive now lugs the suitcase back one more time to Louie. This time, Louie removes his own lock, hustles Olive out of the house, and— at last—receives the prized message. Notice that even though Olive was in reach of the message the whole time, she never could actually gain possession of it. And Frankie and Louie never had to exchange a key; rather, by using their own keys in a fixed sequence, they managed to secure the message throughout the entire course of transmission.

In the solution crafted by Diffie, Hellman, and Merkle, of course, these locks and keys are considerably more complex. They involve a particular kind of arithmetic known as modular arithmetic and use keys comprised of very big numbers. The basic idea, though, is exactly as laid out in the final round of our Frankie-and-Louie saga. In the first stage, the sender encrypts his message by applying a certain number or algorithm to it. Suppose, for example, that the message is BASEBALL and that it has already been converted into a basic numeric code with each letter simply represented by its place in the alphabet:

```
B  A  S   E  B  A  L   L
2  1  19  5  2  1  12  12
```

If the sender's key is MULTIPLY BY 2, then BASEBALL rapidly becomes 4-2-38-10-4-2-24-24. In the next round, the receiver takes this message and further encodes it with his key—say, MULTIPLY BY 3—so that the message now reads 12-6-114-30-12-6-72-72. When the sender gets this garbled message back, he partially decodes it by applying the reverse of his key. In this case, then, he divides by 2 and gets 6-3-57-15-6-3-36-36. Finally, the receiver takes this message and applies the reverse of *his* key: dividing by 3, he would get the original code of 2-1-19-5-2-1-12-12, which he can easily convert into BASEBALL.[25] Note that the same wondrous logic applies: there is no exchange of keys in the process and no likelihood of interception.

Once Diffie, Hellman, and Merkle had worked out the basic logic and math of this system, the real breakthrough came in 1975, when Diffie stumbled onto public key encryption. In the example above, both Frank and Louie have a single key, which they use to code and decode messages. Diffie's breakthrough was to tinker with the math so that Frankie and Louie could theoretically each have two keys, one for coding and one for decoding. This way, they could actually make the initial key public—giving it to Olive, or their parents, or whomever—and then decode any messages with their own private key. Because the keys were mathematically distinct, there was little chance that anyone with access to the public key could decipher the private one. Mathematically, then, it was as if everyone who wanted to send Louie a message had access to his bicycle lock, but only he retained the key. Practically, it meant a vital simplification of very complex, very high-powered encryption. Using Diffie's public keys, any pair of correspondents could communicate quickly, efficiently, and confidently. There was no need to worry about the vulnerable process of key distribution, no need to listen for eavesdroppers or constantly switch codes. Rather, with public keys, encryption became both accessible and secure. Which, in the emerging world of electronic communications, was a very big deal.

In 1977, another team of academics forged the final piece of public key encryption. Motivated by news of Diffie's work, three

---

[25]This is a very simple recounting of key encryption. For a more complete account, see Martin E. Hellman, "The Mathematics of Public-Key Cryptography," *Scientific American,* August 1979, pp. 146–57; Philip R. Zimmermann, "Cryptography for the Internet," *Scientific American,* October 1998, pp. 110–15; and Singh, *The Code Book,* pp. 256–71.

mathematicians at MIT, Ron Rivest, Adi Shamir, and Leonard Adleman, developed an elegant solution to Diffie's theoretical concept. In retrospect, the solution is almost staggeringly simple. A sender employs two very large prime numbers (as large, say, as 10 raised to the 309th power, or 10 followed by 309 zeros). She then multiplies these two together and releases the product as her public, published, key. Anyone can access this key; anyone can use it to encrypt a message to her. When she receives the message—wrapped securely in her own key—she can then unlock it by using her private key, a combination of the prime numbers that only she can use. No keys are transmitted during this process and, so long as the prime numbers remain large enough, it will be almost impossible for any outsider to determine the private keys based on the public one. So powerful was this idea that Rivest, Shamir, and Adleman filed for a patent in 1977, formally creating the world's first practical public key system.[26] In 1982, they established an accompanying firm, RSA Data Security, and began to provide encryption software to a variety of customers: to banks, for example, and government agencies, and eventually satellite television companies such as Sky.

In retrospect, the period from 1974 to around 1994 can be seen as a time of nearly pure innovation. Diffie, Hellman, and Merkle crashed the key barrier because it was there and because, in Diffie's case at least, he believed in the social benefits of enhanced privacy. "I thought," he later recalled, that "having an essential technology of privacy that was a government secret was a bad idea."[27] Rivest, Shamir, and Adleman then perfected the technology as a mathematical puzzle, a solution to the cryptographic challenge that Diffie and his colleagues had posed. When commerce crossed the technological frontier it was almost as an afterthought, something to do with all this math, rather than

---

[26]At the time that the group filed for their patent, the idea of patenting computer software or mathematical algorithms was very new, and thus not particularly well received by the U.S. Patent Office. See James Bell, "Clipper Encryption Must Be Destroyed," *Electronic Engineering Times*, March 14, 1994, p. 32. The group did not receive a patent until September 20, 1983 (U.S. Patent and Trademark Office, Patent #4,405,829).

[27]Quoted in Neal Thompson, "Whitfield Diffie's Crusade for Privacy," *Baltimore Sun*, December 31, 1998, p. 1A.

the motivation behind it. During this stage—the classic first stage of a technological breakthrough—science prevailed over commercial demands and the innovators worked alone.

In the early 1990s, though, the Internet barged into cryptology's silent world. Suddenly, private firms and individuals were clamoring for secure, private encryption. Suddenly, the market demanded something it had never wanted before, something that apparently only private key encryption could provide. It was at this point that public key encryption slipped into the second phase of its evolution, leaving the research laboratories and migrating, loudly and publicly, toward the commercial sector. Technically, of course, this migration was brilliant, since it matched twenty years of cutting-edge thought to a perfectly suited problem. Commercially, it held phenomenal promise. But politically it was tough. For in the United States, at least, certain uses of public key cryptography—including some of the most important and profitable uses—were distinctly illegal.

The Politics of Public Keys

To understand the strange politics of encryption, it is necessary to recall where the technology came from and how it has been applied throughout history. Until the advent of the Net, encryption was almost entirely a military tool. It was what Caesar used to foil his allies, what powered the Germans' inscrutable ENIGMA, and what heightened Cold War battles between western allies and the Soviet bloc. Although outsiders (such as Alberti or Diffie) had always played a key role in advancing the science of cryptography, the power of the technology has customarily resided in the state—in closed and secretive agencies such as the NSA or Britain's GCHQ.

These links grew particularly strong during the Cold War, when both the United States and Great Britain treated cutting-edge cryptography as a sharply restricted science. In the United Kingdom, nearly all work on cryptography was pulled under the auspices of the GCHQ, where it remained wholly confidential. Indeed, according to one recent account, cryptographers at GCHQ actually solved the problem of public keys several years before their American counterparts, but were forbidden under

British law from ever publishing their discoveries.[28] In the United States, meanwhile, encryption fell under the joint purview of the CIA, the FBI, and, especially, the NSA. All encryption technologies were officially classified as "munitions" and regulated by the Department of State. Export of high-powered encryption was strictly forbidden, meaning that no U.S. firm could either sell or use these encryption products outside the United States.[29] Inside the United States, private encryption was permitted, but the small circle of users was encouraged to employ the rather weak, government-approved Data Encryption Standard.

When the Internet went mainstream, it thus thrust the national security and law enforcement agencies into an awkward dilemma. On the one hand, they, like the rest of the U.S. government, saw the inherent potential of the Net and were reluctant to impede its growth. On the other hand, they were terrified by what the Net could do to their own world of law enforcement and national defense. For just as cyberspace opened up vast new spaces for social and commercial contact, so too did it expand the possibilities for criminal exchange. On the Internet, terrorists could easily exchange information about bomb-making or chemical weapons. Hate groups could recruit new members; gangsters could launder money. New-age pirates would be able to manipulate the ever-growing flows of money and information, and hostile governments could extend their reach in subtle and undetected ways. While none of these threats was completely novel—there had always been hostile governments, and money launderers and terrorists—they all loomed larger in cyberspace, where contact was more immediate, communication more expansive, and activity of any sort more difficult to detect. In this context, secure encryption was seen, paradoxically perhaps, as a threat. If everyone could protect their communication, the agencies feared, then law enforcement would be thrust into an unprecedented vacuum and

[28]See Singh, *The Code Book,* pp. 279–92.
[29]In most cases, Canada was an exception to these rules and U.S. firms were permitted to sell encryption products there. In 1993, a change in U.S. regulation allowed the overseas subsidiaries of U.S. firms to employ encryption products that were still banned for sale to foreign firms. See Steve Higgins, "Do Export Laws Threaten On-Line Software Firms?" *Investor's Business Daily,* October 9, 1995, p. A8.

national security would suffer. In a world of perfect information and total privacy, governments would lose out.

Part of this concern, perhaps, was a predictable response to a drastic power shift. For decades, the national security agencies had presided over a very secluded, very controlled realm. They played by a strict set of rules and had full power to enforce them. When hordes of newcomers threatened to rush into this world and upset its balance, the agencies were understandably caught off guard. A larger part of the concern, though, and a more important one, relates to the underlying dynamic of technological change. When the Internet created an insatiable demand for encryption, it pulled the mass market along the technological frontier and changed the underlying rules of the game. In the old world, a world marked by non-electronic communications and Cold War politics, the rules of encryption made perfect sense. They worked for the government; they worked for the public; and they worked well enough for the market, which didn't see any real need for high-powered encryption. In the world of the Net, however, these rules no longer made sense. And thus the familiar lines of battle were set, pitching those who wanted to preserve some semblance of the old order against those who preferred the anarchy of the new.

For the next several years, these two sides waged a nearly constant tug-of-war, pulling back and forth between the dictates of security and the driving momentum of the Net. [30] The first confrontation occurred in 1993, when the Clinton Administration proposed a security solution known as the Clipper chip. Backed by the national security agencies, Clipper was designed as a compromise of sorts, a way to permit private encryption without compromising the traditional tools of law enforcement. Under the Clipper scheme, all computers and telephones were to be equipped with a tiny black computer chip—the Clipper chip—that would contain, in essence, a separate private key. When the computer or telephone was sold, the corresponding key would be deposited in a secure kind of escrow account. It would stay there, untouched and unused, unless a government

[30]For a recent review of this debate, see Stewart A. Baker and Paul R. Hurst, *The Limits of Trust: Cryptography, Governments and Electronic Commerce* (Boston: Kluwer Law International, 1998).

agency received a court-approved wiretap. In that case, the key would be released, enabling the government to track and decode any of the subject's communications.

According to its defenders, Clipper's only purpose was to tug the new world of the Internet back to the traditional world of law enforcement. It was the same, they argued, as old-fashioned wiretapping: an unobtrusive way to catch those who were already suspected of being criminals. Without Clipper or something like it, warned Dorothy Denning, a prominent computer scientist at Georgetown University, "[a]ll communications on the information highway would be immune from lawful interception. In a world threatened by international organized crime, terrorism and rogue governments, this would be folly."[31] Others were equally committed to Clipper's cause, arguing that it was the only way, or at least the best way, to extend law enforcement into cyberspace. "Do we want a digital superhighway where not only the commerce of the nation can take place but where major criminals can operate impervious to the legal process?" queried Jim Kallstrom, a technical specialist with the FBI. "If we don't want that, then we have to look at Clipper."[32]

Within the close-knit world of private encryption, though, the response to Clipper was incendiary. Led in part by Whitfield Diffie, a small band of cryptographers, hackers, and computer engineers charged that Clipper was an unnatural intrusion, an early sign of a "cyber police state." They formed a group known as the Cypherpunks and began to rally against the chip. "The war is upon us," proclaimed Tim May, a co-founder of the group. "Clinton and Gore folks have shown themselves to be enthusiastic supporters of Big Brother."[33] Other newly founded groups, such as the Electronic Frontier Foundation and the Electronic Privacy Information Center, rushed to add their support for the cypherpunks and to protest against what they too saw as an undue invasion of privacy. "The idea that the Government holds the keys to all our locks," argued another opponent, "doesn't parse with the public. It's not American."[34]

---

[31]Quoted in Levy, "Battle of the Clipper Chip," p. 47.
[32]Quoted in Levy, "Battle of the Clipper Chip," p. 48.
[33]Quoted in Levy, "Battle of the Clipper Chip," p. 46.
[34]Quoted in Levy, "Battle of the Clipper Chip," p. 70.

"Clipper," argued another, "is a last ditch attempt . . . to establish imperial control over cyberspace."[35] On the Net and in a score of Net-related publications, opposition to Clipper grew at a furious pace. Critics charged that it was technically faulty, politically foolish, and inherently unconstitutional. They claimed that the chip was almost absurdly easy to avoid, and that the real burden would thereby fall on U.S. telephone and computer manufacturers, who would see their sales migrate to foreign competitors. Most critically, perhaps, they also argued that Clipper was already too late. By 1994, argued May, the era of "crypto anarchy" had arrived. "And just," he wrote, "as a seemingly minor invention like barbed wire made possible the fencing-off of vast ranches and farms . . . so too will the seemingly minor discovery out of an arcane branch of mathematics come to be the wire clippers which will dismantle the barbed wire around intellectual property."[36] Crypto, he and the cypherpunks insisted, was here to stay.

Under a barrage of criticism, the White House eventually retreated from the Clipper chip proposal. By the middle of 1995, the Clinton Administration had quietly backed down from the idea, conceding that more public debate on the topic would be necessary. Yet the notion behind Clipper and the concerns of the national security establishment remained unchanged. Set against the cypherpunks' call for anarchy, they were to dominate the debate for the next several years.

Between 1994 and 1998, the U.S. political system was awash with encryption proposals, bills, and cases. There were a series of completely contradictory legal decisions and a raft of legislation. In 1997 alone, for example, seven bills were proposed, each supported by its own cluster of interest groups and many bearing acronyms that ostensibly revealed their larger purpose: SAFE, for example, the Safety and Freedom through Encryption bill; and Pro-CODE, the Promotion of Commerce Online in the Digital Era. Beneath all this clamor, meanwhile, was a constant and common thread. There was a surging demand for private encryption and a growing supply of encryption prod-

---

[35]John Perry Barlow, "Jackboots on the Infobahn," *Wired*, April 2, 1994, p. 40.
[36]Quoted in Levy, "Battle of the Clipper Chip," pp. 51, 60.

ucts, and there were absolutely no rules in place to order or regulate this trade.

Two issues in particular were troubling, both of which were addressed in various bills and cases but neither of which was fully resolved. First, in the wake of Clipper's demise, the national security community continued to press for some regulation of high-powered encryption and some way to regain the kind of controlled access that had existed in the world of telephones and wiretapping. Mustering support from across the federal agencies, they urged Congress and the Clinton Administration to limit the strength of private encryption products, either by inserting some kind of access feature (à la Clipper) or by capping the size (or "bit length") of the underlying key. The Internet community, though, scoffed at even the mention of control. Led by the cypherpunks and their allies, they argued vehemently against any sort of backdoor access or any attempt to restrict the complexity of private encryption. Or, as one activist proclaimed in Senate testimony, "Trying to stop encryption is like trying to legislate the tides and the weather. It's like the buggy whip manufacturers trying to stop the cars— even with the NSA and the FBI on their side, it's still impossible."[37] "The mind boggles," wrote another. "Does the President really believe that spies and drug dealers will use compromised technology? Is he also confident that our own spies will not justify cracking citizens' conversations by claiming reasons of national security? . . . The issue here is whether you want this government (and all future U.S. governments) to be able to decode anything it thinks might breach public safety, national security, or (just possibly) acceptable moral behavior."[38]

The second issue was subtler, but no less contentious. Ever since encryption had become a product rather than just an art, it had been formally classified as a weapon. Officially, this meant that encryption exports were severely constrained and under the formal authority of the U.S. State Department. Commercially, it meant that selling encryption products, or even *using*

[37]Testimony of Philip R. Zimmermann to the Subcommittee on Science, Technology and Space of the U.S. Senate Committee on Commerce, Science, and Transportation, June 26, 1996.
[38]Richard Morin, "The Other Boot," *Unix Review* 15, no. 8 (July 1997), p. 83.

them in an international transaction, was enormously complicated for multinational firms, or indeed for any firm whose
business slipped naturally into the fluid realm of cyberspace.
Accordingly, as the Internet market expanded, a host of business groups joined the cypherpunks in lobbying for relaxed export controls. Such restrictions, they argued, were an ill-fitting
relic of a wholly different time. They made no sense in cyberspace, where borders essentially disappeared; and they meant
that U.S. firms would eventually cede the lucrative encryption
market to less-constrained foreign competitors from abroad.
"The only effect of export controls," argued an official with the
Software Publishers Association, "is to cripple our ability to
compete."[39] "Why are customers who don't want to give advanced access to a third party going to buy American products
when they can buy non-key escrow products abroad?" queried
another opponent. "The administration's continuing misguided
policies on encryption threaten the continued preeminence of
the American software industry in the twenty-first century."[40]
If the old rules were not changed, critics charged, U.S. firms
would lose their lead in the Net's soaring economy.

To the national security agencies, however, there was more
at stake than market share or profits. There was instead the very
real fear that hostile governments or foreign terrorists would be
able to hide behind an impenetrable wall of secrecy, immune to
any of the tracking technologies that U.S. agencies had employed in the past. Only recently, the agencies urged, these technologies had played a crucial role in national security: Ramsey
Yousef, for example, one of the terrorists involved in the World
Trade Center bombing, kept plans for future attacks encrypted
on his laptop; Aldrich Ames, the CIA spy, relied on encryption
as well. If either of these individuals had been able to encrypt
his communications more securely, he might very well have
proceeded with his plans. And so the old rules, lawmakers argued, still worked. Or as Stewart Baker, general counsel of the
NSA, put it:

---

[39]Quoted in Levy, "Battle of the Clipper Chip," p. 48.
[40]Bruce Heiman, outside legal counsel for the Business Software Alliance, quoted in
Daniela Cimino, "Escrow? Hell No!" *Software Magazine*, February 1997.

[Private encryption,] they say, is out there to protect free-dom fighters in Latvia. But the fact is, the only use that has come to the attention of law enforcement agencies is a guy who was using [encryption] so the police could not tell what little boys he had seduced over the net. Now that's what people will use this for—it's not the only thing people will use it for, but they will use it for that—and by insisting on having a claim to privacy that is be-yond social regulation, we are creating a world in which people like that will flourish and be able to do more than they can do today...

To say that NSA shouldn't be involved in this issue is to say that Government should try to solve this difficult technical and social problem with both hands tied behind its back.[41]

For several years these debates raged inconclusively. A 1996 initiative leaned toward the cypherpunks by taking encryption off the munitions list, but then leaned back toward the national security community by maintaining limits on the strength of private encryption; a bill introduced in the House of Represen-tatives by Bob Goodlatte (R-Virginia) called for a full removal of export controls; another, put forth in the Senate by John McCain (R-Arizona) and Robert Kerrey (D-Nebraska) vowed to maintain controls on high-powered exports. There was little room for compromise between the two sides and only the dim-mest prospect for consensus. "Never in peacetime has our government attempted so completely to monopolize a single form of communication," wrote a group of law professors in re-sponse to one congressional proposal. "[N]ever has it required, in effect, a license to exercise the right to speak."[42] By this stage, both sides of the debate clearly understood that a new market had developed and that a new game was being played upon it. Where they differed critically, though, was in determining how the rules of this game should be shaped and toward what end they might aspire.

---

[41]Quoted in Levy, "Battle of the Clipper Chip," p. 70.
[42]See David Braun, "It's D-Day for Encryption," *TechWeb*, September 24, 1997. Avail-able at www.techweb.com/wire/news/1997/09/0924crypto.html.

Return to the Market

If all the confusion over encryption had created a legislative mess in Washington, the impact of that mess was even greater in Silicon Valley, where firms were simultaneously dealing with a rapidly evolving market and a constantly shifting set of laws. In the early days of the commercial Net, several firms got caught in this web, bearing the burden of rules that they either hadn't considered or chose to ignore. In 1997, for example, Sun Microsystems announced that it would license 128-bit encryption software from Elvis+, a Russian company, and then redistribute the technology to its global customers. Sun claimed that it had spent two years working through the legal ramifications of the deal, and that it was confident that the Elvis plan was "in full compliance with the letter of the law."[43] Yet Sun pulled back from the project just months later, once the Commerce Department launched an investigation. Likewise, RSA Data Security slowed plans in 1997 to fund the development of a Chinese software company before Commerce could investigate.[44] In both of these cases, the firms insisted that they were in full compliance with existing export laws. But the growing rift between the technology and the law was clearly causing problems for the executives and bureaucrats alike.

The most spectacular victim of these changing rules was Philip Zimmermann, an amateur cryptographer who came to symbolize the growing chaos of encryption policy. Unlike Diffie or Hellman or members of the RSA team, Zimmermann was not a formal mathematician or a research scientist. He was instead a sort of itinerant programmer and activist, a soft-spoken man who believed passionately in individual rights and worried about the long arm of government intervention. In the 1980s, Zimmermann had focused his concern on U.S.–Soviet relations and become an active member of the Nuclear Weapons Freeze Campaign. In the 1990s, with the Cold War subsiding, he returned to a childhood fascination with cryptography and con-

[43]Quoted in John Fontana, "Feds Review Sun's Crypto," *Communications Week*, May 26, 1997.
[44]See David Bank, "Encryption Firms Plan Challenge to U.S. Controls," *Wall Street Journal Europe*, March 20, 1998, p. 5A.

centrated instead on the social implications of code. To Zimmermann, cryptography in the digital age was about more than military power or mathematical wizardry. It was about power—about "the power relationship between a government and its people." It was about "the right to privacy, freedom of speech, freedom of political association, freedom of the press, freedom from unreasonable search and seizure, freedom to be left alone."[45] Cryptography, he stressed, was a "surprisingly political technology." With it, people could shield their communications from the prying eyes of government. They could exchange information freely and without fear of reproach. They could learn things that governments didn't want them to learn, and they could press for actions that governments wanted to stop. All of which, Zimmermann believed, would lead to the toppling of dictatorships and the spread of democracy.

Motivated by this concern, the erstwhile peace activist began in the late 1980s to craft an accessible version of the RSA system, something that would allow ordinary citizens to encrypt their communications securely and painlessly. Using freely available technical literature, Zimmermann cobbled together a software package in 1991, which he then distributed for free. He called this package PGP, for Pretty Good Privacy. As the Internet developed over the next few years, PGP spread, Zimmermann recalls, "like a prairie fire."[46] It was widely used by individuals, by human rights groups, by fledgling e-businesses—by anyone, indeed, who wanted online privacy and had no other way to achieve it. Grateful users posted PGP to web sites and bulletin boards, where other users were then able to download the software and spread it along. By the mid-1990s, with millions of copies installed, PGP had become the global standard for private encryption software.

Because PGP relied on a version of the RSA system, though, it too employed extremely long, extremely complex keys—keys that were 128 bits in length, or double the official DES standard. And because these keys had made their way around the world, Zimmermann, according to the U.S. Customs Service, was

---

[45]Quoted in Singh, *The Code Book*, p. 296.
[46]Testimony of Philip R. Zimmermann.

technically in violation of U.S. export control law. In 1993, government investigators informed Zimmermann that he was under investigation for having exported illegal weapons across the Internet. Several years later, RSA also brought suit, claiming that Zimmermann had effectively given away their property. As these suits dragged on, Zimmermann became a virtual icon, a symbol, simultaneously, of how the Internet expanded individual rights and how both big business and big government threatened to squash these emerging rights. Newly formed privacy groups sprang to Zimmermann's defense and human rights organizations from around the world rallied to defend PGP and its beleaguered creator. Zimmermann, they cried, was a freedom fighter, not a criminal. He was a pioneer in the purest sense, a man who had used technology to fight abuse and protect human dignity, who had "done more to open up the Internet than all of Al Gore's minions combined."[47] In the end, both RSA and the U.S. government backed down. RSA negotiated a settlement in 1998, extending new licenses to its technology, and the government simply dropped its investigation in 1996. Technically, Zimmermann probably had violated both patent rights and export controls. Yet as his case so vibrantly demonstrated, it was becoming harder and harder in the mid-1990s to apply old rules to this new space. And so, in a rapid replay of the Clipper controversy, the U.S. government yielded to the force of the market and the protests of the pioneers.

A final blow came in 1998, when Network Associates Incorporated (NAI), a newly formed Internet security firm, blatantly and publicly circumvented U.S. export control law.[48] In 1997, Network Associates had purchased Pretty Good Privacy, Inc., the firm that Zimmermann had hastily assembled after the government dropped its suit against him. Armed with the software, and with Zimmermann himself on board as a consultant, NAI had then wrestled with the rather thorny issue of what to do with this high-profile product. The good news was that the firm

---

[47]David Post, "Encryption—It's Not Just for Spies Anymore," *American Lawyer*, December 1994, p. 106.

[48]This section draws heavily on earlier work by the author. See Jennifer L. Burns and Debora L. Spar, *Network Associates: Securing the Internet*, Harvard Business School case 9-799-087, March 15, 1999.

now controlled the world's most popular private encryption software. The bad news was that none of its customers was used to paying for this software and its export was still—technically at least—illegal. Somewhat to their surprise, executives at NAI were able to solve the first problem almost at once. They simply phoned PGP users and asked them to pay a royalty fee. Since over 90 percent of these users were now legitimate businesses, most were happy to pay the small fee and sign on as NAI customers; piracy, in this instance, had slipped seamlessly into commerce. The second problem, though, was more complex. For it was the same problem that had plagued Elvis, and Zimmermann, and all previous attempts to sell encryption software in a global marketplace. If NAI sold PGP outside the United States, or if it even allowed its U.S. customers to use PGP outside the United States, it was formally in violation of U.S. export law. But if it restricted its sales to U.S. customers operating within the United States, then the product was effectively useless: no one wanted encryption that could be applied only at one end of a conversation.

In the spring of 1998, Network Associates attacked this legal dilemma by running directly into it. On March 20, the company announced that it would soon begin selling Pretty Good Privacy through cn Labs, an independent Swiss company. According to NAI, the Swiss researchers had recently developed encryption software that was "functionally equivalent" to PGP; so similar was it, in fact, that NAI customers outside the United States could purchase the Swiss software and expect it to be fully compatible with NAI's own U.S. version. In a blatantly public announcement, NAI declared that all its customers would finally be able to "securely communicate from Japan or Australia to the US or Germany or Switzerland, or wherever it may be."[49]

It was a brilliant move. For officially, Network Associates had broken no laws: it hadn't exported anything from the United States; it hadn't transferred technology; it hadn't provided technical assistance. All the company did, it turns out, was to publish the source code for PGP—a simple expression of

---

[49]"Network Associates Ships 128-bit Encryption without License," *Newsbytes News Network*, March 20, 1998.

free speech—and then contract with a perfectly legitimate foreign firm. "Essentially," argued a top NAI official, "any third
party could have contracted with any non-U.S. entity for any
product . . . and offer it worldwide."[50] The U.S. government,
however, was not amused. Charging that "This [action] flies in
the face of our policy," Undersecretary of Commerce William
Reinsch opened an immediate investigation into the company's
activities.[51]

Executives at NAI were well aware that they were skating
along the gray edges of the law. As Peter Watkins, a top executive with NAI, recalls: "I asked myself, 'Am I going to go to jail?
Well, probably not. OK.'"[52] They were betting that the law in
this case would bend once again toward the market and that the
government would refrain from any blatant attempt to punish
NAI or halt its thinly disguised export strategy. It was a big bet,
but given the outcome of Clipper and Zimmermann, and given
the weight of public opinion and the leakiness of existing law,
also a relatively safe one. In the end, no formal charges were
ever filed against Network Associates.[53]

Instead, over the course of 1998, the Clinton Administration
began to fumble toward a new kind of encryption policy, something that, in the words of Vice President Albert Gore Jr., would
"enable electronic commerce to grow and to thrive," while still
giving law enforcement agencies "the ability to fight 21st century crimes with 21st century technology."[54] In September, the
details of this policy were announced, representing a breathtak-
ing change. Henceforth, 56-bit encryption could be exported
anywhere in the world except seven "terrorist" nations (Cuba,

---

[50]Quoted in "Network Associates Ships 128-bit Encryption."

[51]Quoted in John Shinal, "Network Associates' Encryption Plans Draw Criticism,"
Bloomberg Financial Newswire, March 20, 1998.

[52]Interview with author, Santa Clara, CA, November 11, 1998.

[53]At around the same time, though, NAI did encounter other legal difficulties. In 1999,
for example, the company was compelled by the SEC to reverse $214 million in write-
offs that it had posted for 1997 and 1998. In the aftermath, disgruntled investors filed a
class action suit against the company, citing accounting and revenue recognition fraud.
See "Analysts Are Unsure if Network Associates Can Reclaim Old Glory," *Dow Jones
Business News,* May 25, 1999; and Edward Iwata, "More Firms Falsify Revenue to Boost
Stocks," *USA Today,* March 29, 2000, p. 5A.

[54]Press Briefing by the Vice President et al., The White House, Office of the Press Secretary, September 16, 1998.

Iran, Iraq, North Korea, Libya, Syria, and Sudan), following a one-time technical review and with no requirement for key recovery. Encryption above 56-bit (that is, 128-bit, or "strong" encryption) could also be exported now, but only with some kind of a key recovery system. In the most startling departure from past policy, the government also announced: "Very strong encryption (with or without key recovery) will now be permitted for export under license exemption, to several industry sectors. For example, U.S. companies will be able to export very strong encryption for use between their headquarters and their foreign subsidiaries worldwide except the seven terrorist countries."[55] With these rules in place, companies such as Network Associates would be free to sell their encryption software to any firms in the insurance, health, online retail, or banking businesses— to the very firms, in fact, that represented the most crucial sectors of their business.

In describing their new policy, U.S. officials made careful reference to the dictates of national security. Officials at the Commerce Department, for example, argued that they had already succeeded in changing the emerging architecture of encryption, and that strong encryption sales were still being permitted only in industries where there was already an established level of trust and government regulation.[56] Yet clearly, the policy was also an acknowledgment of the government's own limitations in this field. Indeed, just days before the new regulations were announced, a high-ranking government advisory committee had released a report citing the problems inherent in export controls and admitting that "the adverse impact of controls on U.S. industry is palpable." Noting that U.S. firms were losing ground in the international market for encryption software, the report warned ominously of a "weakening of the U.S. position as a leader in electronic commerce generally."[57] In the calculus of national security versus national profitability, the market appeared once again to have won.

[55]The White House, Office of the Press Secretary, "Administration Updates Encryption Policy," press release, September 16, 1998.
[56]Interview with author, Washington, D.C., October 6, 1998.
[57]The President's Export Council Subcommittee on Encryption, "Findings of the President's Export Council Subcommittee on Encryption," September 18, 1998.

The Puzzles of Encryption

In some ways, the battle of encryption thus ended in 1998. It ended when the U.S. government began to chip away at its own restrictive policies, confessing—in spirit, if not in word—that it could no longer control how people cloaked their secrets or where in the world they sent them. It ended, it appeared, with a resounding victory for the free market and perhaps even for freedom. In cyberspace, U.S. policy seemed to admit, it was simply too difficult to regulate the flow of information or the confidentiality that protected it. On the Net, secrets would be preserved and governments would have to shed at last their ancient habit of eavesdropping.

In the broader sweep of history, then, the arcane battles over encryption seem poised to assume a much greater significance. If governments will truly be unable to track communications in cyberspace, and if private citizens can route their communication along secure and confidential routes, then the balance of power between citizens and the state will begin to shift—noticeably in countries such as the United States, Canada, and France; more sharply in China, Burma, and Saudi Arabia. The political importance of this shift cannot be denied. For as Phil Zimmermann explained, private encryption "changes the power relationship between a government and its people. For better or worse, that's what it does."[58]

Yet at the same time, it is also important to keep the encryption debate in perspective and to plot its likely evolution. Although the debate over encryption simmered to a draw in 1998, the issues left on the table still have not been resolved. There remain serious concerns, both public and private, about how to deal with new-age criminals and how to protect national security in a more fluid age. The United States still retains some controls over encryption exports (restricted sectors, restricted countries, required documentation), and the European Union is slowly considering encryption policies of its own. In 1997, members of the Organization for Economic Cooperation and

---

[58]Quoted in Jeff Ubois, "Hero or Villain?" *Internet World*, August 1995, p. 78.

Development announced the adoption of joint cryptography guidelines[59]; in 1998, thirty-three developed countries, including the United States and most of its allies, jointly agreed to limit the export of encryption software above 64 bits in length.[60] Elsewhere, of course, governments maintain substantially stronger formal controls: France, for example, controls both exports of encryption products and the domestic use of encryption; so do Israel, Russia, China, and a slew of other countries.[61]

As a policy, then, encryption still has a slightly uncertain future. We don't know for sure whether governments will fully cede their power to control it or, more important, whether they'll be able to scratch back doors into even the most apparently secure system. Recall, after all, that nearly every encryption method eventually has been cracked: the monks' enigmatic nomenclators, the Germans' ENIGMA, the infamous *chiffre indéchiffrable*. Already, experts are poking into 128- and even 512-bit encryption, testing their limits against the ever-increasing force of computers. In 1999, for example, a group of European cryptographers managed to factor both a 456-bit and then a 512-bit number[62]; several months later, one of the creators of the RSA algorithm described a computer that was theoretically capable of cracking 512-bit keys in only a few days.[63] Over time, it is eminently possible that governments, either publicly or in the shadow of their own national security agencies, will gain the means to crack even what is currently regarded as uncrackable encryption. And if they do, the new world of the Internet will slowly evolve into the older world of telephones and radio and

[59]See "OECD Adopts Guidelines for Cryptography Policy," OECD News Release, Paris, March 27, 1997. For the full recommendations, see OECD, "Cryptography Policy Guidelines," Recommendation of the Council Concerning Guidelines for Cryptography Policy, OECD, Paris, March 27, 1997.

[60]See John Markoff, "International Group Reaches Agreement on Encryption," *New York Times,* December 4, 1998, p. C4.

[61]For more on these policies, see Richard C. Barth and Clint N. Smith, "International Regulation of Encryption: Technology Will Drive Policy," in Brian Kahin and Charles Nesson, eds., *Borders in Cyberspace* (Cambridge, Mass.: MIT Press, 1997), pp. 283–99.

[62]See Bruce Schneier, "The 1999 Crypto Year-in-Review," *Information Security,* December 1999, p. 20.

[63]See "Code Experts See Threat to E-Commerce Privacy," *The Wall Street Journal,* August 16, 1999, p. B11.

telegraph: there will be privacy when governments permit it, and eavesdropping when they choose—either legally or clandestinely—to listen in.

An even more interesting possibility concerns the role of private firms. Publicly, most firms (at least in the United States) have positioned themselves squarely on the libertarian side of the encryption debate. They have argued against restrictions on encryption exports; against the Clipper chip; and strongly in favor of an expanded private market for encryption. They have been, not surprisingly, arrayed on the side of the market and against government involvement in it. As the technology evolves, however, and as encryption becomes more prevalent, it's quite likely that private-sector firms will embrace a kind of new-age eavesdropping, adopting technologies that favor backdoor access instead of complete security. Firms such as Nissan and Charles Schwab, for example, have already announced that they monitor employees' e-mail and restrict "inappropriate" use.[64] Several Internet service providers, and most notably AOL, explicitly review the material that passes through their sites, scanning for signs of illegal or illicit activity. In several high-profile cases, AOL clearly cooperated with federal law enforcement agencies, allowing them to track the identity and activities of suspected criminals.[65] While such events do not constitute an easy substitution for wiretapping and envelope-opening, they do suggest that private firms may choose to work with government agencies in the future, stopping the kinds of behavior that threaten both national security and private commerce. Firms such as AOL, after all, don't want child molesters or drug dealers running around their sites; they don't want hackers stealing

[64]Hooman Bassirian, "Reining in the E-mail Beast," *Computer Weekly,* May 20, 1999, p. 10; and "Strategy: Stop Your Employees from Abusing the Net," *Computing,* September 7, 2000, p. 31.

[65]AOL provided the FBI, for example, with information that led to the arrest of the alleged author of the Melissa virus. It also contacted the FBI after uncovering web pages that seemed related to the Columbine massacre; and it disclosed to a navy investigator the identity of a sailor who had revealed himself as gay online. See Leslie Helm et al., "Alleged Spreader of Melissa Virus Arrested in N.J.," *Los Angeles Times,* April 3, 1999, p. A1; Kevin McCoy, "Internet Targeted in Probe," *Daily News (New York),* April 22, 1999, p. 34; and "You're a Pioneer of the Electronic Frontier . . . ," *Houston Chronicle,* June 27, 1999, p. 1.

credit card numbers from unwary shoppers or criminals infiltrating their banking networks. And to prevent or punish such crimes, private firms may well turn back to public authorities, relying on agencies such as the FBI to provide the legal mandate and physical muscle that not even the most successful Internet start-up can supply. In the end, private firms may actually want public surveillance—or at least public prosecution for any crimes that private surveillance has revealed.

And yet, despite these possibilities, despite encryption's ancient bonds to wars and spies and governments, the Internet has categorically changed the rules that govern encryption. It has taken what was once a restricted technology and tool of the state and thrust it into private hands. It has created a market where one didn't exist and has denied states the power they once had to control the flow of secrets. Why has this change been so dramatic? According to the prophets and cypherpunks, it is because the Internet itself is so radical—because it inherently shifts power and erodes the basis of state control. It is because, as one author wrote in a 1993 *Harvard Business Review* article, "information technology is the greatest democratizer the world has ever seen"; because "high technology can put unequal human beings on equal footing."[66] Perhaps. But encryption also seems to have some special qualities, characteristics that make it particularly susceptible to this kind of a power shift and particularly immune to governmental involvement.

Consider first the process of standardization. In other fields and other technologies, this is frequently a very messy business. There is generally a rash of competing standards, as we saw for example in the telegraph industry, and a rush to dominate the market and create an installed base. Often, standards supported by competing firms don't physically work with one another, as was the case with early railroad gauges; or they operate slowly and painfully, as was again the case with telegraphs. If firms combine to create a common standard, they encounter all the problems that customarily haunt cartels; and if one firm dominates, then the familiar issues of monopoly arise. In each of these

---

[66]Sam Pitroda, "Development, Democracy, and the Village Telephone," *Harvard Business Review,* November–December 1993, p. 66.

cases, government intervention offers—potentially at least—a way out of the problem. By setting a standard or regulating a standards battle, governments can often pave the way to a smoother and less chaotic market. This is what happened, for example, when the European powers agreed to a common telegraph system in the 1850s, or when national governments adopted common rules for ship-to-shore radio communications.

In the field of encryption, by contrast, the process of standardization feels oddly like a natural phenomenon. When innovation occurs—as with Alberti's disk, Vigenère's cipher, or Whitfield and Diffie's algorithm—it tends to flow rapidly across the code-making community, creating a common plateau and undermining whatever technology came before it. There are none of the bloody battles here that mark other technological breakthroughs; there is no frantic scramble to establish a particular standard or impose a certain system upon an emerging market. Instead there is just a quietly meritocratic procession, as each technology falls before whatever else can break it and then supplant it. There is no need in this process for an impartial referee and thus no demand—from either business or society—for governmental involvement.

A second characteristic concerns congestion. In many other emerging technologies, problems of congestion quickly plagued the frontier. In the radio industry, for example, broadcasting ground to a halt in mid-1920s, when hundreds of new stations clambered onto the airwaves and then drowned in the din that quickly ensued. In television and then satellite TV, broadcasters knew from the start that they would be competing for space not only with each other, but also with the radio stations that already filled the airwaves. The only way to solve the problem was to have someone parcel out the spectrum, creating property rights and preventing a rush that would otherwise destroy them all. This is a typical problem of the frontier, and a typical solution: at some point, private interests generally want a marshal to come in and solve the traffic problems. But with encryption, there simply are no traffic problems: no scarce resource to be divided, no issues around congestion. And thus there is again no inherent demand for governmental involvement.

Finally, until recently at least, the world of encryption was marked by a distinct disregard for property rights. Encryption was the stuff of spies and generals, of mathematicians and puzzle solvers and, occasionally, frustrated lovers. But it was hardly ever a commercial tool and rarely considered a form of property. People created codes, they broke codes, but they didn't *own* codes. On the contrary, codes were either shared freely (with colleagues and allies) or guarded ferociously (from enemies and eavesdroppers). There was no reason for anyone other than a government to own a code until the Internet came along, and thus no established system, or even norm, of property rights. As a result, there was no demand outside of government for the enforcement of property rights, something that in other industries has driven even the most hardened pioneer back to the services of the state. Over time, of course, this community-minded view is nearly certain to change. Indeed, it has been changing ever since 1977, when Rivest, Shamir, and Adleman filed for their patent and brought private property into the secluded world of cryptography. As the market for encryption edges ever closer to the commercial mainstream, researchers may come to value profits over pooling and shift their efforts away from the tightly knit cloister of research. But in political terms, this movement has already come too late. During the formative years of the encryption industry, governments forestalled a demand for property by holding all the property—the codes and code-breaking techniques—for themselves. There were, as a result, no commercial pirates in this realm and no pioneers scrambling to stop them.

In many ways, therefore, the technology of encryption has been unique. It has encountered no problems of congestion or coordination, no need for external standards, and little immediate pressure for property rights. It hasn't really needed governance, then, and so it is not surprising that private forces have banished governments from their fold. In the area of online encryption, markets have truly been able to develop on their own: regulating themselves, coordinating their actions, and seizing a function—security—that once adhered solely to the state. The result of this shift is dramatic, since it promises (or threatens, depending on one's view) to explode the limits of privacy, giving individuals an unprecedented ability to keep their communications—their thoughts, their news, their personal data—shielded from the prying eyes of the state. This is a crucial shift that cannot be denied. Indeed, it may be one of the most crucial political developments to emerge from the Internet, a technically driven political shift that ranks with Gutenberg's printing press as a source of individual freedom and an obstacle to authority.

Yet even with this most private of technologies, public issues still arise. Privacy itself, of course, is a public issue, and as the Net economy expands, individuals may want to bolster their technical means of privacy with a set of legal rights and protections. In the European Union, for example, privacy is considered a basic right, guaranteed by law and written most recently into the 1995 EU Data Protection Directive.[67] In the United States, similar concerns may well drive citizens back to the state, lobbying for some kind of legal assurance about what privacy means in an online world and how those who violate it will be punished. Or as one Net executive acknowledged: "Standards are good, but they need some teeth, and this is where government becomes a good partner."[68] Meanwhile, the old security concerns of the state—the concerns that led to the formation of the NSA, the creation of Clipper, and the imposition of export controls—could well become public concerns over time, depending again on how the technology evolves and who dominates its use. We may see a market here that remains truly free of government intervention—a market, in fact, that shrinks government authority and widens the range of personal liberty. Or we may see the pendulum swing slowly back, as public concern over legal rights and physical security drives both individuals and private firms back toward the arms of the state.

---

[67]Directive 95/46/EC of the European Parliament and of the Council of 24 October 1995 on the Protection of the Individuals with Regard to the Processing of Personal Data and on the Free Movement of Such Data.

[68]Quoted in Heather Green et al., "It's Time for Rules in Wonderland," *Business Week* March 20, 2000, p. 86.

CHAPTER 6

Trusting Microsoft

We set the standard.

MICROSOFT SLOGAN

On November 5, 1999, the U.S. District Court for the District of Columbia handed down its preliminary judgment in the case of *United States v. Microsoft*.[1] Eagerly awaited and instantly devoured, it was a harsher finding than many analysts had predicted. According to Judge Thomas Penfield Jackson, Microsoft, the world's largest and most profitable computer software company, had violated U.S. antitrust law. It had blackmailed competitors, reduced competition, and used its "prodigious market power" to "harm any firm" that might

---

[1] *United States v. Microsoft Corporation*, Civil Action No. 98-1232 and Civil Action No. 98-1233. Formally, the November 5 decision was not a final verdict but a preliminary "finding of fact." The final verdict in the case, which largely echoed this initial view, was handed down on June 7, 2000.

threaten its hold.[2] It had hurt consumers and stifled innovation, and its chairman, the brilliant and mercurial Bill Gates, had watched it happen.

When word of the decision was announced, a cacophony of voices rose to analyze and explain it. Some saw the court's decision as a rightful judgment about clearly injurious behavior, "a victory for consumers and for companies trying to compete."[3] Others described a gross misunderstanding about one of the world's most profitable markets and most productive companies. Some saw Jackson as a bumbling and ill-informed meddler, a government agent with no comprehension of how business operated along technology's cutting edge. Others saw him as a savior of free markets and painted the case as a simple imposition of justice—"a statement," one commentator wrote, "about basic fairness, truth-telling and the rule of law."[4] Microsoft, of course, maintained that it had committed no wrong and deserved no punishment. On July 26, lawyers for Microsoft filed a brief with the Supreme Court, arguing that their case needed to be reheard by a federal appeals court.[5]

Even before the appeal was filed, however, it was clear that the Microsoft case would not end with Judge Jackson's decision. For this was a huge case, one of the most spectacular of the late twentieth century and a harbinger of both business and government in the twenty-first. Together with his high school buddy Paul Allen, Bill Gates had seen the revolution of personal computing as early as 1968 and rushed to meet it. Gates and Allen had developed software for the emerging personal computer in 1975; they had worked with IBM to create the first great mass market for personal computing in the 1980s, and they had designed wave after wave of wildly successful pro-

[2]See Joel Brinkley, "U.S. Judge Declares Microsoft a Monopoly Stifling a Market," *New York Times,* November 6, 1999, p. 1.

[3]Michael Cowpland, president of Corel Corporation, quoted in Bert Hill, "Victory for Consumers: Cowpland," *Ottawa Citizen,* November 6, 1999, p. D1.

[4]Thomas L. Friedman, "The Way We Are," *New York Times,* November 14, 1999, p. 15.

[5]Formally, Microsoft requested that the case not be sent directly to the Supreme Court for appeal, which was what lawyers for the government had proposed. See Steve Lohr, "Microsoft Files Brief Asking Supreme Court to Send Antitrust Case to Appeals Court," *New York Times,* July 27, 2000, p. C2; and James V. Grimaldi, "Microsoft Seeks Appellate Review," *Washington Post,* July 27, 2000, p. E1.

grams. By 1999 Microsoft presided over an exploding global software market and had earned a staggering $19.7 billion in revenues. Now chairman of the company (Allen had resigned his operating role in 1983), Gates was arguably the world's most famous executive and certainly its richest. Under his prodding, Microsoft was racing now toward the next revolution, trying to catapult its massive stock of assets (its people, its brand name, its capital) into the Internet age.

Jackson's decision, however, called much of this plan into question. According to the court's ruling, Microsoft's Internet strategy was characterized by monopolistic intent and marked by anti-competitive behavior. Using its massive clout in the software market, Microsoft hadn't just embraced the Internet; it had smothered it, wiping out competitors and forcing its own system and standards upon some unwilling hardware manufacturers. If the court's decision were upheld, Microsoft would be broken into separate and competing companies and the U.S. Justice Department would have played a pivotal role in shaping Microsoft's move to the Net.

In responding to the court's claims, lawyers for Microsoft insisted that their company had done nothing wrong; that competition in cyberspace demanded aggressive tactics, and that Microsoft was merely following the accepted rules of the game. "None of these snippets," argued one company lawyer, "none of this rhetoric even approaches proof of anti-competitive behavior."[6] "Saying you're going to compete like heck," echoed another, "is not a violation of the nation's antitrust laws."[7] Moreover, the lawyers suggested, even if Microsoft's style of competition was rough and ready, it still brought benefits that few outside the company could match. Because Microsoft already produced the world's most popular operating system, for example, efficiency demanded that consumers receive a Net browser that was fully integrated with this system. Because Microsoft already had a standard language for computing, it made sense to design Internet products and connections based on this

[6]William H. Neukom, quoted in Joel Brinkley, "As Microsoft Trial Gets Started, Gates's Credibility Is Questioned," *New York Times,* October 20, 1998, p. A1.
[7]Charles Rule, quoted in Steve Lohr, "Disarming a Giant," *New York Times,* November 9, 1998, p. C6.

same standard. And so forth. Most passionately, though, Microsoft's supporters also insisted that in a market like the Internet, in a space that was vast and untested and raw, it was foolish to impose complicated rules from an earlier age. John Herrington, for example, a former U.S. secretary of energy, argued in one heated editorial that the Department of Justice was being fueled in this case by "an outdated, industrial-age philosophy ill-suited to the rigorous, fluid economic environment of the Digital Age." Such misguided actions, he continued, "imperil our economy and our position as a technological world leader. While innovations created by the American technology industry are the linchpins of America's phenomenal economic expansion, federal actions can choke the geese laying the golden eggs of economic growth."[8]

Lawyers for the U.S. Justice Department, however, were unmoved. As the agency responsible for bringing the Microsoft case to court, the Department of Justice felt equally passionate about defending its own position and taking Microsoft to task. The rules for competition in cyberspace, they maintained, were the same as the rules in steel or autos or oil—the same for Bill Gates as they had been for John D. Rockefeller.[9] If Gates and Microsoft had engaged in anti-competitive practices—if, as Judge Jackson apparently agreed, they had used their immense power in one market to harm customers in another—then they needed to be punished and stopped.

One of the few things that both sides agreed upon was that the stakes of *United States v. Microsoft* were huge. For the case would determine not only how Microsoft operated and behaved, but also how other firms—big or small, established or starting up—would need to approach and think about the Internet. Was the Net, as Judge Jackson implied, simply a new space in which to apply old rules? Was it subject to the same kinds of competition law and government authorities that ap-

---

[8]John Herrington, "Microsoft Ruling Serves Rivals, Not Consumers," Editorial, *San Francisco Chronicle*, November 30, 1999, p. A29.
[9]In a landmark 1911 case, the U.S. Supreme Court agreed that Rockefeller's Standard Oil Company had attempted to monopolize the U.S. oil industry. The remedy in this case consisted of dissolving the massive company into thirty-three geographically separate units. See *Standard Oil Co. of New Jersey v. United States,* 221 U.S. 1 (1911).

plied in other areas of commerce? Or was it, as Microsoft and its supporters argued, a new form of activity, one too big and unruly to be subject to competition laws that dated from the last century? Was there really something different about the Net, something that uprooted traditional notions of competition and changed the contours of monopoly? Or was Microsoft just borrowing the prophets' language to cloak an old-fashioned form of bullying? Most critically, what was at stake in the Microsoft case was the question of commercial governance in the Internet age: What were the rules that defined competition in this new market? And who would get to set them?

Land Grab:
Pioneer Days at Microsoft

Once upon a time, Bill Gates was a rebel. Born to wealthy parents in a comfortable section of Seattle, Washington, William Henry Gates III discovered the world of computers in 1968, in the earliest days of the computer revolution and long before most people had even seen it coming. He was thirteen years old. Together with Paul Allen (a positively ancient fifteen-year-old), Gates began to use the facilities at his all-boys prep school to write basic programs for the PDP-10, a refrigerator-sized minicomputer that DEC (the Digital Equipment Corporation) had recently begun to market.[10] Within several months, Gates and Allen had launched their first business venture, helping a local company find "bugs" (or errors) in programs running on their PDP-10.

Working in exchange for time on the company's machines, the boys rapidly became expert programmers. They became intimate with the inner workings of the DEC machine, sold a class-scheduling program to their high school, and launched a firm called Traf-o-Data in 1971. They also became quite skilled in the emerging art of breaking into computer systems. In 1969,

---

[10]For a full account of Gates and Allen's early ventures, see James Wallace and Jim Erickson, *Hard Drive: Bill Gates and the Making of the Microsoft Empire* (New York: John Wiley & Sons, Inc., 1992), pp. 19–51.

Gates was reprimanded after company officials realized that he had bypassed their internal security system, gained access to confidential information, and then crashed the firm's entire computer system. Shortly thereafter, he used machines at the University of Washington to gain access to Cybernet, a national computer network, and crashed that as well.[11]

By this point, Gates and Allen had graduated from the PDP-10 and were experimenting with the 8008, a microprocessor introduced by the fledgling Intel Corporation in 1972. It was a critical development. For before the 8008, no computer was small and simple enough to exist outside a business or laboratory. Early machines had been mostly mainframes, vast and clunky behemoths constructed of tens of thousands of discrete components soldered together and large enough to fill an entire room. They also tended to be very expensive, and to demand a dedicated group of technicians to maintain and operate them. In the 1960s, DEC had begun to shrink these massive machines to a more manageable size, using transistors and then integrated circuits to build machines that smaller businesses and laboratories—and hence more of them—could use. But even these "minicomputers" were made out of multiple-printed circuit boards assembled in racks, and were too large, expensive, and complex to be owned by individual consumers. The 8008, by comparison, had the whole central processing unit of a computer on a single chip—a piece of silicon etched with several layers of circuits. Like many young computer buffs, Gates and Allen were awed by the potential of this new device, by the ability to plant a tiny chip of silicon in a small computer box and make a new kind of machine—a personal computer. Unlike most of the other buffs, though, Gates and Allen were lured as much by the commercial as the technological potential. They didn't just want to tinker with the new technology or imagine its myriad possibilities; they wanted to create a mass market on top of it.

In the fall of 1973, Gates left Seattle for Harvard. Allen followed close behind, taking a job with the computer manufac-

---

[11]Although this story is well repeated, it may be apocryphal. Wallace and Erickson, for example, insist that Gates was not involved in the Cybernet crash. See *Hard Drive,* p. 42.

turer Honeywell and constantly pressing Gates to drop out of school and start another business. In December 1974, Allen saw his chance. That month, the cover of *Popular Electronics* carried a headline proclaiming "World's First Microcomputer Kit to Rival Commercial Models." In the accompanying article, the magazine described how Ed Roberts, founder of a small electronics company, had assembled a kit that allowed hobbyists to build their own microcomputers.[12] Reportedly, Allen saw a copy of the magazine in Harvard Square, grabbed it off the stand, and ran to Gates's dormitory. "Look," he yelled, "it's going to happen! I told you it was going to happen! And we're going to miss it!"[13] Gates was persuaded. Just days after seeing the article in *Popular Electronics,* he called Ed Roberts in Albuquerque and offered to sell him some software.

When Gates made his rather presumptuous offer, hundreds of other computer buffs were also knocking on Roberts's door. Indeed, during the early months of 1975 there was a veritable pilgrimage to Albuquerque, a parade of enthusiasts who saw Roberts's machine—the Altair—as the stuff of high-tech dreams. As the editor of *Popular Electronics* later recalled, "The only word which could come to mind was 'magic.' . . . About two-thousand people, sight unseen, sent checks, money orders, three, four, five hundred dollars apiece, to an unknown company in a relatively unknown city, in a technologically unknown state. These people were different. They were adventurers in a new land. They were the same people who went West in the early days of America. The weirdos who decided they were going to California, or Oregon or Christ knows where."[14] One determined hacker drove clear across the country to be near Roberts's plant, and slept for weeks in a trailer, waiting for his Altair.[15]

[12]These machines were still far from user-friendly. They required hours of delicate assembly and looked like metal boxes once complete. There was no keyboard or monitor, and users had to do all of their own programming. See Daniel Ichbiah and Susan L. Knepper, *The Making of Microsoft: How Bill Gates and His Team Created the World's Most Successful Software Company* (Rocklin, Calif.: Prima Publishing, 1991, pp. 19–20.
[13]Ichbiah and Knepper, *The Making of Microsoft,* p. 21.
[14]Quoted in Steven Levy, *Hackers: Heroes of the Computer Revolution* (New York: Anchor Press/Doubleday, 1984).
[15]Wallace and Erickson, *Hard Drive,* p. 87.

If Gates and Allen were not alone in their enthusiasm, however, they were unique in their ambition. While hordes of hobbyists wanted to own an Altair, there were only a few commercial pioneers who actually dreamed of profiting from it. When Gates called Roberts, he offered to take a standard programming language and create a version of it specifically for the Altair, a version that would enable users to program and operate the machine without having to go through a daunting series of switches. To convince Roberts, Gates swore that he and Allen had already written a similarly modified language for Intel's 8080 chip (the microprocessor that powered the Altair) and that they could easily make it work on this new machine. It was a lie. Gates and Allen had never even worked with an 8080 chip, much less written a programming language for it. But they saw the vast potential of the Altair and were determined to play some role in its development. And so when Roberts agreed to look at their supposed language, the two worked feverishly for weeks, creating an entirely new program for a machine and a microprocessor they still had never seen. They pulled off this feat by programming the powerful Harvard PDP-10 to make it behave like the much simpler 8080 processor, and then writing 8080 code, using the tried and true programming methods they had learned years earlier.

Remarkably, it worked. In February 1975, Paul Allen flew to Albuquerque and successfully demonstrated the new software—which ran successfully the first time—thousands of miles away from the machine on which it had been developed, and the only machine he could have used to fix any bugs that might have come up. Five months later, Gates and Allen formed a new company, Micro-Soft, to license their programming language.

From these awkward beginnings a giant was born. Gates and Allen licensed their software to Roberts's company, developed a disk-based version of their programming language, and adjusted the underlying code to serve a growing list of corporate clients. By 1977, their slightly renamed Microsoft Corporation had nine employees and $381,715 in revenues.[16] Critically, Microsoft also retained full legal rights to all of the company's software.

---

[16]The Microsoft Timeline, www.microsoft.com/MSCCorp/Museum/timelines/microsoft/timeline.asp.

The decision to own software—to license its use rather than sell programs or code directly—was simultaneously the secret to Microsoft's commercial success and the source of intense envy, even hatred, within the burgeoning computer community. From the very start of their operations, Gates and Allen had decided to break from standard practice in the computing world, which was all about sharing ideas and eschewing profits. Rather than acceding to this model, they instead decided to make a business from the software they created, copyrighting the code or program and earning revenues by licensing the right to use it. Commercially, this licensing strategy was a brilliant move, since it allowed the young company to maximize revenues and increase control. The more copies of its software that Microsoft licensed, the more money it made and the more entrenched its program became across the community of computer users. The more it sold, in other words, the more its software became the standard—the language that everyone else wanted to program for and use. The problem, however, was that the very core of this strategy ran over the traditional soul of computing. Until Gates and Microsoft came along, software hadn't really been conceived of as property. It was a code of sorts, a common good, a thing to be shared and developed and grown. It was what defined a certain group of like-minded tinkerers and cemented their common bond. In fact, when Gates and Allen "created" a programming language for the Altair, all they really did was to take an existing language known as BASIC—a language developed at Dartmouth University by John Kemeny and Thomas Kurtz—and build a modified version to fit the Altair's needs. In other words, like legions of other enthusiasts, they employed a common knowledge of programming, a knowledge that had already been spread around the community and tweaked and expanded by all. Unlike these other programmers, though, and unlike even the original creators of BASIC, Gates and Allen decided to license and own *their* particular version of BASIC. And in doing so, they created a tempest that was to linger for decades.

The seat of the controversy was the Homebrew Computing Club, an amateur computing club formed in a Menlo Park garage. With early members such as Steve Wozniak and Steve Jobs (who would go on to found Apple Computers), the

Homebrew Club was a group of ardent computer enthusiasts—engineers, gadget buffs, hobbyists, tinkerers, and anyone
else entranced by the prospects of computing. Among the
members of this community, the ethic of cooperation was paramount. "By sharing our experience and exchanging tips,"
wrote its founder, "we advance the state of the art and make
low-cost computing possible for more folks."[17] As the Altair
and other microcomputers became commercially available,
groups such as Homebrew multiplied around the country, fueled by their members' belief that the era of personal computing—the era when individuals could seize the power that once
belonged only to governments and corporations—had at last
arrived.[18] Thus, when members of the Homebrew Club found
a copy of Gates's BASIC code for the Altair, they liberally
copied and distributed it. Gates saw the impact on his licensing
revenues and was incensed. In a blistering open letter to the
Homebrew Club, he raged against what he defined as theft.
"Who can work for nothing?" he rhetorically asked. "What
hobbyist is ready to spend three years of his life programming,
finding bugs, and documenting his program, only to have it
freely distributed?"[19] After three hundred hobbyists responded
to Gates's letter with their own passionate words, the twenty-
one-year-old chairman of Microsoft relented somewhat, ac-
knowledging in a second open letter that not all hobbyists were
thieves and proposing some modest technical fixes for piracy.
But the battle itself remained fiery. To members of the Homebrew Club, Microsoft was a pirate. It had taken something that
belonged to the public, something scratched out by hundreds
of people over several years, and had grabbed this common
property for its private use. From Microsoft's perspective, however, the Homebrewers were the pirates, stealing property that
rightfully and legally belonged to their firm. It was a classic
standoff along the technological frontier; a moment when
property rights were defined and pirates morphed smoothly
into pioneers.

---

[17]Quoted in Levy, *Hackers*, pp. 207–208.
[18]The spirit and ethic of this time are beautifully captured in Levy, *Hackers*.
[19]Quoted in Ichbiah and Knepper, *The Making of Microsoft*, p. 31.

Matters came to a head in 1977, when Ed Roberts sold his now-flourishing business to a larger California firm. Realizing the value of the BASIC embedded in each Altair, this new firm claimed that they owned the programs that Gates and Allen had created and refused to license them to any other manufacturer. Gates, predictably, raged again. Arguing that the Altair's version of BASIC belonged to Microsoft, he fought the company in a messy six-month legal battle and won. Under the terms of settlement, Microsoft was allowed to market its BASIC however it liked, while the California company, Pertec, was found guilty of "business piracy." From this point on, the legal lines of battle were clear. Software was private property and Microsoft controlled the BASIC programs that Gates and Allen had written.

For the next several years, Microsoft flourished in the market it had helped to create. Between 1977 and 1981, the company licensed its software to a growing legion of personal computer firms: Apple, Commodore, Tandy. Each of these firms used a different version of Microsoft software, a version written specifically for its machine and then included as an integral part of its computer package. By the beginning of 1979, Microsoft had made its first million dollars and moved its corporate headquarters back to Seattle. The company's major breakthrough, though, came in 1980, when it signed a now-legendary deal with IBM, the world's largest manufacturer of mainframe computers. In the early days of the computer revolution, IBM had remained confidently aloof from the personal computer market, disdaining these "mini" machines as the cheap plaything of hobbyists. Once the market had shown its incumbent strength, however, and firms such as Apple and Tandy were beginning to clamber into this space, IBM realized that there was a major market for PCs. In the summer of 1980, therefore, the giant computer firm approached Microsoft and asked them to develop some software.[20]

The IBM project was bigger than anything Microsoft had ever contemplated. For IBM executives didn't just want another

[20]The story of IBM's relationship with Microsoft and making of the IBM PC has spawned a significant literature. See for example Paul Carroll, *Big Blues: The Unmaking of IBM* (New York: Crown Publishers, 1993).

version of BASIC or a particular software application; they wanted a full-fledged system, an underlying computer code that would sit at the heart of their new line of personal computers and allow users to run a whole range of programs atop of it. They wanted something seamless and invisible, something for people who had no desire to program their own computers. In short, they wanted something that would take the personal computer out of the hobbyist and engineer market and put it right into the mainstream. It was a strategy that meshed perfectly with Microsoft's own ambitions and intent. The only problem, though, was that Microsoft didn't have the kind of operating system that IBM needed[21]; it had access to programming languages, such as BASIC, that translated keyboard commands into computer responses, but it didn't have the inner code, the part that directly controls the workings of the microprocessor.[22] And so to meet IBM's terms, Microsoft arranged to purchase QDOS, a "Quick and Dirty Operating System" designed by an outside programmer named Tim Paterson. Although full details of the transaction remain a hotly contested secret, it appears that Microsoft paid Paterson's company about $100,000 for QDOS.[23]

[21]At the time, Microsoft was producing some operating systems, but it didn't have the rights to sell or license the CP/M system, which it used for some Apple products; and the XENIX system, which it could license, wasn't appropriate for IBM. See the description in Wallace and Erickson, *Hard Drive*, pp. 170–82; and Stephen Manes and Paul Andrews, *Gates: How Microsoft's Mogul Reinvented the Industry and Made Himself the Richest Man in America* (New York: Doubleday, 1993), pp. 156–57.

[22]There are three basic kinds, or layers, of computer software. The innermost layer, the operating system, controls the most basic tasks, such as establishing how data are handled and stored. The second layer is the programming language, which enables the user to communicate with the computer using a nonmathematical language. The final layer is the application, the specific program that the user chooses to perform a task such as word processing or bookkeeping. For most users, the only real interaction comes at the application level, since the first two levels are already deeply embedded in the machine. For more on these distinctions and the technology behind them see Tom Forester, *High Tech Society: The Story of the Information Technology Revolution* (Cambridge, Mass.: The MIT Press, 1987), pp. 146–51; and Dennis Longley and Michael Shain, *The Dictionary of Information Technology* (New York: Oxford University Press, 1986).

[23]This figure is from Ichbiah and Knepper, *The Making of Microsoft*, p. 76. In *Hard Drive*, Wallace and Erickson contend that the total figure was closer to $50,000. (See *Hard Drive*, pp. 194–95, 202–204.)

The IBM PC made its commercial debut in October of 1981. Although critics had argued that the machine would never sell—that its corporate origins wouldn't mesh with the free-wheeling spirit of personal computing—the PC was an almost instant success. In the first year of production, IBM sold an estimated 250 thousand PCs, each one linked to Microsoft's recently named MS-DOS system.[24] There were hundreds, and then thousands, of other sales too, from "IBM compatibles" made by smaller computer companies and from a growing range of applications (for word processing, data processing, and so forth) that Microsoft programmers had created to sit on top of MS-DOS. As the computer revolution churned and expanded, therefore, MS-DOS became the brains behind a wide range of personal computers, the code that powered even competing machines. And Microsoft owned that code. In 1986, Microsoft earned $197 million in revenue, half of which came from MS-DOS.[25]

Setting Standards, Fighting Suits

After the birth of the IBM PC, Microsoft enjoyed nearly two decades of spectacular prosperity. With DOS firmly established as the industry's standard operating system, the company's revenues grew at a nearly geometric pace, increasing from $24 million in 1982 to over $1 billion by 1990. Its market capitalization increased accordingly, so that the company was valued in 1991 at $21.9 billion. Secretaries who had joined Microsoft in the 1980s became fabulously rich from their Microsoft

[24]The original IBM PC came with a choice of three operating systems: CP/M-86, a PASCAL-based system from the University of California, and Microsoft's MS-DOS. Because CP/M-86 was not fully available at the time of launch and few customers were interested in the PASCAL version, DOS quickly became the dominant operating system. See Paul Ceruzzi, *A History of Modern Computing* (Cambridge, Mass.: The MIT Press, 1998), p. 268; and Sol Libes, "Bytelines," *Byte*, November 1982, p. 540.
[25]Ichbiah and Knepper, *The Making of Microsoft*, p. 93; Michael A. Cusumano and Richard W. Selby, *Microsoft Secrets: How the World's Most Powerful Software Company Creates Technology, Shapes Markets, and Manages People* (New York: The Free Press, 1995), p. 3.

stock, and Bill Gates alone was worth a staggering $7 billion by 1992.[26] "Microsoft millionaires" cluttered the offices of Seattle and Microsoft—the firm started on a whim and a lie—had become a household word.

The history of Microsoft in the 1980s and 1990s reads like a history of the personal computing revolution. In September of 1983, Microsoft launched Word, the first word processing program that allowed users to change the style and font of their documents and the first to be controlled with a hand-held mouse. In 1984, the company helped to create the user-friendly applications software that would distinguish the Apple Macintosh (the "computer for the rest of us") from the still more awkward PC models.[27] And then in 1985 it released Windows, software that replicated the look and feel of the friendlier Apple for the lower-priced and more pervasive PC. By 1987, Windows had shipped more than a million copies and earned revenues of over $345 million.

Throughout this period, Microsoft's power and revenues drew from the unique relationship that licensing had created. Once the IBM PC became the world's best-selling computer, MS-DOS became, almost by default, the standard operating system for personal computers. This meant that every computer manufacturer who wanted to make an IBM-compatible machine had to license from Microsoft, and that everyone who bought an IBM or IBM-compatible was eligible to purchase applications software from Microsoft as well. When Microsoft teamed up with Apple in 1985, the net grew even denser, giving Microsoft a desktop presence in nearly every personal computer produced in the United States. It was this overwhelming pres-

---

[26]Since much of Gates's wealth is based on his holdings of Microsoft stock, his net worth clearly varies along with Microsoft's stock price. The $7 billion figure is from Wallace and Erickson, *Hard Drive,* p. 410.

[27]For more on this story and Microsoft's often-stormy relationship with Apple, see Jim Carlton, *Apple: The Inside Story of Intrigue, Egomania and Business Blunders* (New York: Times Business/Random House, 1997), pp. 40–61, 131–36, 273–79; Robert Cringely, *Accidental Empire: How the Boys of Silicon Valley Make Their Millions, Battle Foreign Competition, and Still Can't Get a Date* (New York: Addison-Wesley, 1992), pp. 226–29; and Owen Linzmayer, *Apple Confidential: The Real Story of Apple Computer, Inc.* (San Francisco: No Starch Press, 1999), pp. 133–42, 193–98, 238–39.

ence in the industry—this ability to carve out and control a
single standard—that made Microsoft such a phenomenally
successful company. But at the same time, it also caused resentment, especially among those who saw Microsoft as a pirate
rather than a pioneer, and who felt that their own fortunes had
been compromised by Gates's standard-setting ways.

The first to bristle at Microsoft's power was Apple, the upstart computer firm launched by former members of the Homebrew Club. Between 1982 and 1988, Apple and Microsoft had
enjoyed a profitable though stormy relationship. Microsoft provided Apple with most of its most popular software—primarily
Word and Excel—while sales of Apple-based products, in turn,
helped to drive Microsoft's revenues. In 1987, software for
Apple's Macintosh computer accounted for nearly half of
Microsoft's total revenues.[28] But in 1985, Microsoft also introduced Windows, the path-breaking software that allowed PC
users to interact much more freely with their machines. With
Windows, users didn't have to give their computers obscure
keyboard codes; they didn't have to deal with prompt signs or
commands that hearkened from the earlier days of computing.
They simply had to point and click, using a mouse and a visual
display of icons to navigate around the machine. Which, of
course, was precisely what Apple was already offering. Indeed,
the whole breakthrough appeal of the Apple—the "machine for
the rest of us"—was that it eliminated the high-tech clunkiness
of the PC and its clones. What Microsoft did with Windows was
precisely what Apple had already done with its entire line of
computers. And Apple, not surprisingly, was not pleased.

In March of 1988, lawyers for Apple Computer filed a copyright suit in San Jose federal court accusing Microsoft of stealing visual display features from Apple's Macintosh line.
Although Apple executives acknowledged that they had granted
Microsoft a limited license to employ some "Macintosh-like
features" in their own software, they argued that Microsoft's most
recent version of Windows—version 2.03—took this borrowing

---

[28]Martin Campbell-Kelly and William Aspray, *Computer: A History of the Information
Machine* (New York: Basic Books, 1996), p. 276.

way too far. Windows 2.03, Apple maintained, was an illegal copy of the "look and feel" of the Macintosh interface. It wasn't innovation; it was piracy.

Apple's suit against Microsoft stirred up a rash of issues: issues about the ownership of design elements, the risks of licensing, and the applicability of copyright law to a field for which it was never intended. But at the heart of the suit was the soul of the old Homebrew controversy. According to Apple, Microsoft had once again appropriated intellectual property and claimed it as its own. The company had stolen the essence of the Apple interface and then used an arsenal of rough tactics— "lies and threats," according to the suit—to push the market toward its new standard. It was the BASIC argument all over again. Only this time the stakes were substantially bigger and Gates was considerably more feared.

In July of 1989, the judge in San Jose reduced the scope of Apple's complaint, removing all but ten of the 189 violations that Apple had claimed.[29] Three years later, the U.S. District Court for Northern California ruled in favor of Microsoft, dismissing all copyright infringement claims against the company. It was a major victory for Microsoft and a painful blow in what by this point appeared to be the slow decline of Apple. But the underlying concerns about Microsoft's conduct remained alive, and they continued to dog the company.

On March 12, 1991, for example, Microsoft admitted that it was under investigation by the U.S. Federal Trade Commission (FTC) for possible violation of antitrust. Although the probe began with a relatively narrow question (whether IBM and Microsoft were colluding to create a new operating system standard), it rapidly widened to include a much broader stretch of Microsoft's business practice. Bolstered by tips and requests from across the computer industry, the FTC began to examine whether Microsoft was monopolizing entire pockets of the industry: the market for operating systems, for operating environments (such as visual interfaces), for computer software, and for consumer peripherals. At the same time, Britain's Office of Fair

---

[29]Apple eventually asked the Supreme Court to review the case, but on February 21, 1995, the Court declined.

Trading launched its own investigation, following local complaints about how Microsoft priced its licensing deals.[30]

As is often the case with antitrust suits, all of these allegations were backed and even prodded by other computing and software firms. By the time that the FTC and EU launched their investigations, they were responding to what had become a virtual cacophony of anti-Microsoft sentiment. Inside the computer industry, hatred—or at least resentment, of Microsoft had become a sort of fraternal bond. Quietly, often anonymously, executives were venting their anger at the computer giant, accusing Microsoft of using its massive clout to squeeze customers and crush competitors. Critics were particularly harsh on Gates, lambasting him for a record of unfair tactics and a now-legendary habit of yelling, screaming, and outwitting others until he eventually got his way. In March of 1991, *Business Month* magazine ran a cover story on Gates, headlined "The Silicon Muscleman." "[I]s Gates really the kind of guy," it asked, "we can trust to carry on our revolution?"[31] At a software industry dinner, Gates was actually booed by his peers; and several competitors had begun to grumble publicly about illegal, or at least unethical, behavior. Although few industry insiders were willing to speak on the record about either Gates or Microsoft, those who did were blunt. "Gates has clearly won," proclaimed Mitch Kapor, founder of competing software firm Lotus Development Corporation. "[T]he revolution is over, and freewheeling innovation in the software industry has ground to a halt. For me it's the Kingdom of the Dead."[32]

Was it just jealousy driving these reactions? Perhaps. In 1990, Microsoft had become the first personal computer software company to exceed $1 billion of sales in a single year. The company was considerably more profitable than any of its competitors and its software dominated the market. Yet there was more to Microsoft than money, it appeared, and a more deeply seated hostility toward Gates. People resented the firm not because it came up with successful products (which it did) and

[30]Philip Robinson, "OFT to Study Microsoft Licensing Deals in UK," *The Times,* December 22, 1992, p. 18.
[31]James S. Henry, "The Silicon Muscleman," *Business Month,* November 1990, p. 31.
[32]Quoted in Henry, "The Silicon Muscleman," p. 31.

not necessarily because it charged high prices (which it arguably did not). They resented it because Microsoft had become far too powerfully established along the technological frontier, far too influential in setting standards for a market that, admittedly, it had helped to create. It was a classic political struggle along the frontier, and Microsoft was at the forefront.

Eventually, this round of the battle ground to a draw. In 1994, the U.S. Justice Department asked Microsoft to sign a consent decree—a legal agreement that settled the case but bound the company to a prescribed list of obligations. Under the decree's rather mild terms, Microsoft was prohibited from "tying" separate products to the distribution of Windows. This meant that the company could no longer sell its operating system "tied" or "bundled" with other pieces of software, a practice that industry critics had long labeled as anti-competitive. It could, however, continue to develop what the decree labeled "integrated products." Microsoft also pledged not to enter into any licensing agreements that were conditioned upon either the licensing of any other Microsoft product or a guarantee that the customer would *not* license or purchase non-Microsoft products.[33] In other words, the company promised (implicitly at least) not to use its licensing agreements as anti-competitive threats or weapons. On the same day that it agreed to the Justice Department's terms, Microsoft signed a separate agreement with the European Union, consenting to an identical set of conditions.

Microsoft Enters the Net

Microsoft might have been excused in 1995 for assuming that its legal troubles were over. The firm, after all, had been engaged in suits and investigations since 1988 and had never actually been indicted. With the signed consent decrees, it looked as if Microsoft's position in the computer industry—its standard-setting role, its dominant market share, its rough-and-

[33]In a related provision, Microsoft also agreed to stop tying its royalties to the number of computers that a licensee shipped, and to base fees instead on the number of copies of software sold or shipped. See *United States v. Microsoft Corporation* 312 U.S. App. D.C. Civ. No. 94-1564; 56 FR 42845 (1994).

tumble style of business—might simply have been accepted as a natural part of the landscape. Yet nothing, as it turns out, could have been further from the truth.

Sometime in the middle of 1995, Gates and senior management at Microsoft stumbled upon the Internet. After several years of nodding vaguely toward the Net, several years of focusing the bulk of their energies on more pressing concerns such as the launch of Windows 95, executives slowly began to realize that the web was there and that it was big. Until this point, the company had basically adopted a wait-and-see attitude. As the Net tumbled into prominence, they concentrated their energies where they had always been—on software—and treated the Net much as IBM had once treated the PC: it was cool, it was new, but it wasn't a commercial space. Even when customers began to demand that Microsoft include TCP/IP (the Net communication format) in their more advanced software products, the company's response, recalls one executive, was, "I don't know what it is. I don't want to know what it is. My customers are screaming about it. Make the pain go away."[34] So the company hired a few young engineers and allowed them to tinker with the technologies. But there was no real excitement about the Net and precious little support from upper management.

In 1994, however, the birth of a firm called Netscape grabbed Microsoft by surprise. Founded by Jim Clark and the twenty-two-year-old Marc Andreessen, Netscape was a classic start-up. It had no money, no reputation, and one truly ingenious product: a web browser formerly known as Mosaic. Within a year of its creation, Netscape had stormed across the emerging Internet, concocting a company valued at $7 billion and creating a market for browsers that it quickly and easily dominated.[35] It also scared the smirk out of Microsoft.

To understand the depth of this fear, it is necessary to consider how a browser operates and what Mosaic means. When Marc Andreessen was at the University of Illinois, he and some

---

[34]J. Allard, quoted in Kathy Rebello, "Inside Microsoft: The Untold Story of How the Internet Forced Bill Gates to Reverse Course," *BusinessWeek,* July 15, 1996, p. 59.
[35]See the discussion in Michael A. Cusumano and David B. Yoffie, *Competing on Internet Time: Lessons from Netscape and Its Battle with Microsoft* (New York: The Free Press, 1998), p. 7.

classmates spent their days tinkering with what was about to become the Net. In 1989, Tim Berners-Lee, a British researcher at CERN (the European Laboratory for Particle Physics), had developed three breakthrough techniques for digital communication. He created HTML (Hypertext Markup Language), a simple language for laying out pages of text; HTTP (Hypertext Transfer Protocol), a system for linking documents; and URLs (Uniform Resource Locators), a scheme for addressing and locating specific nodes of information. Together, these three innovations transformed the landscape of communication, allowing information to flow much more easily across computer networks and raising the possibility that a new crop of users— people beyond the scientific or academic community—might be tempted online as well. Intrigued by this potential, Andreessen and some friends began to think about the next stage: how to hook graphics and multimedia into this emerging network and run it across a much wider set of computers. Their answer was Mosaic, the first Internet browser.[36]

The beauty of Mosaic, and indeed of any Net browser, was that it enabled web users to move seamlessly between far-flung corners of the Net. It was the browser that linked online sites into a connected mesh of information, that sifted through the information and presented it in a friendly, logical, visual map. The browser, in short, let people surf. And it transformed the web.

In April of 1994, Andreessen left his first job in Silicon Valley to launch Netscape Communications Corporation.[37] After teaming up with Jim Clark, the founder of Silicon Graphics, Andreessen released a revised version of Mosaic called Netscape Navigator in December 1994. It was an instant success. By April 1995, the company had distributed six million versions of Navigator, capturing over 75 percent of the now-exploding market.[38] The following year, its market share hit 90 percent and revenues climbed to $346 million.

[36]For more on this period of innovation, see Joshua Quittner and Michelle Slatalla, *Speeding the Net: The Inside Story of Netscape and How It Challenged Microsoft* (New York: Atlantic Monthly Press, 1998), pp. 23–63; and Robert H. Reid, *Architects of the Web* (New York: John Wiley & Sons, Inc., 1997) pp. 1–68.
[37]Briefly called Electric Media and then Mosaic Communications Corporation.
[38]Quittner and Slatalla, *Speeding the Net*, p. 174.

By this stage, the folks at Microsoft had begun to notice. To be sure, individual voices within the firm had been pointing excitedly at the Internet for years. In 1994, for example, one top-level executive argued that Microsoft needed to "'embrace' Internet standards and 'extend' Windows to the Net"; several months later, Gates himself had outlined his commitment to the Net and pledged to invest in it.[39] But the great company moved slowly in this case, and it wasn't until Netscape stormed onto the market that Microsoft truly took note.[40] When they did begin to move, however, things happened quickly.

On December 7, 1995, Bill Gates convened an all-day presentation for analysts, journalists, and leading Microsoft customers. Proclaiming that Microsoft was now "hard core" about the Net, he announced that the company was planning to do whatever it took to dominate the Internet market.[41] Microsoft, Gates promised, would make web browsers and web servers an integral part of its business; it would "webize" existing Microsoft programs and license Java, a new web-based programming language created by Sun. In a direct blow to Netscape, Gates also pledged that Microsoft would distribute all of its web browsers and Internet servers for free. (Netscape, by comparison, offered free distribution only to noncommercial users.) The symbolism of this announcement, coming as it did on Pearl Harbor Day, was not lost on the event's attendees. From this point on, Microsoft was primed for attack.

Over the next several months, the outline of the battle plan grew clearer. In February 1996, Microsoft created an Internet Platform and Tools Division, a twenty-five-hundred–person group charged with overseeing Net projects. This group also took charge of Internet Explorer, a web browser designed to compete with Netscape. When version 3.0 of Explorer debuted in August, it looked and felt remarkably like Netscape's Navigator, complete with bright colors, easy-to-use buttons, and high-quality

---

[39]See Rebello, "Inside Microsoft," pp. 56–67.
[40]There is a considerable literature exploring why Microsoft was relatively slow to the come to the Net, and how they responded to the sudden emergence of Netscape. See for example James Wallace, *Overdrive: Bill Gates and the Race to Control Cyberspace* (New York: John Wiley & Sons, 1997); and Reid, *Architects of the Web*, pp. 60–68.
[41]Rebello, "Inside Microsoft," p. 56.

sound.[42] "The current path," wrote one senior Microsoft executive to another, "is simply to copy everything Netscape does."[43] Microsoft announced plans to integrate Explorer into future versions of Windows and to develop its own cache of online content. At a gathering of software developers in the spring of 1996, a Microsoft representative made his company's position clear: "Our intent," he reportedly announced, "is to flood the market with free Internet software and squeeze Netscape until they run out of cash."[44]

More controversially, Microsoft also allegedly used its own comfortable position to tighten the reins on Netscape. During the course of 1995, Microsoft tried to take the easiest course into Internet browsers. It tried, that is, to invest in Netscape, or at least to reach a commercial understanding with the upstart firm. When these talks floundered, Microsoft launched an all-out assault on Netscape and the browser market. In 1996, for example, an internal memo directed Microsoft negotiators to "break most of Netscape's licensing deals, and return them to our advantage." Company executives then went after some of Netscape's most influential and profitable customers, convincing them, as the memo suggested, "that it does not make any sense to buy Netscape Navigator."[45] To accentuate this argument, they used a combination of threats and inducements, taking back an account with KPMG, for instance, in exchange for a basket of services and $10 million, and suggesting to Compaq Computer that its Windows license would be revoked if Compaq were to replace Internet Explorer with Netscape Navigator.[46] In all these situations, Microsoft's intent was presumably the same: to make its browser the industry standard and wipe Netscape off the web. Could Microsoft have adopted instead a

[42]See Joshua Cooper Ramo, "Winner Take All," *Time*, September 16, 1996, pp. 57–64.
[43]Quoted in Steve Lohr and John Markoff, "Why Microsoft Is Taking a Hard Line with the Government," *New York Times*, January 12, 1998, p. D1.
[44]This statement was originally repeated in a letter sent by Netscape to the Department of Justice, so it may bear some of the bias of its source. See Joshua Cooper Ramo, "Winner Take All," p. 64.
[45]Cited in Cusumano and Yoffie, *Competing on Internet Time*, p. 146.
[46]See Steve Lohr and John Markoff, "How Software's Giant Played a Hardball Game," *New York Times*, October 8, 1998, p. A1; and Cusumano and Yoffie, *Competing on Internet Time*, pp. 145–46.

live-and-let-live attitude? Could it have coexisted more peaceably with Netscape? It's not clear. For if Netscape or any other browser became the default platform for the Net, there was a real possibility that the kind of software that Microsoft produced—operating systems, interfaces, and applications—would eventually be replaced by a menu of web-based options. If a user, for example, could boot up a different kind of computer, access the web via Netscape, and then download a particular bit of software, he or she might never need a Microsoft-type program. This was the fear that drove Microsoft and that made the market for browsers, potentially at least, such an all-or-nothing game. Or as a senior executive at Microsoft put it, "if there was ever a bullet with Microsoft's name on it, [Navigator] is it."[47]

On March 12, 1996, Microsoft stole a major round of this struggle. It offered America Online what a lawyer for the Justice Department later described as a "bribe": a prominent place on the Windows desktop in exchange for making Internet Explorer the preferred browser for AOL users.[48] Eventually, then, every AOL subscriber (and there were five million in 1996) became a de facto Explorer, and Microsoft took a giant step toward establishing yet another industry standard. For Netscape, it was devastating. As Microsoft became increasingly dominant in the market, Andreessen was caught in a downward spiral of deals. Would-be partners began to shy away from the firm, and customers who had lapped up Netscape's Navigator just a year earlier began migrating slowly back to Microsoft. By 1998, Netscape's share of the browser market had dropped by 50 percent[49] and the firm began to distribute all its browsers for free, "killing," as one article acknowledged, "what had been a $180 million business in 1996."[50] Summarizing this state of affairs, a disgruntled James Barksdale, CEO of Netscape, complained, "I

[47]Quoted in Cusumano and Yoffie, *Competing on Internet Time*, p. 145. For more on this argument, see also Jonathan Zittrain, "The Un-Microsoft Un-Remedy: Law Can Prevent the Problem that It Can't Patch Later," *Connecticut Law Review* 31, no. 4 (Summer 1999), pp. 1361–74.
[48]Correspondence with author, October 16, 2000.
[49]Cusumano and Yoffie, *Competing on Internet Time*, p. 10.
[50]Quoted in Andrew Pollack, "Netscape Plays it Cool as Rival is Sued," *New York Times*, May 25, 199, p. D1.

don't have any browser business . . . It was half of my revenues a year ago; it's none now."[51]

Meanwhile, other forces in the computing and software industries were also beginning to rally against what they perceived as a Microsoft attack. Together, companies such as IBM, Sun, and Oracle, all of whom had felt besieged in the past by Microsoft's aggressive tactics, cobbled together a quiet but powerful alliance explicitly determined *not* to allow the next generation of computing standards to rely on Microsoft products. Instead, these firms wanted to use the innate capacity of the Internet to move software away from the personal computer and toward the more open and less proprietary web. They wanted the old model of software—housed in the computer, licensed by Microsoft—to be replaced by a new and more dynamic one, based this time on downloads from the Net and coded in Java, Sun's new programming language. In this model, there was no shrink-wrapped software, no bundled computers and, most important, no Microsoft. Describing the group's ambition, Oracle chairman Lawrence Ellison was particularly blunt. "The whole Internet," he asserted, "is part of a conspiracy to get Microsoft."[52] In 1996, members of this group contributed $100,000 to a fund that would help start-up firms challenge Microsoft; in 1998, they helped the Software Publishers Association to draft a set of "competition principles" that clearly targeted Microsoft's behavior.[53]

By 1998, then, teams of feuding giants had already descended upon the infant Net. Microsoft was driving to reestablish the dominance it had already achieved in the world of personal computing; Sun and its allies were working to create a new, common standard; and start-ups such as Netscape were determined to wrest power from all of the existing players. It

---

[51]Ibid.
[52]Quoted in David Bank, "Uneasy Bedfellows: An Alliance of Rivals Sets Out to Contain Microsoft Hegemony," *Wall Street Journal Europe*, November 1, 1998, p. 1. For more on the alliance, see David Bank, "Noise Inhibitor: Rivals of Microsoft Find Collaboration Is Easier Said than Done," *Wall Street Journal*, November 19, 1998, p. A1.
[53]Both examples are from Cusumano and Yoffie, *Competing on Internet Time*, p. 135. See also John Markoff, "Plan Offered to Cut Power of Microsoft," *New York Times*, April 7, 1998, p. D1.

was a tumultuous and chaotic time, a classic movement along the technological frontier. And then the U.S. Justice Department came along.

Mr. Gates Takes the Stand

On May 18, 1998, the U.S. Department of Justice joined twenty state governments in launching one of the widest antitrust cases of the late twentieth century. According to U.S. Attorney General Janet Reno, Microsoft had used its monopoly power in the market for operating systems to control the emerging market for web browsers and other Internet products.[54] It had "restricted the choices available for consumers in America,"[55] behavior that was patently illegal under the 1890 Sherman Antitrust Act.

Although the breadth of the suit was undoubtedly unique, *United States v. Microsoft* was in many respects just the logical culmination of the legal and societal issues that had long dogged Microsoft. Indeed, the central contention of the government's suit—that Microsoft was a monopolist that used its power unfairly—was achingly similar to what Apple had alleged in its 1988 suit, what Sun and Oracle were allied against, and what members of the Homebrew Club had asserted more than twenty years before. Throughout the industry, Microsoft was perceived as playing far too great a role in setting the standards upon which everyone else relied. The legal question, however, was whether Microsoft's behavior constituted a legal violation: was Microsoft a pirate firm that employed illegitimate means, or a hard-working pioneer whose success inspired envy? Had it violated the law, or simply operated in an area where the law had not yet caught up to the market? As with many such distinctions, the devil lay in the definition.

The first issue at stake dealt with the practice of bundling. In 1997, the U.S. Justice Department had accused Microsoft of

---

[54]Joel Brinkley, "U.S. and 20 States File Suits Claiming Microsoft Blocks Competition over Internet," *New York Times,* May 19, 1998, p. A1.
[55]U.S. Attorney General Janet Reno, quoted in Brinkley, "U.S. and 20 States File Suits," p. A1.

violating the 1994 consent decree by tying its web browser, Internet Explorer, into its Windows operating system. According to the Justice Department, once Microsoft sold the browser as an integral part of Windows it was using its position in one market (operating systems) to muscle its way into a new one (web browsers). Commercially, they argued, bundling was unnecessary, since any computer manufacturer could combine Windows and Internet Explorer by themselves. Legally, it was an explicit violation of the 1994 consent decree, which prohibited Microsoft from:

> enter[ing] into any License Agreement in which the terms of that agreement are expressly or impliedly conditioned upon . . . the licensing of any other Covered Product [or] Operating System Software product.[56]

Lawyers for Microsoft, however, pointed to the next sentence of the consent decree, inserted in 1995 at the personal insistence of Bill Gates, which added that "this provision in and of itself shall not be construed to prohibit Microsoft from developing integrated products."[57] So which one was it? Was Microsoft's bundling an illegal attempt to extend its market dominance, or a legitimate piece of marketing? It depended on how the practice and the markets were defined. And this, in 1997, was unclear. In an initial ruling, the soon-to-be-famous Judge Jackson agreed that bundling was inappropriate under the circumstances and ordered Microsoft to separate Internet Explorer from Windows. But Microsoft insisted that if it removed Explorer, it would disable Windows. What the government had asked for, argued Microsoft's lead lawyer, "doesn't work. The reason it doesn't work is that this is an integrated product and it performs as an integrated product."[58] Stubbornly, Microsoft refused to comply with Judge Jackson's request.

Two days later, in a highly publicized standoff, Jackson brought two new computers into his chambers and booted them up. He and a technical assistant then proceeded to use the

[56]See *United States of America v. Microsoft Corporation* 980 F. Supp. 537 (1997).
[57]*United States of America v. Microsoft Corporation* 980 F. Supp. 537 (1997). See also Lohr and Markoff, "Why Microsoft is Taking a Hard Line," p. D1.
[58]Stephen Labaton, "U.S. Assails Microsoft and Seeks New Oversight Role," *New York Times,* December 18, 1997, p. D1.

"add/delete" function to remove Explorer from Windows, causing no apparent harm to the operating system. Case closed, he implied. However, although Microsoft's lawyers remained quiet after Jackson announced the results of his test, company representatives later explained that the judge hadn't really removed the program, but had only hidden it, leaving "97 percent of the code still on the machine."[59] So once again, the crucial definitions were cloudy. If Judge Jackson had only meant for Microsoft to hide their icon, then the company was in contempt of the court. But if he was asking them to re-engineer the basic code in order to remove Explorer fully, he might have been demanding something that was either technically infeasible or commercially unfair. At the end of this initial round, therefore, the key issues around bundling were left ambiguous. The federal court ruled against Microsoft, an appeals court subsequently reversed their decision, and the issue of bundling was then rolled into the broader antitrust suit.

A second issue concerned Microsoft's alleged attempts to divide the booming Internet market between itself and Netscape. According to the Justice Department, Microsoft had approached Netscape early on in their rivalry, trying to persuade the newborn company to produce its browsers only for non-Windows machines. Their initial approach, as Marc Andreessen recalled it, was "like a visit by Don Corleone . . . I expected to find a bloody computer monitor in my bed the next day."[60] If Netscape refused to cooperate with Microsoft, Andreessen felt, the larger company would refuse to give his firm the access they needed to develop Netscape Navigator for Windows. Microsoft, on the other hand, denied anything but a technical interest in Navigator and insisted that it was Netscape that actually proposed their meeting. "I never instructed any Microsoft personnel to seek a 'division of the market,'" testified Paul Maritz, Microsoft's vice-president for software development and marketing, "nor do I believe any such proposal was ever made."[61]

[59]Stephen Labaton, "A Few Clicks, and Microsoft Has a Problem," *New York Times,* December 20, 1997, p. A1.
[60]John R. Wilke, "Microsoft Subject of New Antitrust Probe," *Wall Street Journal,* April 24, 1998, p. A2.
[61]Quoted in Steve Lohr, "Microsoft Executive's Testimony Attacks Accusers," *New York Times,* January 23, 1999, p. C1.

Part of what defined this issue, then, was a simple matter of truth: who was telling it, and who was lying. But part of its also dwelled in another area of newfound ambiguity. If Microsoft had directly asked Netscape to divide the market for browsers into Windows and non-Windows machines, it had clearly violated U.S. laws that prohibited market allocation. However, if it had only refused to give Netscape the technical information it needed to build a Windows-friendly browser, the legal implications were less clear. Could Microsoft violate antitrust law simply by refusing to license property that it legally controlled? And if so, then what did this mean for a whole industry based on networked standards and proprietary technology?

A third issue was implicit in the Don Corleone story and explicit in the wording of the case. Over and over again, Microsoft had been accused of playing meanly—of squeezing suppliers, pressuring customers, and killing its competition. Now these whispered allegations had been gathered into a formal complaint, an argument that Microsoft had crossed the boundary of acceptable behavior into the realm of illegality.[62] "What cannot be tolerated," asserted Assistant U.S. Attorney Joel Klein, "and what the antitrust laws forbid—is the barrage of illegal, anticompetitive practices that Microsoft uses to destroy its rivals and to avoid competition."[63] Microsoft maintained, however, that its conduct was wholly ordinary and the case a travesty of justice. "This is a step backwards for America," stormed Bill Gates, "for consumers, and for the personal-computer industry that is leading our nation's economy into the 21st century. How ironic that in the United States—where freedom and innovation are core values—that these regulators are trying to punish an American company that has worked hard and successfully to deliver on these values."[64] Microsoft wasn't about anti-competitive behavior, Gates and his lieutenants stressed,

---

[62]In the original case, filed on May 18, 1998, the suit was focused primarily on the complaint that Microsoft had used its position in the market for operating systems to affect the market for browsers. It was gradually extended, though, to include the substantially broader charge of anticompetitive behavior. See Steve Lohr, "Microsoft On Trial: The Overview," *New York Times,* October 19, 1998, p. C1.

[63]Quoted in Joel Brinkley, "U.S. and 20 States File Suits," p. A1.

[64]Ibid.

but rather about the very epitome of *competitive* behavior; about seizing new technologies, building new products, and creating mass markets that enhanced consumer welfare. If this were illegal, they queried, how could commerce ever grow?

Hearings in the case of *U.S. v. Microsoft* began on October 19, 1998.[65] The lead lawyer for the U.S. government was David Boies, a celebrated courtroom litigator who had spent thirteen years defending IBM in the government's last great antitrust battle. Microsoft's team was headed by William Neukom, the company's long-time legal counsel. During the course of the proceedings, which stretched for nearly a year, a veritable parade of high-tech heavyweights took the stand in Washington, including James Barksdale of Netscape, Intuit CEO William Harris, and Steven McGeady, vice president of Intel. Gates himself gave a videotaped deposition, fidgeting through twenty hours of questioning and cross-examination.

Although the case was closely watched and meticulously analyzed, its outcome was never clear. Lawyers lined up on both sides of the debate, economists argued vociferously about the market implications of Microsoft's behavior, and prophets from across the high-tech spectrum were loath to reach any sort of consensus. For example, according to Richard Schmalensee, an MIT economist who testified for the defense, the government's inability to demonstrate harm meant that Microsoft couldn't possibly have violated U.S. antitrust law. "Proper economic inquiry into a whether a company is engaged in anticompetitive conduct," he urged, "should end if it concludes that consumers have not been harmed by the conduct at issue."[66] Yet his colleague Franklin Fisher, also an economist at MIT, made precisely the opposite argument in his rebuttal testimony. Calling Schmalensee's testimony "peculiar," "ridiculous," and "credulous," Fisher argued that Microsoft's behavior confirmed its monopolistic intent. "Microsoft has shown," he insisted, "that it will decide the ways in which innovation takes place in this

---

[65]In order to speed up the case, Judge Jackson had ordered that direct testimony for both sides would be taken beforehand in written form. Therefore, the case began directly with cross-examinations.

[66]Quoted in Steve Lohr, "Microsoft Puts Its First Witness on the Stand," *New York Times*, January 12, 1999, C1.

industry, and that any innovation which threatens Microsoft's platform monopoly will be squashed. . . . We will live in a Microsoft world."[67]

Part of this confusion, of course, was the normal give-and-take of any trial, the arguments over what happened when and who did what to whom. But part—indeed a great part—was also a reflection of how new these issues were and how difficult they were to assess. Did Microsoft bundle its new web browser with its existing operating system? Incontrovertibly. Did this bundling help the company to establish a strong position in the emerging market for browsers? Yes. But is this behavior illegal or anti-competitive? It depends on how one defines things—on whether the market for browsing software is fundamentally separate from the market for operating system software, and whether consumer welfare is helped or harmed by having the two products provided by the same firm. During the course of the trial, Microsoft insisted that "browsing capability" had always been considered an essential part of Windows and thus that the linking of Windows and Internet Explorer was a technical and commercial move, rather than an attack on Netscape. In testimony before the court, for example, a senior Microsoft official pointed to numerous "capabilities that are only available through the deep integration of Internet technologies into the Windows operating system."[68] Another witness quoted from a published technology review, which had found that "[t]he incorporation of browsing and other Internet functions into Windows is a powerful innovation . . . [and] a big gain for consumers."[69] Lawyers for the Justice Department, however, countered by pointing to Microsoft's late Internet entry and its rather transparent fear of Netscape. If browsers were so important to Microsoft, they argued, why hadn't the firm moved earlier?

[67]United States District Court for the District of Columbia, Franklin Fisher's rebuttal testimony in *United States vs. Microsoft Corporation,* January 12, 1999. (http://www.microsoft.com/presspass/trial/transcripts/jan99/01–12-am.asp)

[68]James Allchin, quoted in Joel Brinkley, "U.S. Pushes to Get a Microsoft Defense to Boomerang," *New York Times,* February 2, 1999, p. C2.

[69]From Stephen H. Wildstrom, "Why I'm Rooting for Microsoft," *BusinessWeek,* February 23, 1998, p. 30. Cited by Richard Schmalensee and quoted in David S. Evans, "All the Facts that Fit: Square Pegs and Round Holes in *U.S. v. Microsoft,*" *Regulation* 22, no. 4 (1999), p. 56.

Returning to the earlier debate in Judge Jackson's court, they also stressed that browsers and operating systems were technically distinct. If they were inseparable in the most recent version of Windows, it was only because Microsoft had engineered things that way.

A similar uncertainty surrounded the entire debate over Microsoft's allegedly anti-competitive behavior. Throughout more than two months of testimony, most of the assembled executives repeated the Justice Department's refrain. A high-level executive at AOL, for example, claimed that Gates had once bluntly asked him, "How much do we need to pay you to screw Netscape?"[70] Sun's chief scientist complained that Microsoft had tailored its version of Java for Windows so that it became incompatible with Sun's standard Java language,[71] and executives from AOL acknowledged that they had switched from Netscape Navigator to Internet Explorer after Microsoft promised to give AOL a featured spot on the Windows desktop.[72] Yet beneath all the gory detail, the same basic question remained: were these the actions of a pirate or a pioneer? A firm that was creating markets or killing them?

In the end, or at least the beginning of the end, Judge Jackson sided vigorously with the Justice Department. On November 5, 1999, he issued his finding of facts, an initial review that would provide the framework for the court's eventual ruling. The conclusions were stark. First of all, Jackson determined that Microsoft did indeed enjoy monopoly power in the market for operating systems and that, "if it wished to exercise this power solely in terms of price, it could charge a price for Windows substantially above that which could be charged in a competitive market." Microsoft, in other words, was a monopolist. Second, despite Microsoft's argument that web browsing was just a feature of its operating system, the judge found unequivocally

[70]Quoted in Joel Brinkley, "As Microsoft Trial Gets Started, Gates's Credibility Is Questioned," *New York Times,* October 20, 1998, p. A1.
[71]Steve Lohr and Joel Brinkley, "Testimony on Microsoft's League of Its Own," *New York Times,* December 2, 1998, p. C2.
[72]See Steve Lohr, "The Prosecution Almost Rests in the Microsoft Trial," *New York Times,* January 8, 1999, p. C1. For more on the debate between competing fiercely and competing unfairly, see Steve Hamm, "Microsoft: No Letup—and No Apologies," *BusinessWeek,* October 26, 1998, p. 58–64.

that "Web browsers and operating systems are separate products." He also concluded that Microsoft had indeed used its
market power to affect at least one customer's decision about
whether or not to distribute Netscape. The only slight concession to Microsoft's position came with regard to its now-
infamous meeting with Netscape. While the government had
alleged that Microsoft called the meeting in order to divide the
browser market (illegally) between the two firms, Judge Jackson
was more opaque in his findings. "Although the discussions
ended before Microsoft was compelled to demarcate precisely
where the boundary between its platform and Netscape's applications would lie," he found, "it is unclear whether Netscape's
acceptance of Microsoft's proposal would have left the firm
with even the ability to survive as an independent business."[73]

Seven months later, in June of 2000, Judge Jackson issued
his final ruling. Echoing the conclusions from his earlier finding
of facts, Jackson found that Microsoft had indeed violated the
antitrust laws of the United States. He ordered Microsoft to
split its existing operations into two separate companies, one
dealing with operating systems and the other with applications.
Arguing that Microsoft had been "untrustworthy in the past,"
Jackson also imposed a long list of restrictions that would govern Microsoft's behavior until the breakup took place. The ruling, predicted a beaming Janet Reno, would have "a profound
impact, not only by promoting competition in the software industry, but also by reaffirming the importance of antitrust law
enforcement in the 21st century."[74]

On July 26, lawyers for Microsoft filed an argument with the
Supreme Court, claiming that the case "went badly awry from
the outset," and requesting a review by the federal appeals
court.[75] At the time of this writing, the appeal was still pending
and no remedy had been imposed. In the year 2000, Microsoft

[73]All quotations in this section are from "How Judge Jackson Ruled," *New York Times,*
November 6, 1999, p. C1.
[74]Quoted in Joel Brinkley, "Microsoft Breakup is Ordered for Antitrust Law Violations,"
New York Times, June 8, 2000, p. A1.
[75]The Justice Department had asked the Supreme Court to hear the appeal directly, a
request that the court denied in September 2000. See Steve Lohr, "Microsoft Files Brief
Asking Supreme Court to Send Antitrust Case to Appeals Court," *New York Times,* July
27, 2000, p. C2; and Stephen Labaton, "Justices Decline a Direct Appeal in Microsoft
Case," *New York Times,* September 27, 2000, p. A1.

reported revenues of nearly $23 billion and net income of $9.4 million.[76] Its Internet Explorer accounted for 86 percent of the U.S. browser market.[77]

The Meaning of Microsoft

In a case as big as Microsoft, there are many meanings to be found. The story is so dense and all-encompassing that, like some giant Rorschach test, it allows analysts to find whatever they may be seeking: evidence of government bungling or corporate mismanagement or changing business paradigms.[78] For the purposes of this book, however, the meaning of *U.S. v. Microsoft* is clear. It demonstrates that even at the peak of technological change, even as markets are booming and erupting, formal governance is entirely possible. This doesn't mean that it will always happen. Indeed, as the case of encryption demonstrates, there may well be times and areas where governments are pushed back from the technological edge, where they want to regulate and try to regulate, but a combination of technical and commercial impulses forces them to desist. In the case of Microsoft, however, the picture is radically different. Here, governments in both the United States and Europe were perfectly able to take old laws and impose them on a new market. They were fully able in this case to rule the Net and to make even one of the world's most powerful companies comply.

The question, then, is why. Why were governments able to

---

[76]Microsoft 2000 Annual Report: Financial Highlights, www.microsoft.com.

[77]Based on data reported by WebSideStory's StatMarket and reported in "Microsoft Dominates Browser Battle," PCWorld.com, September 28, 2000.

[78]Some of the most important contributions to this debate include Stan J. Liebowitz, "Breaking Windows: Estimating the Cost of Breaking Up Microsoft Windows," Mimeo, Management School of the University of Texas at Dallas, April 30, 1999; Robert J. Levinson, R. Craig Romaine, and Steven C. Salop, "The Flawed Fragmentation Critique of Structural Remedies in the Microsoft Case," Georgetown University Law Center, 1999 Working Paper Series in Business, Economics and Regulatory Law, Working Paper No. 204874; Pierluigi Sabbatini, "The Microsoft Case," Mimeo, Autorità Garante della Concorrenza e del Mercato, 1999; Jeffrey A. Eisenbach and Thomas M. Lenard, eds., *Competition, Innovation and the Microsoft Monopoly: Antitrust in the Digital Marketplace* (Boston: Kluwer Academic Publishers, 1999); and Steven C. Salop and R. Craig Romaine, "Preserving Monopoly: Economic Analysis, Legal Standards and Microsoft," *George Mason Law Review* Vol. 7, no. 3 (1999).

impose their will on Microsoft and not on the tiny, newborn Network Associates? Why did the cypherpunks get away while Bill Gates got caught?

Part of the answer concerns the obvious differences between export law and antitrust. Unlike antitrust, export controls are inherently tied to the physical nature of the state. They are laws about things that cross borders, things that states can see and track and regulate. When these things suddenly lose their shape and migrate to the more fluid world of the web, governments are left with both a definitional quandary and a practical nightmare. How can they regulate borders in areas where they—physically at least—no longer exist? How can they stop the spread of information when increasingly large chunks of their economy depend upon its free flow? They can't—or at least they thus far have chosen not to make the drastic political trade-offs that such a policy, presuming it is even feasible, would demand.

In the case of antitrust, however, the advent of cyberspace has had a less dramatic effect. The Net creates new kinds of markets and new forms of competition, but it doesn't necessarily transform the basic notion of competition law or the desire to enforce it. Moreover, physical and political boundaries remain as clear in this regard as they did with older industries and in earlier centuries. If Microsoft is located in the United States and does business in the United States, it can be subject to U.S. antitrust law. As the European case shows, moreover, it can even be subject to European law if it has any presence in the European Union and the Europeans choose to press their case. In fact, the demand for antitrust law may well expand along with the digital age, as the increased flow of information across national borders raises concerns about the possible monopolization or manipulation of these flows.[79]

---

[79]Already, there is some evidence that this trend is underway. In the year 2000 alone, the European Union launched competition reviews of three information-based deals: EMI's proposed joint venture with Time Warner, AOL's merger with Time Warner, and a three-way merger between Vivendi, Seagram, and Canal Plus. See Deborah Hargreaves et al., "EU to Clear $127bn AOL Deal," *Financial Times,* September 29, 2000, p. 1. For more on the general trend toward antitrust prosecution in Europe, see Deborah Hargreaves, "Monti Shows His Tough Streak," *Financial Times,* March 6, 2000, p. 23; and Peter S. Goodman, "Europe Resists Big U.S. Mergers," *Washington Post,* June 22, 2000, p. E1.

A second distinction concerns the nature of standards. In the case of encryption, standards remained free and open for decades, buoyed by a common philosophy and a widespread norm of academic sharing. In the Microsoft case, by contrast, this openness ended in 1975, when Microsoft created its version of BASIC and insisted on ownership. From this point on, Microsoft essentially set and controlled software standards for the personal computing industry. When Gates and Allen wrote their early software for the Altair, for example, they provided a common platform for what was then a disparate and ragtag band of users. When they purchased QDOS and wrote MS-DOS, they created a common language for what became millions of computer users and a critical mass for subsequent waves of innovation. The beauty of this system, of course, was that it enabled Microsoft to solve a problem of coordination that might otherwise have fragmented the personal computing industry and delayed its commercial development. By setting standards and deploying them across the market, Microsoft undeniably helped the computer industry over a critical stage of the technological frontier.

The rub, however, was that in solving one problem of the frontier, Microsoft ran smack into another. Like other great standard-setting firms—Marconi, Western Union, BSkyB—Microsoft eventually became too big and too powerful. It created the standards; it controlled the standards; and it became so dominant that no other firm could break into what had become its market. To some extent, this kind of behavior is natural. Once a firm like Microsoft or Marconi has created a successful standard and disseminated it across a market, it is very difficult for anyone else to enter the market and convince consumers to switch. Unless standards are truly open, therefore, as they were in the early days of encryption, there is a natural tendency for standard setters to mature into monopolists. This is precisely what happened to Marconi, and to Murdoch, and to Microsoft. In each case, the standard-setting firm grew so dominant that allegations of antitrust became quite plausible and governments were eventually pressured to move in.

With Microsoft, though, this natural march to monopoly was aggravated by the firm's own glaring history of making, breaking, and flaunting the rules. Even when Microsoft became a firmly established corporation, it retained in many respects its pirate culture. It openly employed ideas—BASIC, visual interfaces, browsers—that others had created and then used these ideas to create a massive and impenetrable installed base. It wielded its power like a gleeful sword, threatening enemies and coercing firms in order to maintain its empire. And thus when Microsoft moved into cyberspace and the Justice Department raised questions about its shift, a line of both friends and enemies stretched up to join the case.

In fact, throughout the saga of *U.S. v. Microsoft,* the role of the company's rivals was a thinly veiled secret.[80] Competitors such as Apple and Sun testified openly and critically about Microsoft's behavior, lending vital credence to the Justice Department's allegations. So, too, did the CEO of Intuit, a major Microsoft rival in the financial software market, and a senior vice president of AOL, one of Microsoft's most important partners. Several private firms, moreover, also took a more active role behind the scenes, bringing various bits of evidence directly to the government and helping to push the case forward. As early as 1995, for example, top-level executives at Netscape began talking with lawyers at the Justice Department about Microsoft's behavior. In 1996, Netscape's general counsel, Roberta Katz, followed up on these conversations, striking out on a quiet campaign to understand what Microsoft was doing in the market that might affect Netscape. When she discovered things that "seemed so far over the line . . . things they would do with customers that would harm us," Katz took her information to Gary Reback, a prominent Silicon Valley attorney who had opposed Microsoft on several high-profile cases and was already well aware of Netscape's plight.[81] Described famously as "the only man Bill Gates fears,"[82] Reback worked with Katz to compile a manuscript-size document that laid out a litany of Microsoft's wrongs: predatory pricing, strong-arm tactics, secret side pay-

---

[80]For more on the behind-the-scenes lobbying of Microsoft's rivals, see John Heilemann, "The Truth, The Whole Truth and Nothing But the Truth," *Wired,* November 2000, pp. 261–311. A longer account appears in Heilemann, *Pride Before the Fall: The Trials of Bill Gates and the End of the Microsoft Era* (New York: HarperCollins, 2001).
[81]This story comes from Cusumano and Yoffie, *Competing on Internet Time,* pp. 152–53.
[82]James Daly, "The Robin Hood of the Rich," *Wired,* August 1997.

ments, exclusive contracts, and an underlying obsession with maintaining its monopoly position.[83] Reback then sent the entire manuscript to Joel Klein at the Justice Department, presenting it as evidence of Microsoft's illegal behavior and urging the Department to reopen its investigation of the giant software firm. Several months after receiving Reback's letter, Klein gathered a team in the Department of Justice's San Francisco office and began to collect data.

Some of this data, of course, was the stuff of library research: market shares, price, financial performance, and the like. But most of the data was based on interviews and subpoenas, information gathered from other computer and Internet companies—the same companies, in many cases, that had already complained to the Justice Department about Microsoft and whose executives would later testify during the Microsoft trial. In 1995, for example, AOL had asked the Justice Department to investigate Microsoft's bundling of Windows and Explorer, and in 1998 "concerned" officials from the U.S. computer industry had presented Justice with their own proposals for how Microsoft's power might be constrained.[84] Reback, the man who helped to put the Justice Department back on the scent of Microsoft, had already filed a series of briefs against the software giant on behalf of anonymous clients who were generally believed to include Sun, Novell, and Borland, a major software firm. He also conducted a veritable road show in 1996, taking Netscape's case to a string of influential software customers— large banks, publishers, and content providers such as Walt Disney—and urging them to see the potential virtues of a case against Microsoft.[85] The implications of these connections are not hard to see: Microsoft may have been indicted by the Justice Department, but it was condemned by its peers.

It is this element above all that characterizes the Microsoft case and differentiates it from other political struggles along the high-technology frontier. In approaching the Internet, Microsoft

---

[83]Described in Heilemann, "The Truth," p. 265.
[84]On Apple's request, see Steve Lohr and John Markoff, "How a Giant Software Maker Played the Game of Hardball," *New York Times*, October 8, 1998; on the industry proposals, see John Markoff, "Proposals Made to Shrink Power of Microsoft," *New York Times*, April 7, 1998.
[85]Conversation with author, October 13, 2000.

probably violated antitrust law, depending on how you interpret the law and which judge is presiding. But its real crime came with regard to the informal rules of the game—to the norms and habits that defined the emerging Net. When Microsoft came to the Net, it landed in a frontier peopled by a new generation of pioneers, who had their own view of the world and their own way of doing business. Microsoft tumbled into this world and tried to bend it back towards its own set of rules. It was an excellent commercial strategy, but a political disaster. For these new pioneers had no love for the old pirate and a powerful tool with which to attack him. It was called antitrust law, and they used it fully.

There was, to be sure, a certain irony in this attack. Throughout the Microsoft case, and indeed throughout the entire saga of Microsoft, the distinctions between piracy and pioneering were never fully resolved. Clearly, Bill Gates and his lieutenants had "borrowed" ideas that others had first and released products based upon others' creations. Certainly they attacked their rivals with a vengeance and saw the software and Internet markets as all-or-nothing fights. In their industry, however, such tactics are common. Mark Andreessen, for example, "borrowed" the idea for Netscape Navigator directly from Mosaic—a product that he had created, of course, but one that was clearly owned by the University of Illinois by the time Andreessen launched Netscape. HTML and HTTP were the creations of Tim Berners-Lee, who never really benefited commercially from their exploitation; and the underlying language of most computer software, be it BASIC or FORTRAN or Linux, has generally been hammered out by armies of enthusiasts who receive neither credit nor compensation for their work. If Bill Gates was a pirate, then, so too were Mark Andreessen and Steve Jobs and a host of Silicon Valley's most luminous citizens. The problem with labels in this case is that the law is simply too weak to substantiate them. When fights over patent rights convulsed the telegraph industry in the 1840s, U.S. courts were able to resolve the disputes without much distress, for the law was there already and the new technology of telegraphy fit rather neatly into it. On the Internet, however, and even in the slightly older world of software, property rights remain contentious.

While a series of U.S. cases has set a precedent for considering software as intellectual property and thus covering it with standard principles of copyright law, both law and commercial sentiment in this area are still weak.[86] There is no law, for example, that clearly delineates the boundaries of property in software or on the Net, nor is there a law that differentiates borrowing an idea or business model from stealing one. Consequently, many denizens of the new economy assert that the whole notion of property rights is rendered moot in cyberspace, as is the question of piracy. Others, though (and generally the older and more established firms), insist that traditional concepts of property can and must be extended online.[87] So long as this debate continues to rage, property rights in cyberspace will remain ambiguous and every pioneer can be accused, with some justification, of piracy.

On June 28, 2001, a U.S. federal appeals court issued its ruling on the Microsoft case. Agreeing that Microsoft had indeed acted like a monopolist, the court upheld the basic findings of the lower court: Microsoft had violated U.S. antitrust law and illegally exercised monopoly power in the software industry. The court disagreed, however, with other elements of Penfield's analysis, and reversed his order of a formal breakup. So Microsoft and the Justice Department were sent back once again to the legal system and the vague possibility of a negotiated settlement. Microsoft's share price, meanwhile, had tumbled from a high of $116 in December of 1999 to roughly $70 by the summer of 2001. Netscape had been purchased in 1999 by America Online, and then rolled into AOL's merger with Time Warner, a deal that was itself the subject of antitrust investigation. With several suits still outstanding before the European Commission, some observers predicted that by the time

---

[86]For a summary of some of the precedent-setting cases, see Joseph Menn, "The Cutting Edge," *Los Angeles Times,* June 14, 1999, p. C1. For an overview of law in this area, see Mark A. Lemley, Peter S. Menell, Robert P. Merges, and Pamela Samuelson, *Software and Internet Law* (Gaithersburg, N.Y.: Aspen Law & Business, 2000).

[87]Contributions to this debate include Robert P. Merges, *Who Owns the Charles River Bridge? Intellectual Property and Competition in the Software Industry;* Mimeo, U. C. Berkeley School of Law, 1999; and M. Ethan Katsh, *Law in a Digital World* (New York: Oxford University Press, 1995).

the Microsoft case was finally resolved, technology would have evolved to a point where Microsoft, the world's first great software company, was already obsolete.[88]

And perhaps it will. But regardless of how the formal case against Microsoft is eventually decided, the real decision occurred in 1998, when the U.S. government, echoed by the Europeans and urged by the private sector, formally accused Microsoft of breaking the rules.

[88]See, for example, John Herrington, "Microsoft Ruling Serves Rivals, Not Consumers," Editorial, *San Francisco Chronicle,* November 30, 1999, p. A29; and Jeremiah Caron, "Dollar Bills—Microsoft's Been Doling out Cash to Service Providers," *tele.com,* August 9, 1999.

CHAPTER 7

Space Music

> A change to a new type of music is something to
> beware of as a hazard to all our fortunes. For the
> modes of music are never disturbed without
> unsettling of the most fundamental political and
> social conventions.
>
> PLATO, *The Republic*

Chuck D is an unlikely hero of the digital age. With
hit albums such as *Yo! Bum Rush the Show* and *Fear of a Black
Planet,* the founder of the rap group Public Enemy would seem
to inhabit a world far removed from the more conspicuous pioneers of cyberspace, from the Netscapes and Yahoos! and AOLs.
In 1998, however, Chuck D stormed into cyberspace. Rather
than giving his latest songs to Def Jam, the label that had
produced his music for over a decade, the rap artist instead released his music directly onto the Internet, at www.public-
enemy.com. It shouldn't have been such a big deal, really: one
artist, a handful of songs, and a funky distribution method that
probably reached several thousand fans. But in the music business this was very big news. For Chuck D had taken one of the
industry's most sacred practices and thrown it, quite literally,

into space. With just a couple of songs, he challenged how music was sold and, even more fundamentally, how it was owned. "This is the beginning," proclaimed the rapper, "of the end of domination."[1]

As far as Chuck D was concerned, putting music online was a matter of power, of using new technologies to right old wrongs and give recording artists the influence and money that was rightfully theirs. To the recording industry, however, it was heresy. For decades, companies such as EMI and Polygram had operated under a traditional and lucrative set of rules. They signed long-term contracts with the artists they deemed most attractive, and then managed the business side of the artists' careers—recording the albums, distributing them, handling all marketing and publicity. The artists would receive the fame that came along with their work, of course, plus a prearranged portion of the proceeds. The studio, however, retained both the remainder of the proceeds and the legal rights to the music. Ownership of the property, in other words, rested with the studios rather than with the artists.

This was the system that Chuck D challenged. By putting music directly on his web site, he effectively circumvented the entire legal and commercial structure that the studios had so carefully erected. He took his music, which had traditionally been their property, and made it his again. And the studios were not pleased.

Had Chuck D been an isolated case, the studios most likely could have looked the other way. They could have dismissed Chuck D as a simple renegade, a rapper gone bad, and forgotten him and his web site. But the problem was that Chuck D, potentially, was everywhere. In cyberspace, any recording artist could distribute his or her music online; any musician could become a mini-studio, circumventing the record labels and their complex, clunky rules. Even worse, the advent of digital technologies such as MP3 meant that the entire legal foundation of the old recording system was thrown into confusion. Legally, the record companies owned the right to distribute and repro-

[1]Quoted in Patti Hartigan, "The Prophet Chuck D, on MP3," *The Boston Globe*, February 12, 1999, p. E1.

duce their artists' music; no one else could distribute or perform these works unless he purchased a license directly from the company.[2] But what if an artist wrote a new song and simply posted it to a web site? What if a college student then downloaded this computer file into a hand-held machine or e-mailed it to a group of her friends? Because these practices were so new in the late 1990s, the law was simply silent on them: there was no regulation of MP3 technology, no system of property rights that explicitly applied to online music.

Matters reached a head in 1999, when a nineteen-year-old college dropout named Shawn Fanning joined Chuck D in storming the frontier. Backed by his uncle in Boston, Fanning created Napster, a revolutionary system that allowed thousands—even millions—of users to trade their music online. Within months of its release, Napster had become a social phenomenon and a massive commercial threat. Universities complained that Napster was suddenly consuming huge chunks of their Internet bandwidth, and the music industry condemned it as piracy of the most blatant sort: "STEALING," as one music lawyer described it, "in big letters."[3] Ironic foes such as Prince and the rock band Metallica joined the labels in pursuit of these new pirates, while prophets predicted the death of the recorded music industry. "A revolution has occurred in the way music is distributed," wrote one observer, "and the big record companies are in a state of panic."[4] Chuck D, as usual, was even blunter. "Our whole notion," he recalled, "was to come into the music business and destroy it."[5]

For a while chaos mingled with euphoria, and it seemed as if cyber music might really be able to destroy the established

---

[2] As described in Chapter Three, radio is one important exception to this rule. Under the terms of an agreement signed first in 1922, radio stations agreed to make a single yearly payment to ASCAP (American Society of Composers, Authors and Publishers) which covered their use of all ASCAP-controlled music. Record labels, by contrast, have traditionally used radio as a conduit for promotion rather than sales. For more on this history, see Russell Sanjek, *American Popular Music and Its Business: The First Four Hundred Years, Volume III: From 1900–1984* (New York: Oxford University Press, 1988), pp. 74–90, 159–211.
[3] Felicia Boyd, "The New Pirates," Faegre & Benson Legal Updates, accessed from http://www.faegre.com/articles/article_468.asp.
[4] Frederic Dannen, "Commentary: What Goes around Comes Around," *Los Angeles Times*, June 24, 1999, p. B–9.
[5] Interview with author, Cambridge, Mass., February 25, 2000.

recording companies. But then quietly, inevitably, old questions started to nag. How would cyber music develop if artists and manufacturers used different standards of recording and transmission? How could artists be protected from the pirates who would steal their songs? And how, most basically, could anyone make money if property rights in music disappeared? By the year 2001, industry groups and legal experts had coalesced around these issues, creeping slowly and often painfully towards some kind of more orderly framework. No one wanted government regulation of the music business. No one (except the record companies, of course) wanted to cede power back to the record companies and their established system of property rights and royalties. But neither did any of the players want to perpetuate a truly anarchic state of affairs. They wanted rules, as it turns out, and they wanted property rights, and they wanted someone or something to enforce them. Such is the nature of business along the technological frontier.

Making Music:
The Old-Fashioned Way

For those who grew up in the golden days of rock 'n' roll, it is natural to associate music with rebellion. Music, and especially recorded music, has long been the stuff of sex and indiscretion, of forbidden longings and illicit lust. And so indeed it still is. Yet the music industry itself is a fortress of conservatism, dominated by five long-standing corporations and run, according to Chuck D at least, "by lawyers and accountants who don't give a fuck about the creative process."[6]

Perhaps. In any case, however, the recording industry has always been small and tightly controlled, run by just a handful of firms with an iron grip on the global market. Like the light bulb, the recording industry began with Thomas Edison, who invented the phonograph in 1877. For nearly a decade, Edison was the only manufacturer of his "talking machine," a fragile device that played songs etched onto tin foil and appeared to

---

[6]Brett Atwood, "Chuck D: The Billboard Interview," *Billboard,* July 17, 1999, p. 82.

have no commercial value. In 1887, though, Alexander Graham Bell designed a more sophisticated device and formed the American Gramophone Company to compete with the Edison Speaking Phonograph Company. Edison then responded with his own upgraded machine, and the two firms were soon selling phonographs across the United States—most of which were used either in coin slot machines or to provide the voice for "talking dolls." After Bell's machine faltered in the market, Edison encountered new (albeit limited) competition from Columbia, a former distributor of Edison's own machines, and then from Victor, a firm that pioneered the use of recorded disks. By the turn of the century, these three companies—Edison, Columbia, and Victor—were the only major players in the global phonograph industry. They controlled the key patents, signed exclusive contracts with popular performers, and they pushed "talking machines" out of arcades and into their customers' homes.[7] They also won critical legal victories in 1902 and 1909, when first Great Britain and then the United States carved out formal structures for protecting musical property rights.[8] Henceforth, any company could record copyrighted music so long as it paid a royalty fee (initially two cents per reproduction in the United States) to the original copyright holder. The record companies, in other words, owned rights to the music they recorded, and to any proceeds that might be made from this music. The legal framework of the industry was now in place.

Over the next forty years, control over the recorded music industry remained tightly in the grasp of a tiny group of firms. When the stock market crash of 1929 forced Edison out of business, a new firm, Decca, migrated from the British record market and quickly edged its way into Edison's old position in the troika. Then, in the wake of World War II, two upstarts named

---

[7]This state of affairs was enhanced by the 1908 case of *White, Smith v. Apollo,* which upheld the notion of copyright in music and extended it to cover mechanical reproductions.
[8]In Great Britain, the relevant legislation was the Musical Copyright Act passed by Parliament in 1902. In the United States, it was the Copyright Law of 1909. For a fascinating glimpse of how British songwriters and music companies dealt with piracy before passage of the 1902 act, see Charles C. Mann, "The Heavenly Jukebox from Hell," *Atlantic Monthly,* September 2000, pp. 41–42. See also James Coover, *Music Publishing, Copyright and Piracy in Victorian England: A Twenty-Five Year Chronicle, 1881–1906* (New York: Mansell, 1985).

Mercury and Capitol began to nudge Columbia and Victor[9] from their decades-old perch, riding the rising tide of singers such as Frank Sinatra and Peggy Lee and the newfound popularity of jukeboxes. Together, these firms presided over the music industry, signing exclusive contracts, wooing radio stations, and releasing new formats such as the LP and 45. Until the mid-1950s, just four labels (Capitol, Mercury, Victor, and Columbia) jointly controlled roughly 70 percent of the titles on the *Billboard* charts, the leading indicator of music sales.[10]

In 1955, however, rock 'n' roll suddenly burst onto the popular scene, catching even the record companies by surprise. It began quietly enough, in the jazz clubs and urban radio stations that had long been showcasing music by African American performers—music that the industry had always segregated under the thinly veiled categories of "sepia" or "race" songs. Slowly, though, this music had begun to slip out of its narrow niche and into the mainstream of American music, where it was christened by *Billboard* magazine as rhythm and blues (R&B) and rapidly copied—illegally—by a growing range of white artists. This was the era, then, of hits such as "Bo Diddley" and "Maybelline," of original R&B music that was rerecorded by white performers and sold to an increasingly interested white audience.

Then, just as this trend was getting underway, two events hit the music world. The first was the release of Bill Haley's "Rock Around the Clock," a breakthrough song that soon became, to quote Frank Zappa, "the Teen-Age national anthem."[11] And the second was Elvis Presley, whose hip-rattling southern sound turned the world of music on its head. Though both Haley and Presley released their music with the major labels, it was their *kind* of music—deeply tied to R&B, worlds away from the crooning of Frank Sinatra and Pat Boone—that inspired an ecstatic embrace of what had become rock 'n' roll. Demand for new titles soared, giving independent labels—and particularly small rhythm and blues labels—a rare opportunity to grab sales away from the Big Four. By 1958, independents such as Atlantic

[9]In 1929, the Radio Corporation of America (RCA) bought Victor through a stock deal and the company was reorganized as RCA Victor.

[10]Cited in Geoffrey P. Hull, *The Recording Industry* (Boston: Allyn & Bacon, 1998), p. 29.

[11]Quoted in Sanjek, *American Popular Music and Its Business*, p. 340.

and Imperial had captured 76 percent of the *Billboard* charts, propelled by the success of artists such as Fats Domino, Muddy Waters, and Ray Charles.[12]

In retrospect, though, the first wave of rock was also the peak of power for the independents. For ironically, once rock and R&B blazed across the mainstream of American music, it became harder and harder for these more specialized labels to compete with the major firms. They didn't have the financial clout to offer star salaries; they didn't have the marketing arms to reach a growing international audience; and they didn't have the resources to spread over hundreds of artists and albums, only a small percentage of whom would ever hit it big. And thus even as their music became more successful, many of the independent labels either formed larger firms or were bought by them. In 1964, for example, upstart Warner Brothers Records acquired another independent, Reprise. It then bought still another independent, Atlantic, in 1967, and yet another, Elektra/ Asylum, several years later. In 1979, United Artists, one of the largest indies, merged with Capitol and A&M joined RCA Victor. Arista and Ariola followed suit in 1983, adding their own labels to RCA's burgeoning fold. By the 1980s, the record industry was again down to a handful of giants, each of which now owned a cluster of ostensibly independent labels. Ten years later, the field had shuffled only slightly, with the addition of several non-American firms and a market that now covered the globe. The "Big Six"—BMG, EMI, Polygram, Sony, Universal, and Warner—controlled an estimated 70 to 80 percent of global recorded music sales in 1996 and nearly all of the world's most popular musicians.[13]

Like their earliest counterparts, the record labels that predominated at the turn of the twenty-first century were large and well-diversified firms. Warner Music, for example, was a division of the powerful Time Warner Entertainment Company, owner of twenty-four magazines, theme parks, book publishers, cable concerns, and cinemas, in addition to its several record

[12]Hull, *The Recording Industry,* p. 29. For a full-fledged account of the rise of independent labels, see Rick Kennedy and Randy McNutt, *Little Labels, Big Sound* (Bloomington: Indiana University Press, 1999).
[13]Robert Burnett, *The Global Jukebox* (London: Routledge, 1996), p. 18.

labels. Universal was likewise a division of MCA, a vast conglomerate that includes Universal Studios, Decca, and Geffen
Records, and was itself purchased in 1995 by Seagrams, the
wine and spirits company. BMG was a unit of Bertelsmann
A.G., Germany's largest media company and a powerhouse in
the publishing world. And so on.

Over time, of course, each of the companies—and each of
the "independent" labels within them—developed its own style
and focus. Some specialized in particular genres, such as classical music or rap; most were tied to a handful of particularly
popular artists. EMI, for example, controlled the work of the
Beatles, Beach Boys and Rolling Stones; Warner Music had
Madonna, Led Zeppelin, and the Eagles. In general, though, all
of the companies ran their business along remarkably similar
lines. They signed a large number of promising artists, held
them to fairly restrictive contracts, and then managed the marketing and distribution for the artists' music. In a typical relationship, then, an artist like Chuck D would sign with a label
such as Polygram's Def Jam Records, agreeing to give Def Jam
rights to an album or series of albums in exchange for a royalty
based on the albums' sales. For Smash Mouth or Bruce Springsteen or Luciano Pavarotti, it would be nearly the same: output
over some period of time, with property rights vested in the
label and a certain percentage of the proceeds paid back to
the artist. To the labels, this was the only way of recouping the
bottom-line profits that their own corporate owners demanded,
of realizing returns from a diversified portfolio of young, generally untested and often unproductive talent. To many in the industry, though, it was a kind of musical bondage, ownership of
the artist's soul in exchange for a measly share of the proceeds.

Consider the matter of royalties. In general, mid-level artists
earn between 14 and 16 percent of the suggested retail price
of their albums.[14] If they make a hit, say a "gold album" that
sells a half million copies in the U.S. market, they would then
earn around 15 percent of the total retail sales—or roughly

---

[14]The following section is based on the description provided in Daniel S. Passman,
All You Need to Know About the Music Business (New York: Simon & Schuster, 1994),
pp. 88–115.

$1,198,500, assuming that the album is a standard CD with an average retail price of $15.98. Not bad—especially if the artist happens to be a sixteen-year-old rocker with her very first album. But in practice, the math turns out to be trickier. First, recording companies typically deduct the cost of packaging from the artist's royalty base—approximately 25 percent for CDs and 20 percent for cassettes—arguing that the packaging shouldn't be counted as part of the product's "real" cost. Then, they also deduct any albums that they give away for free, arguing once again that these sales shouldn't be considered as part of the artist's base. Fair enough, except that these promotions often account for up to 15 percent of an album's total "sales." Finally, artists are also expected to cover their own "recoupable" expenses, including recording costs, video production costs, equipment rental, travel, and salaries for a producer and, often, a promoter. In practice, most of these costs are deducted from the album's proceeds, rather than being paid by the artist up front.

Once these various costs are factored into the equation, the artist's payout begins to look very different, as expressed in Exhibit 7.1 below.

At $261,490 our sixteen-year-old ingenue is still not doing too badly, but her proceeds from even a best-selling album are

EXHIBIT 7.1 Breakdown of Typical Royalty Payment

CD, suggested retail price	$15.98
(25% packaging)	$3.99
Royalty base	$11.99
Royalty rate (15% − 3% for producer)	× 12%
	$1.44
Units	× 500,000
	$719,400
(15% free goods)	−$107,910
	$611,490
(Recording costs)	−$200,000
(50% of promotion costs)	−$75,000
(50% of video costs)	−$75,000
	$261,490

Source: Adapted from Passman, p. 114.

336 Ruling the Waves

quite a bit lower than the original figure of nearly $1.2 million. What's more, in practice the record label pays only a fraction of these royalties in the early stages of an album's release, holding on to a reserve of 35 to 50 percent in case the album's sales were suddenly to drop off. The label also retains all rights to the artist's master recordings and to any duplication of these masters.

Not surprisingly, this system of contracts and accounting has generated a fair amount of criticism. When, for example, Warner Brothers refused to let Prince release albums as frequently as he would have liked, the popular singer-songwriter changed his name to an unpronounceable symbol, scrawled "SLAVE" on his face, and declared both himself and his contract deceased. This was in 1993, less than a year after signing a new contract that the artist had claimed could have been worth as much as $100 million. "Warner wanted a record only every 18 months," said The Artist Formerly Known as Prince. "I could release a record every seven months. I could not record when I wanted to."[15] After months of battling with Warner, The Artist began to record for his own label, NPG Records, and to experiment with the CD-ROM format; several years later, back with Arista Records, he was still singing that "Heavy rotation never made my world go 'round / Commercialization of the music is what brought it down."[16] He is echoed by others such as Chuck D, who raps,

> If you don't own the master
> Then the master owns you.
> Dollar a rhyme
> But we barely get a dime.[17]

To a large extent, of course, the tension between artists and recording labels is a natural result of a complicated symbiosis. Throughout the history of their industry, the artists and the labels have relied almost entirely on one another: without the music, the labels had nothing to sell; but without the labels, the

[15]Quoted in Aleksandrs Rozens, "'The Artist' Is with Major Label but Dodges Clauses," *Reuters English News Service*, November 10, 1999.
[16]Ibid.
[17]Quoted in Patti Hartigan, "The Prophet Chuck D," p. E1.

musicians had no efficient way to reach their audience. It was a thorny relationship, but an essential one. Neither partner could make music—or money—without the other. Yet, throughout this relationship, the power had seemed indelibly glued to the labels' side—to the much-maligned "accountants" or "suits" who controlled the artists' gateway to the fame and the fortune and the fans. And because the number of artists (and would-be artists) always exceeded by far the number of labels, the artists were essentially at the mercy of the labels. They could argue about contractual terms, of course, and leverage whatever fame might come their way, but it was still the labels who called the shots; it was the labels who maintained connections with retail stores and radio stations, and thus with the fans who were the ultimate consumers of any musician's product. But then something new hit the scene, a novel technology that was called MP3.

The Birth of MP3

Like many digital technologies, MP3 began life as an esoteric project of scientists and inventors. Ever since computers had become practical devices, researchers had been tinkering with ways to play and transmit music on them—to enable the computer to "read" digitized music just as it could already read text and mathematical notation. In 1983, members of the electronic music community took the first step, devising an interface known as MIDI that stored musical information almost like a player piano and allowed synthesizers, samplers, and specialized computers to communicate with one another. Eight years later, when personal computers had become popular and affordable, IBM and Microsoft built upon this standard to create the Waveform Audio File format (or WAV), which actually recorded sound waves through a microphone and then encoded them into digital signals. In 1992, all versions of the popular Windows 3.1 software were equipped with the ability to read this jointly concocted format.

Despite having slipped so easily into the popular market, though, the WAV (pronounced "wave") file format was too clunky to be practical. A two-minute song used 20 megabytes of disk space and, on the average modem, might take more than two hours to download.[18] As if that were not enough to discourage casual use, the sound quality was not particularly good, and certainly nowhere near the level of traditional CDs or cassettes.

Yet a small band of scientists was still convinced that computers could play music and that users, given the right equipment, would want to listen. One of the leaders of this band was Leonardo Chiariglione, a mild-mannered Italian engineer who was captivated by the prospect for standards. In 1987, while working for Italy's CSELT (Centro Studi e Laboratori Telecomunicazioni SpA) telecommunications laboratory, Chiariglione attended a meeting of the Joint Photographic Expert Group (JPEG), an industry group convened under the auspices of the International Standards Organization (ISO). Like all ISO groups, JPEG was a voluntary consortium of private firms and government officials, a venue for experts from around the world to cobble together common technical or operational standards. Chiariglione, who had never attended this type of meeting before, was hooked. Within months, he had approached the chairman of JPEG and won approval to launch a new ISO-linked group, MPEG, which would devise standards for the digital coding of motion pictures.

On the surface, MPEG looked like a standard standards group. It was convened under the joint auspices of the ISO and the IEC (International Electrotechnical Commission) and had the normal mandate of developing a common standard for a particular branch of technology—in this case, for the way moving pictures and audio could be translated into digital code. Like other ISO or IEC groups, it was composed of delegations from national standards bodies across the world, and shared a common mix of academic experts and corporate engineers. What made MPEG different, though, was Chiariglione. For to Chiariglione, standards weren't just standards, and multimedia wasn't just any technology; they were instead the core of human communication, the very platform on which societies could either progress together or decline alone. Or, as he put it:

[18]Robert Kendall, "MIDI Software: MIDI Goes Mainstream," *PC Magazine*, March 31, 1992, p. 181.

Communications standards are at the basis of civilised life. Human beings can achieve collective goals through sharing a common understanding that certain utterances are associated with certain objects, concepts, and all the way up to certain intellectual values. Civilisation is preserved and enhanced from generation to generation because there is an agreed mapping between certain utterances and certain signs on paper that enable a human being to leave messages to posterity and posterity to revisit the experience of people who have long ago departed.[19]

It's not clear whether other members of the MPEG group shared Chiariglione's grand vision. Yet they apparently were captured by his enthusiasm, and by his remarkable ability to smooth corporate or national squabbles in pursuit of a common goal. By 1994, MPEG had grown from its original twenty-five members to encompass more than 150 delegations from national standards bodies and interested firms, many of whom controlled proprietary, often competing, digital technologies. With so many participants, industry critics worried that MPEG would break down under its own weight, crushed by technical squabbles or disputes over patent rights. Yet somehow, Chiariglione and his vision prevailed. In 1994, MPEG released its MPEG-2 Audio standard, based on technology created by Fraunhofer Schaltungen, a German engineering firm. Though no one really knew it at the time, the era of MP3 had been launched.

Technically, MP3 (an abbreviation of MPEG-1 Layer III) is a standard for coding sound into electronic bits and then compressing those bits into easily manageable files. Using it, or technology based upon it, programmers can take ordinary recordings (music, soundtracks, personalized greetings) and convert them into a standardized digital format—a stream of zeroes and ones recognizable to any other device based on the MPEG-2 format. Essentially, then, MP3 is a code: a mathematized, digitized version of Morse's dots and dashes or the Chappes' system of clanging pots and synchronized clocks. Like

[19]http://www.cselt.it/leonardo/paper/standardisation/html

them, it enables users to take information from one form and translate it, seamlessly and universally, into another.

Practically, MP3 was also a major improvement over the older WAV format. In all of the earlier iterations, programmers had tried to retain and reproduce the exact audio signal, meaning that their files contained huge amounts of data. MP3, by contrast, employed a process known as "auditory masking," which replicated how the signal would sound to a human ear. By stripping away the additional sounds, auditory masking could compress the encoded signal by a factor of 12 to 14, meaning that fourteen hours of music could suddenly fit in a single CD. This was the breakthrough that drove renegade artists to embrace the new format and embed it in their development plans.

Between 1994 and 1996, MP3 slowly made its way to the world of music. There was no great clamor for the technology itself, and certainly no demand from within the traditional recording studios. But on the fringes, MP3 won a growing band of advocates, mostly among those who either saw the prospects for the new technology or bore an outstanding grudge against the labels. In 1994, for example, Public Enemy released a CD entitled *Muse Sick-N-Hour Mess Age*. While the album was released the old-fashioned way—via Def Jam Records and retail outlets—its message paid direct homage to the emerging world of MP3, warning artists that "We're talking about a shift in the way this music is distributed."[20] That same year, the rock band Aerosmith posted an unreleased single directly to a CompuServe web site, announcing, "If our fans are out there driving down that information superhighway, we want to be playing at the truck stop."[21] They were joined a year later by British rocker David Bowie, who posted his unreleased single to another site, and then by the popular Irish band U2, whose tracks were stolen in Hungary and posted instantly to sites around the world. In the wake of the theft, Marc Marot, managing director of U2's label, commented, "There is this sense that they [the thieves] were liberating U2 from the clutches of some monstrous record com-

---

[20]Quoted in Hartigan, "The Prophet Chuck D," p. E1.
[21]Quoted in Marilyn A. Gillen, "Geffen Puts Aerosmith Music on Line," *Billboard,* June 18, 1994, p. 19.

pany regime."[22] To counteract the damage, U2 hastened the release of its new single, and then its new album. "The Internet," mused Marot, "has changed our lives completely and forever."[23]

Pirates, Again

So long as MP3 remained the playground of scientists and the platform for a few disgruntled artists, the recording industry could afford to look the other way. There had always been technological innovations in music delivery, and there had long been pirates of all musical formats. Most of the original radio stations, after all, had been built around "unauthorized" music; most popular artists were regularly available in "bootleg" (illegally recorded) or pirated (illegally copied) form. Yet the record industry had survived all these pirates and thrived among them. MP3, for a while at least, didn't seem particularly threatening; it was just a new way of capturing sound and storing it in an efficient, standardized format. But then it hit the mainstream.

The first takers were mostly college students, who came online in droves in the late 1990s. Like most people of their age, these students were fertile territory for any kind of music, and particularly for music marked with a radical, anti-establishment twist. They were also, though, armed with a high level of technical expertise—this, after all, was a generation raised on PCs—and with the high-powered Internet connections that many colleges and universities were rushing to provide around this time. Quickly, the students found the music and began to play it, downloading favorite cuts, compiling personal collections, and e-mailing songs to their friends. Like the radio boys who preceded them, some began to spend virtually all their free time online, building networks of invisible friends and sharing tips and songs. Some were even more ambitious and, bitten by the Net bug, began to build small communities around their MP3 downloads.

As these sites and services proliferated, a transformation ripped through the world of MP3. What had been an esoteric

[22]Quoted in Chris Marlowe, "The Cyberspace Race: Living with the Threat of the Net," *Music and Media,* June 14, 1997, p. 10.
[23]Quoted in Marlowe, "The Cyberspace Race," p. 10.

technology became in a matter of months a casual plaything—
a way for even relatively unsophisticated computer users to find
music and download it to their hard drives. Seizing this potential, a handful of entrepreneurs built official MP3 sites, places
where avid listeners could download—for a small fee—the particular songs they chose. Nordic Entertainment Worldwide, for
example, debuted in April 1997 with a site described by its
founders as "a record store that happens to be selling music online"—a fully legal compendium of older, relatively obscure
tracks.[24] Global Music Outlet did the same thing for South
African and then alternative music[25]; X-Radio Corp., another
small start-up, focused on a small genre of club music known as
"techno." In 1998, two major commercial sites emerged, Good-
Noise and MP3.com, both of which were devoted to playing
and distributing more mainstream music.

The real action in 1997 and 1998, however, came from the underground sites—from the college students and their supporters
who were letting loose an avalanche of free online music. Initially,
much of this music was like the Net in general. It was alternative
or independent, posted by little-known bands who saw the Net as
a novel form of distribution and didn't have much to lose by giving their music away for free. One of the earliest large-scale free
music sites, for example, was the International Underground
Music Archive, a self-styled "coffeehouse for 20 million people"
that flaunted its alternative leanings and actively supported a network of new and independent bands.[26] As interest in online
music rose, however, the music sites quickly moved from the underground into the mainstream. Students copied their favorite
pop songs onto their hard drives; industry insiders secretly "released" new music directly to the Net; and pirate sites in far-flung
locations such as Bulgaria and China started to post CD-quality
versions of Madonna and Bruce Springsteen recordings. By 1998,
according to one estimate, more than twenty-six thousand illicit

[24]Brett Atwood, "Nordic Bows Sales of Digitally Sent Music," *Billboard,* April 26, 1997,
p. 6.
[25]Brett Atwood, "GMO Vies for Digital Download of Music," *Billboard,* May 17, 1997,
p. 57.
[26]Quoted in James Daly, "Music by Modem," *Rolling Stone,* July 14, 1994, p. 31.

music sites had proliferated across the Net,[27] offering roughly two hundred thousand songs. There were 3,462 sites for Nirvana alone, each offering free versions of the band's popular grunge rock sound. It was Chiariglione's vision, but with a youthful, radical twist: music for the masses, plucked from cyberspace and behind the backs of the established recording labels.

To the students who remained the largest users of MP3, all of this activity was exhilarating and harmless. For what were they actually doing? Just downloading intangible bits of sound and playing or distributing their favorite music. As one nineteen-year-old sophomore with hundreds of songs on his hard drive, argued, "I feel bad sometimes because I know I'm making a mess for the music companies. I know it's technically a crime, but anyone can say copies are being made for back-up purposes."[28] Or, in the blunter words of another online fan, "It may be illegal, but, hey, it's free."[29] To the record companies, however, this was theft pure and simple—digital piracy cloaked in the ill-fitting coat of personal entertainment. "With a touch of a button," complained one executive, "any 13-year-old can make music available to millions of people around the world. We're talking about a whole new dimension of piracy."[30] As word of the music sites seeped out, the record companies began to fight back, working through the Washington-based offices of the RIAA (Recording Industry Association of America), the industry's major trade association. There, in a windowless room, a handful of investigators began to surf for pirates, scouring college servers and music sites in search of illicit recordings. They weren't hard to find. In its first eighteen months of operation, the RIAA investigation found thousands of copyrighted songs floating freely around the Web and sent formal warnings to 350 "pirate" sites. It seized 23,858 unauthorized CD-Rs (recordable

[27]British Music Rights, cited in Carol Midgley, "Musicians Lose £40 Million in 'Theft' on Internet," *The Times,* May 28, 1998.
[28]Quoted in "Student Bootleggers Copy, Trade Music from the Internet," *Pittsburgh Post-Gazette,* April 12, 1998, p. G–5.
[29]Quoted in Benny Evangelista, "Download that Tune—It's Web Music," *The San Francisco Chronicle,* December 3, 1998, p. A1.
[30]Quoted in Hartigan, "The Prophet Chuck D," p. E1.

CDs) and brought suit against three of the most egregious on-line music sites, charging them with violation of copyright law.

The three offenders were not commercial entities; they either offered music free of charge or followed a "ratio" system, asking that visitors upload a certain number of songs in exchange for those that they downloaded. Legally, then, their position was ambiguous, since none of the named sites was actually realizing a commercial benefit from its downloaded music. To the RIAA, however, the principle remained the same: copyright was property, and abrogation of copyright—even without commercial gain—was theft. "We want a decision," said Hilary Rosen, president and Chief Operating Officer of the RIAA, "affirming the rights of copyright owners."[31] According to the RIAA, one of the named sites was already receiving more than twenty-nine thousand visits a month.

Eventually, the recording industry triumphed in court. In each of the three cases, the RIAA won damages of $100,000 from the site operators, and then, in exchange for waiving the fines, won a promise that the defendants would not repeat their bad behavior. It was a fairly straightforward victory, with the court determining that the operators had indeed engaged in illegal copying. Yet it was also an ephemeral one, denounced across the online community and widely described as unenforceable. What made matters even worse—or better, depending on one's perspective—was that the technology of digital compression was evolving as rapidly as the online music industry. The files themselves were becoming easier and easier to download and a new generation of players, such as Nullsoft's Winamp and RealNetwork's Real Jukebox, recreated the feel of a traditional stereo system—only hipper now, and fed by an ever-growing stream of music. By 1998, anyone with a compact disc, a CD-ROM, and a computer was fully able to send CD-quality music over the Net.

And so, even as the record companies were basking in their legal victory, a new wave of artists were gleefully ignoring the legal issues and rushing to deal directly with their fans. Inde-

---

[31]Quoted in Don Jeffrey, "Downloading Songs Subject of RIAA Suit," *Billboard*, June 21, 1997, p. 3.

pendent labels, such as Rykodisc, Sub Pop Records, and Parasol Music, began catering to the MP3 format and releasing new music in it. By March of 1998, the online music retailer Music Boulevard had sold more than four thousand songs from these labels at ninety-nine cents a piece. It didn't add up to all that much money, of course (only $3,960 in sales for Music Boulevard!), but the threat to the music industry was clear.

An even bigger blow, though, came from Napster, the web service launched in 1999 by college freshman Shawn Fanning. Like Chuck D or Chiariglione, Fanning was in many respects an unlikely pioneer. The son of a single mother from the working-class town of Brockton, Massachusetts, Fanning had a tough childhood and an unremarkable academic record. He entered Northeastern University in 1998 as a computer science major but spent most of his time at his uncle's nearby office, fiddling with the computers there and trying to build a cache of MP3 files.[32] Eventually, while trying to find a more efficient way of accessing and downloading music, Fanning stumbled upon what is now known as "peer-to-peer" technology—a way for Net users to copy and exchange files without necessarily being aware of each other's presence. In most of the older models of file exchange, one user would deliberately send her copy of, say, Ricky Martin's "Vuelve" directly to her six best friends. In Fanning's system, by contrast, once this user had logged on to a central site, any MP3 files located on her computer would automatically be available to anyone else logged on to the site. Which meant, of course, that the potential number of "Vuelve" downloads had now become immense—any Ricky Martin fan could simply copy the song, for free, from any other fan who happened to be online.

To Fanning, the goal of the technology he quickly dubbed Napster was just to extend the music community that had already assembled online. It was a way to share more efficiently, to get the songs that he and his buddies craved. "I didn't see us turning into a business," he recalls. "I just did it because I loved

---

[32]Fanning's early history is adapted from Spencer E. Ante, "Inside Napster: Its Struggle to Survive," *Business Week*, August 14, 2000, pp. 113–20. See also Karl Taro Greenfeld, "Meet the Napster," *Time*, October 2, 2000, p. 60.

the technology."[33] To Fanning's uncle John, though, who had already launched an online gaming site, Napster was a commercial breakthrough waiting to happen. And to the industry, of course, it was hell.

In January of 1999, Shawn Fanning dropped out of college with his uncle's approval. He launched Napster on June 1, and had between three and four thousand "customers" in just the first few days. John Fanning then began raising money in earnest, meeting with some of Silicon Valley's most respected venture capitalists and touting Napster's revolutionary potential. "We all knew from the beginning," he remembers, "that this would be huge."[34] Certainly, college students bore out this prediction in full force. Within months of its launch, Napster was raging across U.S. universities, consuming in some cases up to 30 percent of the schools' total bandwidth capacity.[35] And then, because Fanning's technology was so easy to use, Napster slammed even faster than other MP3 technologies into the musical mainstream, reaching everyone from suburban dentists searching for classical tracks to professors with a taste for German folk songs. By the fall of 2000, Napster claimed to have more than thirty-eight million users.[36]

For the labels, of course, this was a nightmare. For it wasn't just that the technology had created a new threat—there had been piracy, after all, ever since the days of sheet music—but rather that the breadth of this technology created whole new kinds of pirates: people who wouldn't dream of stealing but simply didn't equate the use of Napster with theft. The problem, in other words, was that the Fannings' technology had redefined the very notion of property rights in music. Under the existing law, after all, consumers had the right to share their music with friends; this was a provision known in the United States as "fair use" and enshrined in such practices as lending libraries. What Napster had done—or at least what the Fannings and their supporters claimed—was only to expand the

---

[33]Quoted in Ante, "Inside Napster," p. 115.
[34]Ibid.
[35]From Ante, "Inside Napster," p. 115.
[36]Cited in Matt Richtel and David Kirkpatrick, "In a Shift, Internet Service Will Pay for Music Rights," *New York Times*, November 1, 2000, p. 1.

fair use provision over a wider, but still noncommercial, community. So was this a violation of copyright, or a perfectly legal practice that just happened to undermine copyright's intent? As is usual along the technological frontier, the answer depended on one's perspective. And the old rules did not easily apply.

During this period of flux, however, the sheer weight of activity seemed to favor Napster and the host of new services, such as Gnutella, FreeNet, and Napigator, that rapidly aped its technology. For if the record labels could not physically stop these services, then all of their markets would soon be awash in free music, "owned" now by consumers who saw no need to pay. And if music were free, how could anyone make money from it? Accordingly, prophets at the turn of the twenty-first century were quick to paint Napster as the grim reaper of the recording industry: the technological blow that would at last reveal the labels as both stupid and doomed. "The power of the cartel," wrote one observer, "is doomed. And for that, the major labels deserve no sympathy at all."[37]

Return of the Labels

In retrospect, however, the record companies' initial response to MP3 and Napster was at least partially understandable. They did, to be sure, move exceedingly slowly. They were cocky and stubborn and were scared to embrace anything that might challenge their own comfortable position. But they did in the end take action, challenging the new breed of pirates and trying—through whatever means—to impose old rules on what even they were forced to admit had become a new world.

Initially, the record labels clustered their forces and concentrated on their strengths. Banding together behind the RIAA, they brought suit against a growing slate of offenders, expanding their arguments to include other industries that might be similarly threatened by MP3 technologies and explicitly refusing to acknowledge that technology had changed either the structure of property rights or the definition of piracy. Late in

[37]Dannen, "Commentary," p. B–9.

1998, for example, the RIAA brought suit against Diamond Multimedia Systems, a California start-up company that planned to produce a hand-held gadget known as the Rio. Like Sony's breakthrough Walkman, the Rio was designed to be both sleek and functional, a hip, portable device that would play MP3 files at near-CD quality. Potentially, at least, Rio was a big technology: if it could mimic the Walkman's performance in a new format, it could thrust MP3 directly into the mainstream of the music market and create, along the way, a whole new branch of consumer electronics.

To the recording industry, though, Rio hit too close to home. For by taking MP3 out of the underground market, it essentially legitimized it beyond the point of no return. Once Rio hit the shelves of Radio Shack or Circuit City, they feared, there was no going back: MP3 would flow out of the dorms and grunge-rock sites and smack into the musical mainstream. Or, as Hadrian Katz, a lawyer for the RIAA, expressed it: "[T]he expansion of the MP3 market from just those of us who like to spend our lives in front of the computers to people who actually have lives, is really going to create a new level of demand for the product."[38] Jim McDermott, vice-president of new media technology at Polygram, was even blunter: "The Rio," he argued, "is like walking into a head shop and buying a bong, and it says, 'For use with tobacco products only.' They fucking know it's going to be used for piracy."[39]

Accordingly, the RIAA began a formal assault on Rio, charging in an echo of its earlier suits that the company was "facilitating" and "encouraging" the unauthorized copying of copyrighted music. This time, though, the argument fell on less sympathetic ears. In a biting reversal, Judge Audrey Collins denied the RIAA's request for a preliminary injunction against Diamond Multimedia, arguing first that because the Rio had no output capacity (that is, one couldn't copy from it) it did little to facilitate unauthorized use, and second that "[b]ecause the Rio is

---

[38]"Diamond Multimedia Files Response to RIAA's Complaint and Files Nine Counterclaims," Diamond Multimedia Press Release, December 2, 1998.

[39]Quoted in Julian Dibbell, "The Record Industry's Digital Daze," *Rolling Stone*, November 26, 1998, p. 102.

capable of recording legitimate digital music, an injunction would deprive the public of a device with significant beneficial uses."[40] Diamond thus pressed ahead with its production plans, and the RIAA was stunned.

As might be expected, prophets of the new economy were quick to seize Diamond's victory as proof of an impending power shift. "This is a win for consumers and a win for musicians," exclaimed Ken Wirt, Diamond's vice-president of corporate marketing. "The big record companies could do great if they got on this train instead of just standing in front of it trying to stop it."[41] Normally conservative *Business Week* dismissed the RIAA's actions as "Luddite folly," scoffing that "[N]o amount of litigation can put the digital genie back in the bottle."[42] And others were downright nasty. "It's bad enough," suggested one observer, "that the RIAA doesn't get it.... But the RIAA doesn't just not get it—they refuse to get it, in their dogged determination to take cluelessness to new depths."[43]

In the next round, however, the record companies did begin to get it—or at least to address the technology of online music on their own terms. In 1998—and then more boldly in 1999, they transferred their efforts from the courtroom and back towards the market, venturing onto the Net and gingerly testing prospects for their own online distribution. Early in 1998, for instance, Polygram, Sony, and Warner Music launched similar retail web sites, inviting customers to listen to digital clips of new songs and order CDs online. The sites weren't particularly radical; users couldn't download music directly and they couldn't buy music that wasn't already available at standard retail outlets. It was at least a start, though, and a whispered admission that the Net might be more than some squalid pirates' lair. In April, Atlantic Records (a subsidiary of Time Warner) edged slightly further ahead, allowing customers who purchased a new Tori Amos CD from Tower Records to download an additional

[40]Cited in Doug Reece, "Ruling Favors Digital Player Rio," *Billboard*, November 7, 1998, p. 8.

[41]Quoted in Reece, "Ruling Favors Digital Player Rio," p. 8.

[42]Neil Gross, "Target Pirates—Not Technology," *Business Week*, November 2, 1998, p. 40.

[43]Dana J. Parker, "The RIAA and MP3: In Search of a Clue," *Emedia*, April 1999.

song from the artist's web site. Others followed suit: Sony allowed fans to download thirty-second clips from recently released albums from its web site; EMI agreed to make the majority of its catalog available for digital distribution in the Liquid Audio file format, a secure digital download system.

To the hard core of MP3 fans, these tentative steps only proved that the labels still didn't understand the new rules of music; that, as one insider put it, they "had taken their heads out of the sand but still had their feet in the concrete."[44] But then, in a move that recalled their earlier plunges into rock and R&B, the labels got more serious. Rather than denying the new world that was already on their doorstep, and rather than trying to compete with a bunch of younger and hipper new entrants, the "suits" simply went back to their own deep pockets and bought the competition. Time Warner and Sony, for example, each purchased 37 percent of CDnow,[45] a leading online music retailer; EMI purchased 50 percent of MusicMaker, a producer of customized music compilations; and BMG and Universal established their own online commercial site, GetMusic.

So far, so good. By 1999 the labels had begun to establish some kind of online presence and had acknowledged the possibility of new distribution methods. They had not, however, grappled with the central issue of the digital frontier—with the ease of piracy let loose by technology and the pressure this placed on existing rules of property. As the spread of Napster yanked this issue to the fore, the labels realized that they had to become more aggressive. They had to reconstitute some structure of property rights in cyberspace and some means—either technical or legal—of defining pirates and stopping them. Obviously, though, these were exceedingly difficult tasks. And as the record companies gradually realized, the only way to accomplish them was by working together, not only with each other this time, but with all the other players who had some claim to property along the digital frontier.

---

[44]Quoted in Steven V. Brull, "Net Nightmare for the Music Biz," *BusinessWeek,* March 2, 1998, p. 89.
[45]CDnow was subsequently purchased by Bertelsmann in July 2000.

Working Together

In February of 1999, a rather odd meeting occurred. Present at the Los Angeles location was a striking combination of bedfellows. The suits were all there, of course, for it was their meeting. But so, ironically, were Diamond Multimedia (creator of the Rio) and RealNetworks (a leading provider of Internet "jukebox" software), plus the now-legendary Chiariglione. This was the start of the Secure Digital Music Initiative (SDMI), a massive effort by the record companies to attack online music by co-opting it.

On the surface, SDMI looked an awful lot like the original MPEG group. It had a similar mixture of technical experts and industry reps, and a similarly broad agenda hidden in a mass of complex detail. It had Leonardo back at the helm as well, presiding once again over an endless series of meetings. Like MPEG, SDMI was devoted to Chiariglione's original concept: that is, to a world bound together by common standards and pushing at the outer frontier of communications technology. Behind the scenes, though, SDMI was clearly the record industry's baby. It wasn't about common standards for the sake of standards, and it didn't seem particularly interested in pushing the technological frontier. Rather, SDMI was a savvy attempt to use technology and standards to recreate the very rules that MPEG-2 had threatened. It was in fact the counter-revolution to MP3: a cooperative effort that aimed to create a new technology for online music, one more amenable to the old system of controlled distribution and property rights. And it had a great deal of support.

It's not clear just who hatched the concept of SDMI, but the idea behind it was simple. By 1999, the record companies realized they were fighting a losing battle against MP3, and particularly against the flow of unlicensed, illegally downloaded music. It was just too easy for listeners to gain access to online content, and too tempting to play and distribute this content on nifty devices such as RealNetwork's Real Jukebox or the much-touted Rio. Were this flood were to continue, Chuck D and the rest of the prophets would indeed be proven right. The record companies would drown and artists would find other, perhaps more sympathetic, channels for distributing their music. The problem, though, was that by 1999 Chuck D was drowning too. For every time that a fan downloaded, say, "Burn, Hollywood, Burn," he wasn't only depriving the labels of money; he was depriving Chuck D as well. Not to mention the marketers, and promoters, and even commercial sites such as EMusic.com (successor to GoodNoise), which were still about *selling* songs rather than giving them away for free. In the initial flush of excitement, new players to the MP3 game had seen the lack of rules and property rights as the source of freedom. As the game wore on, however, they quickly saw the other side of chaos—the side that meant no control and little cash. And quietly, begrudgingly, they began to embrace the record companies' agenda.

Part of this agenda was about formal, legal rights. In 1997, the major labels had rallied behind passage of the No Electronic Theft Act, which made it explicitly illegal to reproduce or distribute copyrighted works (including digital recordings) in cyberspace. In 1999, they joined forces to promote three new pieces of legislation: a U.S. Digital Millennium Copyright Act, which would actually expand copyright holders' rights in cyberspace; an EU Copyright Directive, which would extend similar rights across the European Union; and a treaty under the auspices of the World Intellectual Property Organization (WIPO), that would—officially at least—stretch these same kind of cyberrights across the global music industry. They also launched another major flurry of suits: against MP3.com; against Scour (an online service that finds music or movies already posted on the web); and, most famously, against Napster.[46] While the RIAA or individual labels were the major movers in all these cases, they were joined increasingly by members of the artistic community, who saw their own property rights reflected in the industry's push. The issue, these artists claimed, was not about music or money—it was about property, and about the legal

[46]For more on these suits, see Stephanie Stoughton, "Online Site Gets Huge Fine for Music Use," *Boston Globe*, September 7, 2000, p. 1; Matt Richtel, "Judge Rules on 2 Issues in Music Case," *New York Times*, May 9, 2000, p. C8; and Amy Harmon and John Sullivan, "Music Industry Wins Ruling in U.S. Court," *New York Times Online*, April 29, 2000.

rights that adhered to this property even when it was flung into cyberspace.

A second part of the artists' agenda, and in many ways a more important one, concerned the informal standards that prevailed within the music industry. During the first wave of MP3, many artists had responded to the technology with either enthusiasm or disinterest; they either ignored it entirely or, like Chuck D and his admirers, embraced it as the wave of a better future. If David Bowie was posting on his web site, people reasoned, and Chuck D was rapping rhapsodic about it, how bad could the thing be? As the online industry evolved, however, many of these same artists began to see the underside of digital distribution and the ill effects it could have on their own careers. Yes, for obscure groups like Slug Oven, the Net offered an unparalleled chance to reach listeners more directly and perhaps, even, to drive sales of the group's music. But for Bowie or Springsteen or any of the more established groups, free downloads were just a hit on legitimate sales. And thus many of these artists began to speak more directly against online music, or at least against the practice of copying and distributing unlicensed tracks. In 1998, Prince (or The Artist Formerly Known as Him) did a remarkable turn-around, threatening to sue his fan sites for illegal distribution only months after announcing that he would distribute music only over the web. Oasis, a popular British band, also threatened to sue fans for offering online music, and Metallica led a high-profile charge against Napster in 1999 and 2000. Led by Lars Ulrich, the band's scruffy and hardly establishment drummer, Metallica filed its own suit against the online site, presenting Napster with a list of 317,377 users that had allegedly violated the band's copyright.[47] Less famous artists also started to complain that the much-touted benefits of Internet distribution had yet to materialize. "I've sold a total of one CD online," griped an independent musician who posted her songs on MP3.com, "and I think that's pretty typical."[48] Indeed, in August of 1999, MP3.com sold a total of

---

[47]In response to Metallica's claim, Napster removed the offending users from its site. See Margaret Kane, "Napster Boots 317,377 Users," *ZDNet News,* May 10, 2000.

[48]Quoted in Emily Vander Veer, "Singing the MP3 Blues," Salon.com, December 2, 1999.

15,600 CDs on behalf of 26,700 artists—or about half a CD per artist and roughly $3 a month.[49] Such desultory results prompted one representative of the American Federation of Musicians to complain, "The bottom line is that these companies want to distribute recordings without even attempting to compensate the artists. We're very concerned."[50]

Meanwhile, a similar commercial impetus was also dragging the new music companies, such as EMusic.com and RealNetworks, closer to the record companies' camp. On the one hand, online distribution gave these companies their entire reason for being: EMusic.com rose to prominence as one of the leading distributors of online tracks, and RealNetworks grew on the back of powerful technology that enabled listeners to play (or "stream") their online music in real time. On the other hand, though, the proliferation of free music and the absence of online property rights also squeezed the profit potential of these firms and many like them. So long as property rights remained fuzzy, it would be difficult for any online music site to gain access to the vast libraries that the labels still controlled, or even to control the newer music that they themselves were rapidly throwing online. In the early stages of e-commerce, when companies such as EMusic.com and RealNetworks were soaring on their own stock valuation, they didn't need to worry so much about revenue models. But as the Net evolved and early business movers looked increasingly like established giants, the absence of property rights and established procedure became more obvious. And in pursuit of some semblance of rights and procedure, many of the "new" music companies began to sidle closer to the old ones, seeking, as one executive with RealNetworks explained, to "reduce chaos" and be in the vanguard of a "legitimate online market."[51]

A final player in this strange coalition were the consumer electronics firms, such as Samsung, Hitachi, and Diamond Multimedia, which made the devices that served the music industry. Even more than the new media companies, their interest lay in standardization, in creating some common platform that all

[49]Ibid.
[50]Quoted in Vander Veer, "Singing the MP3 Blues."
[51]Interview with author, Seattle, Wash., October 1999.

users and distributors of music would eventually share. To some extent, of course, MP3 was already this platform. It was an official standard of the MPEG group and the unofficial standard of online music community. Yet because the record labels loathed MP3 and feared its widening spread, they had refused, of course, to convert their own libraries or cutting-edge music to the MP3 format. And until they did, the market for any kind of MP3-related hardware was going to be severely limited. Yes, there were the hordes of students and underground fans of Slug Oven and such; but the bulk of the market, the core of the conservative, mainstream, music-listening public wasn't there yet. Which meant that it was considerably riskier for any consumer electronics firm to invest in a mass-market gadget—particularly if it feared that the standard it chose would eventually be overturned by something else. This is exactly what had happened in the early days of video cassette recorders, when the reluctance of the movie studios to release movies in any video format had prolonged a battle between VHS and Betamax and hurt, in the process, all companies who rushed to the soon-to-be-obsolete Betamax. Nobody wanted that to happen again, and the best protection was some kind of common standard, some platform that would ease the fears of the record labels and convince them to convert the mainstream content that was still so critical to the entire industry's success. The electronics companies didn't care so much what the standard *was*; they just wanted one that would have some permanence and allay the labels' fears.

And thus, over the course of 1999, Leonardo Chiariglione presided once again at a series of endless meetings and hotly contested debates. From the start, SDMI set itself a swift and ambitious target. The goal was to create a new standard for digital music, something that would enable listeners to access on-line music while still protecting the copyrights that adhered to this music. The standard had to be based on an "open architecture"—that is, it had to allow all sorts of programs and devices to communicate seamlessly with one another. It had to be pirate-proof. And it had to be done by Christmas of 1999, when most analysts predicted that the market for Rio-like devices would explode. To achieve these targets, members of SDMI divided their work into two tranches. First, they planned to create a preliminary specification for portable devices, something that would enable machines such as the Rio to play digital music files in a more secure format. Initially, this specification would still allow users to listen to any kind of digital file, including both legitimate and illicit MP3 files. Then, in Phase II, a new set of specifications would enable music providers to release their songs in an advanced, SDMI-compliant format, one that would prohibit users from copying a song more than a certain number of times. To hear this newly secure music, users would have to upgrade the software in their hand-held devices. And once they did so, the device would refuse to download any music that it recognized as being pirated.

Technically, crafting such specifications was exceedingly tricky, since they entailed a considerable amount of inter-operability and overlap with previous formats. Politically, it was even worse, since the SDMI remained a large coalition of very strange bedfellows, many of whom were competing against each other and nearly all of whom still resented the record companies' dominance. Yet somehow, Chiariglione prevailed once again; and on June 28, just two days before its self-imposed deadline of June 30, SDMI released its Portable Device Specifications. From here on in, proclaimed the SDMI in its official press release, portable devices such as the Rio would "respect the usage rules embedded in music by its creators." "This," they predicted, "will enable new business models that will provide consumers with new ways to enjoy the latest music."[52]

But then, even as the industry was gearing up for the promised Christmas rush, SDMI appeared to slow down. It pushed ahead with its working groups, and selected a firm to create the digital watermarks that would differentiate SDMI-compliant files, but it didn't develop the full Phase II specification. Christmas of 1999 came and went, without either a new set of specifications or the much-promised rush to hand-held players. By the spring of 2000, the music community was abuzz with doubts about SDMI and complaints—once again—about the record industry's motives. "If one thing is certain in the battle over Internet music," declared an editorial in the *Industry Standard*,

[52]Secure Digital Music Initiative, "SDMI Announces Standard for New Portable Devices," Press Release, June 28, 1999.

"it's that the Recording Industry Association of America's new audio format, the Secure Digital Music Initiative, is doomed. The RIAA's attempt to head off the MP3 juggernaut with a standard that no one is using—and a standard from which many of its members are already diverging—will go away quickly, as do most RIAA attempts to bend technological advances to fit its antiquated notions about copyright."[53] Going even further, one of the founders of EMusic griped in a public forum that because SDMI had failed to deliver its proposed specification, the initiative was now "dead as a doornail."[54] SDMI, of course, denied all reports of its demise and disputed any undue allegiance to the established recording industry. "The notion that SDMI is a Satanic conspiracy of special interests is hyperbolic," said one participant. "It's a very inclusive process."[55] The RIAA's Rosen also defended the process, noting that before SDMI, "the music and technology industries had spoken completely different languages." "Now," she commented, "at least they have each other's glossaries."[56]

Meanwhile, as SDMI was being torn from within and battered from without, members of the coalition began to grope for some other kind of standard—something that would achieve the technical objectives of SDMI without being dragged down by its increasingly massive weight. In January of 1999, for example, even as it was participating in SDMI, Liquid Audio, a leading developer of online audio systems, simultaneously created the Genuine Music Coalition, a group of hardware and software vendors, independent record labels, rights societies, and others, who were also trying to cobble together a common system for digital audio. With fewer members on board, and apparently less tension, the Coalition rapidly created its own standard—a sort of interim SDMI that allowed content providers to mark their music with a tamper-proof stamp and thus guarantee that certain songs had been obtained and recorded

[53]Jimmy Guterman, "MP3 Death Watch," *The Standard.com,* April 5, 1999.
[54]"Panel Spars Over Need for SDMI," *Webnoize,* November 17, 1999. (news.webnoize. com)
[55]Talal Shamoon of InterTrust Technologies Corporation, quoted in "Panel Spars over Need for SDMI," *Webnoize,* November 17, 1999 (news. webnoize.com).
[56]Interview with author, Washington, D.C., March 8, 2000.

legally. Unlike the goals of SDMI, though, there was no way for Genuine Music's stamp to prevent future piracy of legitimate songs, or to prohibit users from listening to pirated content. Not surprisingly, then, the major labels refused to endorse the standard. But they did begin to experiment with other forms of watermarking and other kinds of technical solutions: Sony, for example, formed an alliance with IBM and agreed to develop compatible technologies for downloading, storing, and playing digital music; Universal formed a similar alliance with Inter-Trust Technologies; and EMI and BMG each declared their support for Microsoft's newly announced audio format.

Conceptually, the labels clearly understood that their strength lay in numbers, and in a collective response to the impending crush of digital distribution. For despite the vast change that had already roiled through the music industry, the labels still possessed some of the industry's most valuable assets. They had the artists, the distribution muscle, and the rights to hundreds of thousands of songs. Or, as one top executive phrased it: "We may be slow and stupid, but we still own the business."[57] If the labels could join forces with one another, and then with the other facets of the industry that shared their underlying faith in copyright, stability, and profits, they could conceivably establish a new set of rules to protect their old kind of property. This was the brilliance of SDMI and the momentum that drove it forward. Practically, though, SDMI was a treacherous venture to pull off. It involved too many interests, too many animosities, and a staggering level of technological change. If the industry could somehow manage to hold this ragtag coalition together, then the shift to online music might well play out like the shift to rock 'n' roll. There would be an explosion of creativity and a raft of new players, but then, slowly and quietly, the power would return to the established companies—to the deep-pocketed labels who still knew better than anyone else how to make and market recording stars. But if SDMI fell apart, then each of the labels would be forced to rely on its own resources, struggling to stake some claim in the newly chaotic world of music. The old world of music would be forever changed, and the powers and rules replaced by a different set of

---

[57]Interview with author, New York, N.Y., March 8, 2000.

players. And that, of course, was precisely what many observers were predicting. To quote Chuck D again: "MP3 is like an asteroid—it's deep impact. And when it hits, it's going to knock out all the dinosaurs. They may come back in some form, but this time, they're gonna be salamanders."[58]

Fast Forward:
The Possible Paths of Online Music

Unlike the cases of encryption and antitrust, the story of online music does not yet have an ending. As of mid-2001, there has been no clear resolution to the issue of online piracy, to Napster and SDMI, or to the fate of traditional labels in a digitally distributed world. Will the pirates kill the suits for good? Will the music industry shatter into tiny shards of commerce? Will music even be *sold* in an online world? It's simply too early to tell. The industry is still only less than a decade old and the technology is still evolving. By 2002, MP3 may well be an obsolete format, superseded by something more powerful and even less understood.

Amidst all this uncertainty, however, one can still discern the outlines of a pattern—faint lines of the progression we have already seen in telegraphy and radio and other key junctures along the technological frontier. Like these other new markets, online music began with the innovators. It began with the engineers and scientists who crafted MIDI and WAV and then MP3. It began, too, with the work of Chiariglione and the MPEG group and with the conviction that standardization could make computers sing. This was the first phase of online music, when technology preceded commerce and the general public had no idea what was underway. It lasted roughly from 1983, when MIDI was created, to 1994, when MP3 was formally adopted by the MPEG group.

The next phase was dominated by a familiar group of followers: the pirates, prophets, and pioneers who flocked to cyberspace in the late 1990s and reveled in the prospects of online music. As is common along the frontier, the pirates paved the

---

[58]Interview with author, Cambridge, Mass., February 25, 2000.

way to this second phase, exploring the possibilities let loose by technological change and prodding at the shifting limits of the law. They were joined in due course by two waves of pioneers: artists such as Chuck D and David Bowie, who saw the new space as a way to circumvent the established structures of the music industry; and then commercial pioneers such as Mike Robertson of MP3.com and the founders of RealNetworks, who built towering commercial ventures on the back of all this change. And in their wake, of course, came the prophets, who saw the flow of change and predicted a revolution.

At the turn of the twenty-first century, online music was lunging into the third and most chaotic phase of its evolution. This was the phase of creative anarchy, a phase marked by a continued roar of enthusiasm and a notable lack of rules. During this phase, power seemed to veer almost palpably from the record companies to the pirates, creating a vacuum of control and uprooting even some of the industry's most sacred tenets: copyright law, for example, and the basic notion that music was something to be sold rather than shared. This phase began sometime in 1999, when digital distribution moved out of the college dorms and into the mainstream. It is still underway in the year 2001.

In music, however, as in all the other industries we've seen, this period of anarchy seems unlikely to last for long. If the arguments of this book hold true, in fact, we should expect a certain level of order to return to the music industry before long, and a restoration, or perhaps a rewriting, of its underlying rules. To be sure, many observers of the digital music scene claim that technology has already dealt a death blow to order, eliminating the need for property or any kind of intellectual property rights. In a 2000 *Wired* article, for example, John Perry Barlow of the Electronic Frontier Foundation wrote that "the future will win; there will be no property in cyberspace."[59] He is echoed by scholars such as Harvard Law School's Charles Nesson, who suggests, "The whole structure of intellectual property, and the value of it, is changing . . . The technical barriers are dropping almost to zero, and . . . law itself is going to get embarrassed or weakened."[60]

---

[59]John Perry Barlow, "The Next Economy of Ideas," *Wired*, October 2000, p. 241.
[60]Quoted in John Gibeaut, "Facing the Music," *ABA Journal*, October 2000, p. 38.

Such arguments have a lofty, end-of-millennium feel. They are inspiring in many respects, and desirable. What music lover, after all, wouldn't like an endless stream of easily accessible, totally free music? Who could disagree with the vision of some great global meetinghouse of wholly public ideas? The problem, however, is that despite the massive technological change to which Barlow and Nesson refer, and despite the attractions of online listening and the undeniable proliferation of downloadable music, the fundamental drivers of the music industry remain largely as they were in the dark ages of CDs. There are still a relatively small number of people who record music, and a relatively large number who like to listen to them. There are still more artists than the average listener can attend to, and preferences that vary with age, location, and personal taste. Most important, there are still commercial motives lurking only barely beneath the surface of all this art. True, some musicians, like Chuck D, see their work primarily as a social vehicle and don't seem to care about giving it away for free. Others can give it away and still make enough money from T-shirts or concert tickets. Yet both of these groups are relatively small; most artists still make most of their money from the sale of recorded music. If these sales go away, then so too will the artists' proceeds. Moreover, simply producing this music demands a large and hungry infrastructure—not just the lone artist with her guitar, but also the songwriter, the backup musicians, the video producer, the independent promoter, the radio station, and so forth. Each of these players contributes to the final product and none, presumably, is willing to work for free.

If we assume, then, that online music will not necessarily mean the end of commercial music, then some system of property rights must be maintained or recreated. There must be some way for artists to own the property they create, or at least to earn proceeds from it. Accordingly, as the pioneers begin to stretch their muscles and voice their demands, the distinctions between property and piracy are likely to be clarified. The established labels will pressure both courts and politicians to extend the reign of copyright into the realm of cyberspace, and the politicians and judges, it appears, are likely to comply. Already, this trend is underway. In April of 2000, for example, a federal judge ruled that MP3.com had violated copyright law by compiling a huge database of downloadable music[61]; in August, another judge prohibited a web site from publishing programs that allowed users to copy digital video disks (DVDs) by breaking their encryption codes[62]; and in December, the U.S. Copyright Office ruled that radio stations that broadcast over the Internet had to pay licensing fees for the songs they played.[63] In the most high-profile case, meanwhile, the RIAA won a powerful victory against Napster in July of 2000, when Judge Marilyn Patel sided with the recording industry in the case of *A&M Records, et al. v. Napster* and issued a preliminary injunction that essentially ordered Napster to halt its operations. A federal appeals court upheld this ruling in February 2001, adding that Napster was also potentially liable for copyright infringement.[64]

To be sure, critics such as Barlow can still argue that any judgments against digital downloads are both irrelevant and doomed. They can predict that the public at large will simply ignore these provisions or rulings, opening a gap between law and practice that will, in the end, prove unsustainable. And perhaps they're right. Yet the evidence thus far suggests otherwise. When Metallica and two record labels sued Indiana University for copyright infringement, for example, the university immediately banned its students from accessing Napster. Oklahoma State University likewise seized one of its student's computers after the RIAA alleged that he was using the school's server to offer free music; and more than two hundred colleges had followed suit by the fall of 2000, banning Napster use on campus.[65] Similarly, after AOL executives discovered that programmers at their Nullsoft division had created a serverless version of Napster called Gnutella, they instantly pulled the program from their web site[66]; and in April of 2000, one high-speed cable ser-

---

[61]Harmon and Sullivan, "Music Industry Wins Ruling."

[62]Christopher Grimes, "Hollywood Wins Case Over DVD Hacking," *New York Times*, August 18, 2000, p. 4.

[63]Matt Richtel, "Record Labels Win Internet Ruling," *New York Times Online*, December 9, 2000.

[64]Lee Gomes and Anna Wilde Matthews, "Napster Suffers a Rout in Appeals Court," *Wall Street Journal*, February 13, 2001, p. A3.

[65]Nicole St. Pierre, "Why More Schools Are Expelling Napster," *BusinessWeek*, October 9, 2000, p. 94.

[66]Jerome Kuptz, "Gnutella: Unstoppable by Design," *Wired*, October 2000, p. 236.

vice provider told its users that they would lose their cable modem accounts if they continued to use Napster.[67] The implications of such cases are clear: so long as there are some physical choke points on the web, and some mass of essentially law-abiding citizens, the problems of enforcement may not be half as severe as many prophets foresee. For in this case the object of theft is not covert information or banned masterpieces, it is Britney Spears's latest hit, available down the street for $15.99 and soon, in an authorized online format, for even less.

Which brings us to the final force for order on the frontier. In the heady days of the late 1990s, online music ventures were among the darlings of the red-hot NASDAQ. MP3.com, for example, went public in July of 1999, and was soon trading at $48.50 a share. Launch.com, another online music site, was selling for $27.50 a share in February of 2000, and Napster raised $15 million in venture capital in May.[68] By late 2000 and early 2001, however, it was becoming increasingly clear that most of these firms were not making any money and had little prospect of ever doing so. Launch.com fell steadily throughout 2000, hitting $3.12 by November; MP3.com slipped to $3.81, and Scour filed for Chapter 11 bankruptcy protection in October.[69] As the prospects for these companies continued to unravel, the situation in the digital music industry became eerily reminiscent of the old British East India Company. The pirates were important now; they had proven their threat and potential; but they were also in fairly desperate financial straits. And thus, just like the British traders, they turned back to the establishment as the spoils of their craft began to run dry. On October 31, 2000, Napster announced that it had signed a deal with BMG, a subsidiary of media giant Bertelsmann. Under the arrangement, the two newfound partners would use Bertelsmann's money and Napster's technology to develop a fee-based

---

[67]Chris Oakes, "Napster Not at Home with Cable," *Wired News,* April 7, 2000. (http://www.wired.com/news/technology/0,1282,35523,00)

[68]See Erik Gruenwedel, "Launch Adds RealPlayer Platform to Music Site," *MEDIAWEEK,* November 20, 2000, p. 37; and Alex Salkever, "A Faustian Deal for Napster?" *BusinessWeek,* December 4, 2000, p. 126.

[69]Benny Evangelista, "Napster Made Recording Industry Change its Tune," *San Francisco Chronicle,* December 26, 2000, p. D1.

music service and a tracking system that would allow labels and artists to be compensated for copied music files.[70]

In the final analysis, then, the story of digital music harkens back in many ways to Prince Henry and the pirates of yore. Like them, it is a story of adventure and exploration; of pioneers who charted new worlds and trumpeted great promises upon them. Like the old explorers, many of digital music's pioneers were men of great conviction,[71] who saw their mission in religious as well as commercial terms: to enlighten the natives as well as steal their gold. As events developed, however, these missionaries ran into the same problems that confounded their ancestors. There were too many pirates afoot in the new territory and not enough property rights, too many players and not enough profit. During the great age of navigation, these problems of anarchy were solved, slowly but surely, by the creation of property rights: following the chaos of the sixteenth-century conquests, order was eventually established in the New World and pirates were banished from the seas. A similar logic is liable to hold sway for digital music, and indeed for digital content in general. Yes, digital distribution fundamentally shifts the underlying rules of the music industry. It challenges the established system of property rights and threatens the survival of the world's leading record labels. But what the prophets of digital nirvana forget is that unless the music industry abolishes its capitalist urges completely (an unlikely proposition), people who make the music will still need some way to make money as well. And to do so, they, like the trading companies before them, will need to carve out a system of property rights and a way of differentiating themselves from the pirates they may closely resemble.

What distinguishes the digital music industry from some of its ancient forbears, however, are the methods by which property rights are likely to be established and secured. In the seventeenth century, the trading companies turned first to the method they knew best. This was physical force, of course, a primitive (albeit quite successful) mode of private regulation.

[70]Matt Richtel and David D. Kirkpatrick, "In a Shift, Internet Service Will Pay for Music Rights," *New York Times*, November 1, 2000, p. 1; and Amy Harmon, "Napster Users Mourn End of Free Music," *New York Times*, November 1, 2000, p. 1.
[71]With a handful of exceptions, the current pioneers—in online music, at least—were once again predominantly male.

When the force fell short or grew too expensive, they then turned back to the state, and to its growing coffers of self-defense. In digital music, by contrast, both individual efforts and state intervention have been relatively limited. Firms have relied instead on technological solutions to their property concerns, and on a novel blend of cooperation and standard-setting. They have experimented with cutting-edge technologies such as digital watermarks and cryptography, trying to stop piracy by physically impeding it. They have played with wide-reaching standards, such as the proposed second phase of SDMI, and with more narrow proprietary formats. In all of these cases, the firms involved have rebuffed formal regulation in favor of a more private approach. Borrowing a page, perhaps, from BSkyB, they are trying to use one form of technology—encryption again—to outwit the implications of another. Unlike Sky, however, most of the music industry's efforts have been collaborative. And they have been spearheaded not by the newest pioneers, but rather by the more established firms that the pioneers hope to replace.

In many respects, therefore, the digital music industry continues to dance to its own tune. For there is quite a lot of cooperation among the leading music firms—more, perhaps, than we've seen in any other industry—and more reliance on formal *private* solutions such as SDMI. Even among established firms such as EMI and Polygram, there is still a certain aloofness from government and a reluctance to get too deeply involved in state-led legislation. This pattern suggests several things. First, it implies that in the much-trumpeted new economy it may indeed be possible for firms to use technology as an implement of governance, a way to protect themselves without having to resort to either force or the state. If this is so, it may lead over time to a very different configuration of power in the global economy, one that has been feted by prophets such as John Perry Barlow and described in some quarters as "the new medievalism."[72] Second, the pattern of evolution suggests that private firms can, perhaps, cooperate in the creation of new technical

[72]See for example Stephen J. Kobrin, "Back to the Future: Neomedievalism and the Postmodern Digital World Economy," *Journal of International Affairs* 51, no. 2 (1998), pp. 361–86.

standards and that these standards can serve both to advance an industry and to protect it.

Finally, the story to date of the music industry raises interesting questions about the relative power of new and established firms—the pirates and those from whom they would hope to steal. In most of the stories we've seen, it is the pirates who morph over time into establishment figures, shaping the rules of their new game and redefining the terms of engagement. Certainly this was the case with Sir Francis Drake, with the British East India Company, with Bill Gates, and Rupert Murdoch. In the digital music industry, by contrast, matters seem to have run in the opposite direction. If BMG absorbs Napster and converts it to a fee-for-service model, or if EMI and Sony continue to pursue start-ups such as CDnow and mimic their operations, then the established firms may wind up swallowing the pirates before they can do that much damage. To be sure, the ease of online piracy means that illicit downloads will always occur in teenagers' bedrooms and college dorms. Large-scale pirates will flourish in their usual offshore lairs and renegade musicians will flaunt their independence by circumventing the labels. But the labels themselves will also evolve, purchasing the rebel players as their venture capital begins to dissolve and developing new models of online commerce.

In the end, therefore, the online music industry may well be regulated by a combination of private and public forces. Firms will set industry standards; governments will enforce them; and even diehard music fans will eventually accept some revised notion of property rights in a digital age. This isn't the future that Chuck D has touted, of course; it isn't a world where music is free and record labels fade into commercial oblivion. It is, however, a world that aligns with what the political history of technology suggests. Markets need rules if they are to survive, and power—slowly sometimes, ironically perhaps, and often unfairly—flows to those who make the rules.

CHAPTER 8

Surfing the Barbary Coast

In 1993 the Nobel Prize for Economics went to Douglass North, a seventy-three-year-old professor who was then teaching at Washington University in St. Louis. North was in many respects an unlikely candidate for the esteemed Swedish prize. He worked in the field of economic history, something of a backwater for economists, and his work displayed little of the mathematical pyrotechnics that graced many of his colleagues' papers. Instead, North's work was a bold and meticulous account of economic development—of how civilizations had grown over thousands of years and why some, but only some, had prospered. And his conclusions, for an economist at least, were radical. For North suggested—insisted, really—that markets never prospered in the absence of a state. Individuals, he claimed, could regulate their businesses perfectly well in a small or traditional society; larger groups could form guilds or associations that would guide transactions as markets first grew. But eventually, if these early markets were to develop into large, efficient, and impersonal enterprises, then states had to step in, creating the rules and institutions that successful commerce demanded. Or, as North expressed it: "Undergirding . . . markets are secure property rights, which entail a polity and judicial system to permit low cost contracting."[1]

It is tempting at the turn of the twentieth century to dismiss North's findings as irrelevant, to argue that breakthrough technologies have laid the path towards a new social order and severed the historical link between market and state. North, after all, was writing about ancient markets and had only a peripheral interest in technology. He didn't know about cyberspace when he crafted his theories and, like most everyone else, had no idea that it was coming. Perhaps, then, there is no reason to believe that his view of history should still apply. Perhaps, as the cyber-libertarians argue, there really is something different about the digital age, something that reverses North's proffered connection between business and politics. Maybe we really are standing at the edge of history, looking forward to a time when, as one computer columnist predicted, "the exodus of the productive will . . . touch off a revolution that will result in the practical separation of commerce and state, a radical, outrageous, earth shaking development that will be a tremendous threat to the existing social order and will challenge the moral underpinnings of all governments as they are now conceived."[2]

The stories here, however, suggest otherwise. They suggest that despite the latest round of prophecy, despite the euphoria and venture capital and buzz, the advent of cyberspace is nothing but another arc along technology's frontier. This isn't to say that the Net is not radical, for it is. And it's not to argue digital technologies won't change the way we work and play and organize our societies, for they will. What the stories of this book suggest, though, is that the emergence of cyberspace is not an unprecedented event. The development of both the printing

[1]Douglass C. North, "Institutions," *Journal of Economic Perspectives* 5, no. 1 (Winter 1991), pp. 97–112.
[2]Bill Frezza, "The Crucible of Radical Capitalism: How the Information Revolution Will Transform the Politics of Power," *DigitaLiberty,* August 1994. (http://www.digitaliberty.com/crucible.html)

press and the compass, for example, had jarring, long-term effects; so did the invention of telegraphy, railroads, steamships, and radio. In all of these cases, the technology of the time seemed just as awesome as the Net does today. There were pioneers who stole the public's imagination, pirates who stole their loot, and prophets who foresaw utopia in the technology's advance. Yet as these technologies evolved, they traced much the same pattern as North might have predicted. They grew from small communities of like-minded inventors, were shaped for a glorious time by entrepreneurs and private rules, and in the end fell fully—or at least largely—under government's sway. The pirates of the Caribbean, for example, succumbed to the power of Britain's navy while the great telegraph companies were eventually standardized and regulated. The airwaves were ruled by a string of bureaucratic agencies, and even Murdoch's BSkyB was gradually reined in by the government it once scorned. In each of these cases, therefore, the ebb and flow of innovation led over time to a period of rules, a period when governments stepped into the evolving market and created some kind of stable framework around it.

In cyberspace, a very similar process is already underway. On both sides of the Atlantic, for instance, governments have indicated that the basic rules of competition—the same rules that stymied Marconi and dogged Western Union—will apply just as strictly across the new technologies. They have begun, as the case of MP3 demonstrates, to stumble towards new definitions of property and new means of enforcement. They have started to address major social issues such as privacy and to drag a handful of old policies—on consumer fraud, accounting conventions, and so forth—into the new space. In most parts of the world, at least, governments have not actively tried to regulate or police the Net. They haven't imposed censorship (with a few select exceptions) and they have not yet tried to raise taxes.[3] They have, however, begun to establish some basic rules: rules of property, of competition, of consumer protection and criminal behavior.[4] Standard procedure along the final tracks of the frontier.

[3]Taxes may be coming soon. See Howard Gleckman, "The Other Tax Battleground of 2001: The Internet," *BusinessWeek*, February 19, 2001, p. 49.
[4]For an interesting discussion of how these rules are evolving in Asia, see Ang Peng Hwa, "Policing Asia's Internet," *Asian Wall Street Journal*, September 7, 2000, p. 8.

Admittedly, there are some reaches of cyberspace that are likely to resist formal rules forever; there are some ways, as described below, in which cyberspace may prove as radical and ungovernable as the printing press. But in the broadest political sense, the most remarkable thing about cyberspace may be just how unremarkable it actually is. For there will be rules in cyberspace, and governments will help to craft and enforce them. Why? Because even along this wildest of frontiers, pioneers need property rights, and standards, and some basic understanding of what constitutes fair and foul play. And the only entity that can sustain and enforce these rules is the state.

Of Politics and Property

At one level, the connection between state action and property rights is no surprise. Indeed, political philosophers have underscored this link for centuries, maintaining that a fundamental function of the state—perhaps, in fact, *the* fundamental function—was to ensure the sanctity of its citizens' property. Cicero expounded upon it; Machiavelli promoted it; and the U.S. founding fathers took it as the very basis of their constitution.[5] In a provocative new book, Hernando de Soto takes these arguments even further, arguing (with echoes of North) that it is an absence of property rights, rather than economic mismanagement or bad luck, that has caused many parts of the world to languish for centuries in poverty.[6]

What is more surprising, perhaps, or at least less obvious, is

---

[5]There is a vast literature describing the relationship between states and property rights. See, for example, John A. James and Mark Thomas, eds., *Capitalism in Context* (Chicago: University of Chicago Press, 1994); Yoram Barzel, "Property Rights and the Evolution of the State," *Economics of Governance* 1 no. 1 (February 2000), pp. 25–51; John L. Campbell and Leon N. Lindberg, "Property Rights and the Organization of Economic Activity by the State," *American Sociological Review* 55, no. 5 (pp. 634–47); Harry N. Schreiber, "Regulation, Property Rights, and Definition of 'The Market': Law and the American Economy," *Journal of Economic History* XLI, no. 1 (March 1981), pp. 103–109; and Douglass C. North and Robert Thomas, *The Rise of the Western World: A New Economic History* (Cambridge: Cambridge University Press, 1973).

[6]Hernando de Soto, *The Mystery of Capital: Why Capitalism Triumphs in the West and Fails Everywhere Else* (London: Bantam, 2000).

how deeply the demand for property rights affects technology's advance. In the earliest stages of invention, property rights are often moot. Inventors such as Heinrich Hertz or Tim Berners-Lee push the limits of knowledge for its own sake, creating theories or devices that often have no direct commercial use. The fact that there may be no property rights in these theories or devices doesn't slow them, since profit is not generally the driving motive for their work. As the innovators give way to the commercial pioneers, however (and sometimes, to be sure, they are one and the same), the lack of property rights becomes distinctly more problematic. For now the game is about profit rather than knowledge, and to earn and keep the profits that they rightfully consider theirs, entrepreneurs must be able to employ their technologies freely and keep them from the grasp of competitive interlopers. They must have property rights, in other words, and some means of enforcing them. Without property rights, rival claims over the new market can rapidly disintegrate into chaos, stunting commercial development as it did in the early days of telegraphy and broadcast radio.

There are various ways in which these rights can be established. If the new technology is vested most directly in a physical device like a telegraph transmitter, existing patent laws can usually migrate toward the new arena. Such was the case with both telegraphy and radio, where fights among the pioneers were resolved through the normal patent reviews of U.S. and European courts. The procedures for establishing ownership were clear, rivals took their claims to court, and contests over property rights never truly clouded the technology's commercial future.

Matters become more complicated, however, when a technology reveals some previously untouched space. In these cases (think of oceanic exploration, broadcast radio, or satellite television), the property in question is bigger and broader than a single device or invention. It is a whole new realm, one without any clear demarcation or inherent forms of ownership. These are the cases that are most likely to slide into anarchy, as each of the pioneers struggles to maintain his own stake and keep interlopers at bay. Theoretically, it should be possible for the pioneers to resolve their squabbles amongst themselves.

They should be able, one might presume, to negotiate private contracts or protect their bit of turf. And sometimes, indeed, they can.[7] Usually, though, private schemes of either governance or enforcement fall prey to their obvious vulnerabilities. Tensions mount without an impartial arbitrator, and the physical costs of enforcement grow too high. Eventually, the largest of the pioneers (who have, after all, the most at stake) turn to the government, asking the state to provide a service that the market cannot. Such was the case with the U.S. broadcasting industry, for example, and with the early British trading companies. It also applies to the current world of digital music, although here the beseeching parties are the established firms and artists who fear that their existing rights will be trampled in a world of MP3.

However it occurs, the establishment of property rights is a crucial stage along the technological frontier. It clarifies relations that often will have been murky until this point and allows the successful pioneers to build their firms and markets in a more stable, less chaotic environment. It is a stage that is absolutely critical for a technology with commercial intent, and one that should make the Marxists proud. For as property rights are created, the state confers power—intentionally or not—to a discrete set of firms. And more often than not, the ones that receive this power are already the most powerful: the British East India Company, Microsoft (in its copyright fights with Apple), RCA in the 1920s, and the giant music firms today. These are the firms that lobby for property rights; that drag their innovation into patent courts and beseech the government to carve up disputed territories. They are the ones that use government to create a legal and enforceable system of rights, and then employ these rights as the very bedrock of their commercial strategy.

---

[7]An interesting example of private regulation along these lines occurred at the end of the nineteenth century, when diamond miners in the new fields of South Africa concocted an elaborate system for demarcating and protecting their plots. See Debora L. Spar, *The Cooperative Edge* (Ithaca: Cornell University Press, 1994), pp. 43–47. For a discussion of how private groups were able to monitor economic exchange before the emergence of a powerful state, see Avner Grief, "On Institutions that Facilitate Impersonal Exchange: From the Community Responsibility System to Individual Legal Responsibility in Premodern Europe," Stanford University Department of Economics Working Paper #97-016, 1997.

Recall, for example, how Amos Kendall used his telegraph patents to lock new competitors out of the U.S. market, or how the largest radio stations prodded the government to allocate the spectrum so that the smaller and less profit-oriented concerns were essentially thrown out of the market. This is exactly what firms such as Amazon and Priceline.com have done more recently, attempting to bolster their market position by patenting everything from one-click shopping to name-your-price auctions.[8]

In cyberspace, of course, the problem of property is complicated by its intangible nature. It is harder to patent a concept than a device (as Amazon may eventually discover) and harder to establish ownership over things that cannot be touched, or felt, or seen. Problems of nationality also hound cyberspace, since property is essentially a territorial concept and physical boundaries are not necessarily fixed upon the Net. Yet the weight of these problems is balanced by the pressure of private demands. Firms like Amazon need property rights to preserve what might otherwise be a fragile market position; firms like BMG or Sony desperately need to keep the Napsters of this world from traipsing all over their music and artists and rights. To solve these problems, firms will inevitably pressure the state—through the courts, the legislature, and perhaps even international bodies such as the World Intellectual Property Organization—and the state, history suggests, is likely to respond. This isn't to say, of course, that violations will not occur; they will. There will always be people like Chuck D who revel in a world without rules and are liable to defeat whatever property rights eventually settle upon the Net. But the Net isn't only, or even largely, composed of renegades. It is composed of thousands of lawfully minded companies and millions of basically law-abiding people, many of whom enter cyberspace through large and well-regulated portals such as AOL. Will some of these users follow the pirates and ignore whatever property rights are eventually imposed upon the Net? Undoubtedly. But

[8]See, for example, Tim Jackson, "Inside Track: Amazon's Patently Unique Association of Online Ideas," *Financial Times*, November 14, 2000, p. 16; and May Wong, "Patently Protective? Dot-coms Grabbing Legal Title to What Some Consider Just Normal Practices," *San Diego Union-Tribune*, July 13, 2000, p. C1.

will this behavior be blunt enough and big enough to compromise the rights themselves? Probably not. For most people, and certainly most companies, have a vested interest in the kind of economic stability that property rights help to ensure. They may not like the marshal, and may not trust him, but as markets expand, order trumps chaos more often than not. There is little reason to suspect that the Net will prove different in this respect from any of its predecessors.

Piracy and Playing Fair

The establishment of property rights is one of the most crucial events along the technological frontier. It allows the market to unfold in a predictable way, and gives pioneers a hefty dose of ownership and security. Most important, perhaps, the creation of property rights also marks the difference between pioneers and pirates, between those whose claim on the new technology is legitimate and those whose is not. In the earliest days of the frontier, such distinctions are often ambiguous. Was Rupert Murdoch a pirate or a pioneer when he circumvented British law to broadcast from Sky? How about Bill Gates, who "borrowed" the BASIC language and Apple-like graphics? Or Shawn Fanning of Napster? For a while, it doesn't matter, since the distinctions simply have no weight. But once property rights are defined, it becomes easier to separate the pirates from the pioneers, and to punish those who engage in piracy. Which is why the pirates-cum-pioneers often play such a large role in helping the state create an appropriate system of rights. Microsoft, for example, went to court time after time to establish that software could be owned and that copying constituted theft. Drake shed his pirate garb for the cloak of respectability and spent the latter half of his life catching pirates for the Queen. There was a redoubtable turnaround in both cases, and one that proved eminently profitable to the officially established pioneers. Shawn Fanning and Chuck D should pay heed.

Of course, it isn't only property rights that define the boundaries of behavior in new markets. Coordination is also a major aspect of this process, as is competition. In the case of coordination, either governments or firms can carve out the standards that allow different variants of a technology to link up with one another. States can establish standards before markets have even arisen (as happened, for instance, in the case of European telegraphy) or firms can serve as de facto agents, creating and deploying a common set of codes, norms, or protocols.[9] Sometimes, a firm like Marconi or Microsoft can even set standards by itself, creating a technical precedent so powerful and pervasive that anyone else who wants to play in this market will be compelled to follow suit. At a normative level, it's really not clear whether any of these methods are better than the others: markets can grow, and technologies flourish, under any of these circumstances. What is critical is simply that a standard does get set and that the key players in a market comply with it before long.

To be sure, though, there are differences between the various modes of standard-setting, differences that may affect how the digital economy rolls out around the world. For instance, if European governments continue along the path laid out in other technologies, they may well try to establish Internet standards in advance of a fully developed market. They may try—as indeed some already have—to create formal standards for technologies such as digital broadcasting or wireless telephony. They may create technical protocols to deal with issues such as privacy and carve out precise systems for exploiting the digital spectrum. In the United States, by contrast, the task of coordination is falling, as we might expect, to private firms and groups. Rather than turning to any kind of government standard for confidentiality, for example, many firms in the United States have adopted their own system of guarantees, seeking the endorsement of private organizations such as TRUST-e.[10] Standards for

---

[9]For a fascinating account of standard setting in a wholly different context, see Stephen A. Brown, *Revolution at the Checkout Counter* (Cambridge, Mass.: Harvard University Press, 1997).

[10]TRUST-e is a private, nonprofit organization that invites member firms to adopt a common privacy policy. If TRUST-e confirms that these firms are complying with its rules, it rewards them with a "trustmark," which the firms can post on their web site. See Denise Caruso, "An On-line Tug-of-War over Consumers' Personal Information," *New York Times*, April 13, 1998, p. D5; and Justin Matlick, "Profit Motive Promotes Internet Privacy," *Journal of Commerce*, June 4, 1999, p. 8. Since TRUST-e's debut, however, concerns have been raised over its actual ability to police online privacy, and many privacy advocates have begun to lobby for more formal solutions. See Marilyn Geewax, "Online

encryption and digital music are being hatched by industry con-
sortia rather than the state; and even the registration of domain
names has been left in the hands of ICANN, an independent
private group.[11] In places like China, meanwhile, the state has
been far more aggressive in setting explicit rules and standards
for all kinds of digital conduct.[12]

Eventually, all of these systems and standards are bound to
converge into some kind of global norm. Because the Internet is
inherently an international technology, and because communications across it ignore territorial boundaries, it is difficult to
imagine how a range of incompatible standards could coexist for
too long. And so if the Net is to become the global powerhouse
that its proponents foresee, it will need to develop a common set
of standards before long, equivalent to the kind of cross-border
standards that arose in telegraphy, radio, and transoceanic trade.
It is too early to tell just how these standards will be established—whether they will be hammered out by government negotiators, crafted by international agencies, or simply imposed
by the most powerful market players. But they are likely to
emerge in any case, and to undermine the blissful anarchy that
might otherwise rule the web.

In the case of competition, the road forward is somewhat
easier to predict. For if we look at the cases of Microsoft and
BSkyB, two of the most successful companies of the early digital age, it is clear that the old rules of competition have not died
away with the new economy. On the contrary, despite the impact of network effects in the satellite television and software
industries; despite the fact that industries like these need the
kind of standard-setting that Gates and Murdoch provided, the

---

Privacy Legislation Expected," *Atlanta Journal and Constitution*, December 12, 2000, p.
10E, and Michelle Rafter, "Trust or Bust?" *The Industry Standard*, March 6, 2000,
http://www.thestandard.com/article/display/0,1151,12445,00.html.

[11]ICANN, however, has also encountered a fairly predictable set of problems. Since its
creation in 1998, the agency has been beleaguered by outside criticism and by its own
inability to reach consensus. See Chip Bayers, "Mission Impossible," *Wired*, December
2000, pp. 130–50.

[12]See, for example, Bruce Einhorn, "Big Brother May Crush China's Web Dreams,"
BusinessWeek, February 14, 2000, p. 64; Lorien Holland and Trish Saywell, "Plugging a
Sieve," *Far Eastern Economic Review*, February 10, 2000, p. 20; and Bruce Einhorn et al.,
"China's Tangled Web," *BusinessWeek*, July 17, 2000, pp. 56–58.

weight of their presence still proved too heavy for the industry to bear. In the end, a widespread coalition of firms and state authorities determined that both BSkyB and Microsoft had not, in fact, played fair, and that the scope of their activities was simply too wide and too forceful. They determined, in other words, that Sky and Microsoft—like Marconi and Western Union before them—had exceeded the boundaries of permissible competition, even if those boundaries had not been explicitly extended into their sphere of operation. The implications of this movement are clear: in the United States and the European Union, at least, rules of antitrust will apply in cyberspace.

How these rules will be implemented, of course, is more mysterious. In the wake of the Microsoft case and protracted bouts with Sky, governments on both sides of the Atlantic may decide to use competition policy as an ever-present weapon, something to wield against firms that threaten to become too big or too powerful in the new economy. They may decide to join forces in this quest (indeed, early steps in this direction are already underway[13]) or to impose their notions of competition and antitrust upon the rest of the world economy. Once again, it is simply too early to tell just how this development will play out. If we project from the historical record, though, two probabilities stand out: first, that states will use antitrust to regulate the final stages of the technological frontier; and second, that private firms will solidly support this kind of intervention. For antitrust is in many ways the last gasp of the pirates—a way for embattled entrepreneurs to attack in the courts what they have already lost in the market.

Return to Partenia

Since this book began in the dusty sands of the Sahara, it is only fitting that it should end there, amidst the chat rooms and free expression of the renegade Gaillot. Because it is

---

[13]See Deborah Hargreaves, "US, EU Discuss Anti-Trust Cooperation," *Financial Times*, October 7, 1999; Guy de Jonquieres, "Crackdown on Hard Core Cartels Urged," *Financial Times*, June 6, 2000, p. 16; and Deborah Hargreaves, "Brussels and Japan Agree Competition Pact," *Financial Times*, July 20, 2000, p. 2.

critical to see that despite all the evidence of governments and rules, despite the prediction that rights will settle into cyber-space and order will replace chaos, Partenia still stands. Not even the Pope has been able to quash the irrepressible Gaillot or steal him from his outpost on technology's frontier. How can we explain this?

The answer lies with the motive behind Gaillot's message. Many of the other pioneers in this book—most of them in-deed—have come to the frontier in search of profits. The con-quistadores, Marconi, Sarnoff, Murdoch, and Gates—all were men of business. Great men, as it turns out, men who could see opportunities through the haze of change and knew how best to seize them. These were the pioneers who both broke the rules and needed them, the ones who wanted property rights and standards and laws once they had settled their businesses on what before was virgin turf. In all these cases, therefore, what drove the demand for rules were the dictates of commerce.

In Partenia, however, there is no commerce. Gaillot isn't try-ing to make money from his site, or to carve out any sort of competitive position. He is instead quite literally a man with a mission, a man who is using cyberspace as a conduit for ideas rather than a means of commerce. And this is the crucial differ-ence. For when cyberspace, like any new technology, is used to conduct business and make profits, it will need rules. And when it is used simply to transmit ideas, it may not.

This simple distinction belies a much greater difference, one that has run quietly through all the stories told here. Most tech-nologies, especially during their start-up days, have several pos-sible uses. Radio could be used by navies or awestruck boys; it could transmit military messages or gossip or opera. The com-pass was used by both wanderers and conquerors, and satellites can transmit news and information as easily as "The Simp-sons." If, as a technology evolves, its dominant use is noncom-mercial, then many of the rules described in this book will not apply. People don't need property rights, after all, if they're not trying to make a profit or establish ownership. And they don't care about competition policy if they're not in the business of competing. When these conditions apply—that is, when large chunks of a new technology are dedicated to noncommercial uses—then chaos becomes a more acceptable state of affairs.

The potential for chaos is particularly true, it appears, with regard to information. When technologies truly expand the scope of information flows, they enable all kinds of new activity—commercial as well as noncommercial—to thrive. They create new modes of interpersonal activity and, theoretically at least, ways for communication to slip around the established channels of authority. Because such possibilities can threaten the state, governments will often try to nip these new technologies in the bud, to stop the information flows before they become a real threat. We have seen this dynamic repeatedly: when the British smashed Marconi's radio, and the FBI halted encryption, and the Church banned Gutenberg's press. What is most interesting about these seizures, though, is how rarely they work. The British government did not stop the spread of radio; the FBI was not able to control encryption; and the Church could not stop Gutenberg. In a book that's all about rules, therefore, these cases stand as stark exceptions—cases where the pirates did in fact get away.

Which brings us back to Partenia. As stated at the outset, Partenia exists in many spots. It is in Libya and Algeria, in the Sahara and in cyberspace, everywhere and nowhere. Mostly, though, Partenia exists in the world of ideas. It is a place that has nothing to do with profits or commerce and everything to do with Gaillot's own thoughts and those of his visitors. If Gaillot's site were trampling onto the boundaries of someone else's commerce, there is a good chance that, over time, it would be ruled and regulated and constrained. This, for example, is precisely what happened to the radio boys of the 1910s and the independent radio stations of the 1920s. It's what happened to the Homebrew Club and is likely to befall Napster. In all of these cases, the free flow of information was ultimately compromised by problems of property rights and congestion: other people wanted access to a common resource and, armed with profits and greater political clout, they tended to get it.

If, however, the evolution of technology leaves distinct niches of noncommercial activity, these niches may be able to live in a relatively free, relatively unregulated environment. They may be able to disdain rules of property or competition and flourish quite happily under chaos. Most important, as the case of Gaillot demonstrates, they may be able to challenge the power of existing authorities, using information to sneak around the state.

And this, indeed, may be the central legacy of the Net. For like the printing press, the Net has both commercial and non-commercial uses. There are thousands of ways in which people can make money in cyberspace, and thus thousands of reasons why these people—at some stage, at least—are going to want the state to provide rules that protect their commerce. At the same time, though, there are also thousands of ways in which information can flow freely in cyberspace: in e-mails and electronic newsletters; in dissident reports and political critiques; in Chuck D's online music and Gaillot's Partenia. So long as the purveyors of this information do not harbor commercial intent, they may be able to avoid the rules that are likely to settle elsewhere in cyberspace. They may be able to stream their information around the authorities who would otherwise constrain them and create what could truly be a digital revolution.

The impact of this revolution—ironically perhaps—is likely to hit hardest away from the technological centers of the West. It will hit in China and the former Soviet Union, in Burma and Iraq, in Saudi Arabia and anywhere else that governments have tried to control the free flow of information. It will be, in this sense, more like the printing press than any of the other technologies described in this book—a way for ideas to escape the traditional bounds of power. The power of the printing press in this regard, though, came largely from the context in which it was hatched. In Europe of the Middle Ages, information was a tightly controlled commodity, tied to the Church and available to only a tiny scholarly elite. When the printing press arrived, therefore, it carved a path of circumvention that had never before been possible. A similar dynamic is likely to surround the Internet in places where governments—like the medieval Church—still try to control what their citizens read and know and think. In these places, it should act over time like an amplified version of satellite television, enabling information to seep around older methods of containment. To be sure, governments may well try to impose whatever technological or legal fixes they can muster; they may try, as Singapore and China already have, to ring their nations with proprietary networks or track and censor communications that flow along them. Such efforts, however, are almost certainly doomed to fail. For technologically, the Internet is a porous web, and if people in China or Iraq want to get information from sites in Silicon Valley, even the most omnipotent of governments will be hard pressed to stop them.

Notice, though, that the anarchy inherent in this relationship does not extend to all corners of the Net, or even to all corners of the world. Sometimes, as in the case of Partenia, people will see and use the Net as a way of distributing information that has no price. Like the lovers who used telegraphy to send encoded notes, and the radio boys who used the ether to gossip and tell tales, these people will be able to seize the new technology's anarchic side. They will be the ones who truly can revel in anarchy and who can use cyberspace to get around the tired laws of nation states. It is critical to realize, however, that these people are not the only denizens of cyberspace. Instead, as we have seen, cyberspace is populated mostly by the same folks who live in the real world: by consumers who want to shop, and parents who want to protect their children, and artists who have no desire to give away their work for free. As these citizens make their claims heard, the state is likely to respond. And its response, more often than not, will take the form of rules.

In cyberspace, though, the greatest pressure for rules is not likely to come from consumers, or parents, or even artists. Instead, it will come from the same direction in which it usually hails, and from the same groups that tend to dominate politics along most technological frontiers. It will come, that is, from private firms, and from those who have cast their commercial lot with the Net. Some of these people, particularly in the early phases of the frontier, may have sounded an awful lot like Gaillot. They may have trumpeted their freedom from government's yoke or prophesied a world of itinerant ideas and untaxed trade. As the technological cycle unwinds, however, and the pirates turn bourgeois, the joys of anarchy pale before the demands of profit. The pioneers want property rights, as we have seen; they want standards and competition laws, and they turn to the state as private regimes fail. It is this commercial pressure that compels the state to regulate what was once untamed turf, and to create rules on top of chaos.[14]

There is a delicious irony in all this, a twist that would have made Marx smile. For it turns out that even as capitalism shifts and evolves, even as technologies push both business and society far beyond what was once even barely conceivable, the patterns of power remain unchanged. There are exceptions, of course, and people like Martin Luther and Gaillot who can use new technologies to break the binds of power without hoping to profit by them. But once commerce enters the game, it appears, the players demand both rules and a state to enforce them.

It is impossible to predict, of course, how the technological cycle will play out over time. It may continue in an endless wave of rules, with each successive technology eliciting its own political struggle and custom-made regime. The entire cycle could migrate, theoretically at least, to the international level, or it could pit firms and governments against a newly emboldened band of consumers.[15] Its duration will shrink, almost certainly, as the pace of technology increases, and it could disappear entirely.

Already, though, there are signs that some of the impending revolutions will unfold much like the present ones. In biotechnology, for example, an initial round of innovation rapidly gave way to a surge of commercialization, led by pioneers such as

[14]This phenomenon has been described somewhat more formally as "regulatory capture," a process in which firms or industries demand regulation in order to advance their own economic interest. The standard treatment of this process is George J. Stigler, "The Theory of Economic Regulation," *Bell Journal of Economics and Management Science* 2 (Spring 1971), pp. 3–21. See also Sam Peltzman, "Toward a More General Theory of Regulation," *Journal of Law and Economics* 19 (August 1976), pp. 211–40; Gary S. Becker, "A Theory of Competition Among Pressure Groups for Political Influence," *Quarterly Journal of Economics* 98 (August 1983), pp. 371–400; and Richard Posner, "Taxation by Regulation," *Bell Journal of Economics and Management Science* 2 (Spring 1971), pp. 22–50.

[15]For a provocative argument along these lines, see Noreena Hertz, *The Silent Takeover: Global Capitalism and the Death of Democracy* (New York: Random House, 2001).

Craig Ventner of Celera and companies like DoubleTwist and Incyte.[16] As one might imagine, these pioneers clamored almost immediately for some system of genetic rights or patents; and they were met in the halls of power by an equally committed set of lobbyists, arguing against property rights and in favor of a strict regulatory regime. Different rules on both sides, to be sure, but rules all the same. A similar process seems likely to emerge in space, once technology's advance clears the way for the next round of exploration, expansion, and eventually, commerce.

Standing at the start of the twenty-first century, though, it's difficult to imagine what revolutions the coming decades will bring. Perhaps politics will, in the end, fall prey to technology. Perhaps nation states will disintegrate like feudal lords, leaving some new form of authority—or even anarchy—in their wake. Maybe innovation is itself the key to governance, full of possibilities for new structures and more efficient forms. Or maybe not. At this moment in time, it's simply too early to tell, for any long-term predictions will bear the stain of our preconceptions, of our own view of what the past means and the future is certain to bring. And thus we are left like Prince Henry of Portugal, sitting alone at the edge of world and dreaming of MP3.

[16]See Tom Abate, "Seeing Green in Genes," *San Francisco Chronicle*, August 28, 2000, p. B1; and Esther Dyson, "DoubleTwist's Program: The Human Genome's Equivalent of Napster," *Los Angeles Times*, August 7, 2000, p. C2.

SELECTED BIBLIOGRAPHY

CHAPTER I: THE FIRST WAVE

Beazley, C. Raymond. *Prince Henry the Navigator.* New York: G. P. Putnam's Sons, 1894.

Boorstin, Daniel J. *The Discoverers.* New York: Random House, 1983.

Brown, Lloyd A. *The Story of Maps.* Boston: Little, Brown and Company, 1950.

Gosse, Philip. *The History of Piracy.* New York: Tudor Publishing Company, 1934.

Hebb, David Delison. *Piracy and the English Government 1616–1642.* Brookfield, Vt.: Ashgate Publishing Company, 1994.

Keay, John. *The Honourable Company: A History of the East English Company.* New York: Macmillan, 1991.

Kimble, George H. T. *Geography in the Middle Ages.* London: Methuen & Co. Ltd., 1938.

Lane, Frederic C. *Venice: A Maritime Republic.* Baltimore: The Johns Hopkins University Press, 1973.

Miskimin, Harry A. *The Economy of Later Renaissance Europe 1460–1600.* London: Cambridge University Press, 1977.

Mokyr, Joel. *The Lever of Riches: Technological Creativity and Economic Progress.* New York: Oxford University Press, 1990.

Parry, J. H. *The Discovery of the Sea.* New York: The Dial Press, 1974.

Prestage, Edgar. *The Portuguese Pioneers.* London: A. & C. Black Ltd., 1933.

Pirenne, Henri. *Economic and Social History of Medieval Europe,* trans. by I. E. Clegg. New York: Harcourt, Brace & Company, 1956.

Rankin, Hugh F. *The Golden Age of Piracy.* New York: Holt, Rhinehart & Winston, Inc., 1969.

Ritchie, Robert C. *Captain Kidd and the War against the Pirates.* Cambridge: Harvard University Press, 1986.

Sherry, Frank. *Raiders and Rebels: The Golden Age of Piracy.* New York: William Morrow, 1986.

Thomson, Janice E. *Mercenaries, Pirates, and Sovereigns: State-Building and Extraterritorial Violence in Early Modern Europe.* Princeton: Princeton University Press, 1994.

Williams, Neville. *The Sea Dogs: Privateers, Plunder and Piracy in the Elizabethan Age.* London: Weidenfeld & Nicolson, 1975.

CHAPTER 2: THE CODEMAKERS

Barty-King, Hugh. *Girdle Round the Earth.* London: Heinemann, 1979.

Blondheim, Menahem. *News over the Wires: The Telegraph and the Flow of Public Information in America, 1844–1897.* Cambridge, Mass.: Harvard University Press, 1994.

Briggs, Charles F., and Augustus Maverick. *The Story of the Telegraph.* New York: Rudd & Carleton, 1863.

Cain, Robert Jasper. "Telegraph Cables in the British Empire, 1850–1900." Ph.D. dissertation, Duke University, 1971.

Coe, Lewis. *The Telegraph: A History of Morse's Invention and Its Predecessors in the United States.* Jefferson, N.C.: McFarland & Company, 1993.

DuBoff, Richard B. "Business Demand and the Development of the Telegraph in the United States, 1844–1860," *Business History Review* LIV, no. 4 (Winter 1980), pp. 459–79.

Field, Henry M. *History of the Atlantic Telegraph.* New York: Charles Scribner & Co., 1866.

Harlow, Alvin F. *Old Wires and New Waves.* New York: D. Appleton-Century Company, 1936.

Headrick, Daniel R. *The Invisible Weapon: Telecommunications and International Politics, 1851–1945.* Oxford: Oxford University Press, 1991.

Hubbard, Geoffrey. *Cooke and Wheatstone and the Invention of the Electric Telegraph.* London: Routledge & Kegan Paul, 1965.

International Telecommunication Union. *From Semaphore to Satellite.* Geneva: International Telecommunication Union, 1965.

Lindley, Lester G. "The Constitution Faces Technology: The Relationship of the National Government to the Telegraph, 1866–1884." Ph.D dissertation, Rice University, 1971.

Marland, E. A. *Early Electrical Communication.* London: Abelard-Schuman Ltd., 1964.

Morse, Edward Lind. *Samuel F. B. Morse: His Letters and Journals.* Boston: Houghton Mifflin Company, 1914.

Oslin, George P. *The Story of Telecommunications.* Macon, Ga.: Mercer University Press, 1992.

Parkinson, J. C. *The Ocean Telegraph to India.* London: William Blackwood & Sons, 1870.

Parsons, Frank. *The Telegraph Monopoly.* Philadelphia: C. F. Taylor, 1899.

Sauer, George. *The Telegraph in Europe.* Paris: 1869.

Shaffner, Taliaferro P. *The Telegraph Manual.* New York: Pudley & Russell, 1859.

Standage, Tom. *The Victorian Internet.* New York: Walker and Company, 1998.

Thompson, Robert Luther. *Wiring a Continent: The History of the Telegraph Industry in the United States, 1844–1866.* Princeton, N.J.: Princeton University Press, 1947.

Towers, Walter Kellogg. *Masters of Space.* New York: Harper & Brothers Publishers, 1917.

Wells, David A. *The Relation of the Government to the Telegraph.* New York: 1873.

CHAPTER 3: RADIO DAYS

Aitken, Hugh G. J. *Syntony and Spark: The Origins of Radio.* New York: John Wiley & Sons, 1976.

———. *The Continuous Wave: Technology and American Radio, 1900–1932.* Princeton: Princeton University Press, 1985.

Archer, Gleason L. *History of Radio to 1926.* New York: The American Historical Society, Inc., 1938.

Baker, W. J. *A History of the Marconi Company.* London: Methuen & Co., 1970.

Barnouw, Erik. *A Tower in Babel: A History of Broadcasting in the United States to 1933.* New York: Oxford University Press, 1966.

Bilby, Kenneth. *The General: David Sarnoff and the Rise of the Communications Industry.* New York: Harper & Row, 1986.

Douglas, George H. *The Early Days of Radio Broadcasting.* Jefferson, N.C.: McFarland & Company, 1987.

Douglas, Susan J. *Inventing American Broadcasting, 1899–1922.* Baltimore: The Johns Hopkins University Press, 1987.

Dunlap, Orrin E. *Marconi: The Man and His Wireless.* New York: The Macmillan Company, 1937.

Hancock, H. E. *Wireless at Sea.* Chelmsford, England: Marconi International Marine Communication Company, 1950.

Headrick, Daniel R. *The Invisible Weapon: Telecommunications and International Politics, 1851–1945.* New York: Oxford University Press, 1991.

Lewis, Tom. *Empire of the Air: The Men Who Made Radio.* New York: HarperCollins Publishers, 1991.

Jolly, W. P. *Marconi.* London: Constable & Company, 1972.

Pocock, Rowland F. *The Early British Radio Industry.* Manchester: Manchester University Press, 1988.

Reith, J. C. W. *Broadcast Over Britain.* London: Hodder & Stoughton, 1924.

Sobel, Robert. *RCA.* New York: Stein & Day, 1986.

Tomlinson, John D. *The International Control of Radiocommunications.* Geneva: University of Geneva, 1938.

CHAPTER 5: LAST STAND OF THE CYPHERPUNKS

Bamford, James. *The Puzzle Palace: A Report on America's Most Secret Agency.* Boston: Houghton Mifflin Company, 1982.

Barth, Richard C., and Smith, Clint N. "International Regulation of Encryption: Technology Will Drive Policy." In *Borders in Cyberspace,* edited by Brian Kahin and Charles Nesson. Cambridge, Mass.: MIT Press, 1997, pp. 283–99.

Kahn, David. *The Codebreakers: The Story of Secret Writing.* London: Weidenfeld & Nicolson, 1967.

Singh, Simon. *The Code Book.* New York: Doubleday, 1999.

CHAPTER 6: TRUSTING MICROSOFT

Cusumano, Michael A., and Selby, Richard W. *Microsoft Secrets: How the World's Most Powerful Software Company Creates Technology, Shapes Markets, and Manages People.* New York: The Free Press, 1995.

Cusumano, Michael A., and Yoffie, David B. *Competing on Internet Time: Lessons from Netscape and Its Battle with Microsoft.* New York: The Free Press, 1998.

Ichbiah, Daniel, and Knepper, Susan L. *The Making of Microsoft: How Bill Gates and His Team Created the World's Most Successful Software Company.* Rocklin, Calif.: Prima Publishing, 1991.

Levy, Steven. *Hackers: Heroes of the Computer Revolution.* New York: Anchor Press/Doubleday, 1984.

Quittner, Joshua, and Slatalla, Michelle. *Speeding the Net: The Inside Story of Netscape and How It Challenged Microsoft.* New York: Atlantic Monthly Press, 1998.

Wallace, James, and Erickson, Jim. *Hard Drive: Bill Gates and the Making of the Microsoft Empire.* New York: John Wiley & Sons, 1992.

CHAPTER 7: SPACE MUSIC

Burnett, Robert. *The Global Jukebox.* London: Routledge, 1996.
Hull, Geoffrey P. *The Recording Industry.* Boston: Allyn & Bacon, 1998.
Kennedy, Rick, and McNutt, Randy. *Little Labels, Big Sound.* Bloomington: Indiana University Press, 1999.
Sanjek, Russell. *American Popular Music and Its Business: The First Four Hundred Years.* New York: Oxford University Press, 1988.
Shemel, Sidney, and Krasilovsky, M. William. *This Business of Music.* Sidney Shemel and M. William Krasilovsky, 1977.

ACKNOWLEDGMENTS

In a book that spans seven centuries and a dozen technologies, it seems only natural that the author's gratitude should be similarly expansive. And indeed it is. Scores of people have helped me with the research required for this book; with the intellectual framework that sustains it; with the funding that was so critical; and with generous doses of enthusiasm and support. I am grateful to them all.

At Harvard, many of my colleagues read successive drafts of the manuscript and added their wisdom to it. I deeply appreciate the efforts of Rawi Abdelal, Sven Beckert, Alexander Dyck, Walter Friedman, Geoffrey Jones, Rosabeth Moss Kanter, Thomas McCraw, Julio Rotemberg, Richard Tedlow, Richard Vietor, Louis Wells and David Yoffie. Dick Rosenbloom was particularly helpful in clarifying my work on the radio and telegraph industries; and Tony Oettinger and Harry Lewis provided masterful help with the intricacies of software and computing. Kalypso Nicolaidis of Oxford University helped me struggle through the articulation of the book's overall formulation, as did Jessica Korn of the Gallup Management Journal. Stephen Barden, Len Schoppa, Christopher Marsden and Gordon Silverstein each intervened at critical points in the book's journey, and Allegra Young and Terence Mulligan plowed through final versions of the chapters. Along the way, the book also benefited from a long and illustrious stream of research assistants: Jennifer Burns, Joshua Friedman, Larry Hamlet, Brianna Huntsberger, Lane LaMure, and Elizabeth Stein. Laura Bures was invaluable in the earliest stages, compiling a monumental survey of piracy and serving as an ever-patient sounding board. Michelle Neve did a wonderful job of handling all administrative details and Christopher Albanese advised on matters both technical and musical.

I am also grateful to Mari Sako and her colleagues at Oxford University for offering me a congenial atmosphere in which to write during the spring of 2000; and to Prof. Dr. Friedhelm Neidhardt and Dr. Georg Thurn at the Wissenschaftszentrum Berlin for providing a similarly conducive arrangement in the fall. Throughout, research funding for this project was generously provided by the Division of Research at the Harvard Business School.

As usual, though, my greatest debts lie closest to home. My sons, Daniel and Andrew Catomeris, have essentially grown up with this book and helped make it fun. They found me pictures of satellites and conquistadores; kept my music current; and listened eagerly to pirate tales. My husband Miltos was a joy and a comfort throughout. And Maria Araujo and Irene Saavedra took care of all the other things while I was busy writing.

Finally, many of the most important contributors to this book are not mentioned here. They are the true pioneers of our era— the men and women working along the technological frontier and helping, in some cases, to rein it in. To protect their confidentiality, I am not mentioning these people by name. But I am grateful to all of them and thankful for their time. I hope that this book captures some of their frontier spirit and helps to establish these pirates, prophets, and pioneers along technology's arc.

D.L.S.
Cambridge, June 2001

INDEX